Dear Michael:

Thank you for all your wisdom, support & friendship

You are such a great mentor and inspiration

Lurdes Del Rosso 7/13/2021.

Sleep Neurology

Lourdes M. DelRosso • Raffaele Ferri
Editors

Sleep Neurology

A Comprehensive Guide to Basic and Clinical Aspects

Springer

Editors
Lourdes M. DelRosso
Sleep Disorders Center
UCSF Benioff Children's Hospital
San Francisco
CA
USA

Raffaele Ferri
Sleep Research Centre
Oasi Maria SS
TROINA
Enna
Italy

ISBN 978-3-030-54358-7 ISBN 978-3-030-54359-4 (eBook)
https://doi.org/10.1007/978-3-030-54359-4

This Springer imprint is published by the registered company Springer Nature Switzerland AG
The registered company address is: Gewerbestrasse 11, 6330 Cham, Switzerland

Foreword

Long ago, when I started learning about Sleep Medicine, the only clinical text available was Christian Guilleminault's landmark book *Sleeping and Waking Disorders: Indications and Techniques*. Wow, has that limited publication availability ever changed over the ensuing years! Currently there so many more books it is sometimes overwhelming for someone seeking good texts. Available formats include encyclopedic approaches, broad coverage–limited depth approaches, update books, small scope–high depth of subject, and board review–focused publications, just to name a few. Now, when I see a new book I ask, "Is this one useful?" and "For what audiences?" As a career educator, I also look for applicability for both learners and teachers as well. I believe the answer for *Sleep Neurology: A Comprehensive Guide to Basic and Clinical Aspects*, edited by Lourdes M. DelRosso and Raffaele Ferri, is a clear yes; it is useful to the sleep medicine field.

Those looking for a solid coverage with updated information about relevant clinical aspects of the field of Sleep Medicine as related to Neurology, but less material than wading through an encyclopedia, will find a good clinical focus and solid coverage of major underlying facts and findings. As an added bonus, many of the authors are experienced faculty, having focused their career on teaching, both at their academic institutions as well as for the American Academy of Sleep Medicine committees, courses, and publications. That combination results in a recipe for success with their subject coverage.

Cross discipline students, residents and fellows, sleep fellows, mid-level practitioners from neurology-based programs, neurologists, and sleep medicine practitioners who want to learn more about sleep and neurology interrelation, and those looking for solid coverage of the indexed subjects, especially relevant for teaching, are all likely beneficiaries.

The book's overall organizational format is well designed to be helpful to the reader. The chapters typically provide an efficient focus for the presented topic with an abstract, keywords, and/or an introduction to indicate the scope. The chapter organization includes integrating the neurological topic with the relevant sleep medicine topic(s) making it useful whether the reader is looking for specific areas of sleep or neurology focus, or for providing a framework for the integration of the fuller topic scope of that chapter. References are provided liberally for readers wanting greater depth in specific areas but who do not wish to get bogged down in excessive detail while reading. The coverage of neurology topics related to sleep appears to have been well thought out.

I believe the reader will likely find their time well spent as they read this book and that it will be a useful addition to the educational literature on Neurology and Sleep Medicine.

Andrew L. Chesson Jr, MD
Dean Emeritus, School of Medicine
Professor Emeritus Neurology and Division of Sleep Medicine
LSU Health Sciences Center – Shreveport
Shreveport, LA, USA

Preface

Sleep science has undergone incredible development during the last decades, probably more pronounced than other areas of science. This rapidly evolving field is still demonstrating a fast growth which will probably continue in future years. As a result of this astonishing increase in knowledge and pieces of information coming from a wide range of studies, experts dealing with sleep disorders may find it difficult to integrate this amount of information in their clinical practice.

It is with great hope we present *Sleep Neurology: A Comprehensive Guide to Basic and Clinical Aspects* as a useful aid to sleep clinicians, researchers, and the neurology and medical community in general. With this book, we want to present easy access to a comprehensive view of the neurologic basis of sleep disorders. We have gathered together experts from all over the world to contribute to this book, with a goal to connect the neurophysiology of sleep with sleep disorders seen in common neurologic conditions. We want the information to be relevant to the understanding of sleep processes, practical to highlight important clinical information, and concise to summarize all the key facts in sleep and neurology. The review of the list of pertinent citations and references at the end of each chapter provides the opportunity to expand the knowledge beyond the current volume. The group of authors are experts in their field and represent colleagues from various regions of the world, including Europe and North and South America. The contributors bring various perspectives and expertise to this volume.

In this first edition, we organize the material into two main parts. The first part covers the neurobiology of normal sleep and sleep functions. We start with a general description of sleep and wakefulness, discussing the circadian and homeostatic control of sleep, expanding into the neurophysiology of dreams, memory consolidation, and sleep deprivation. The second part is dedicated to the understanding of sleep symptomatology in various neurologic conditions. Here, we discuss sleep changes in patients with dementia, seizures, headaches, and stroke among other common neurologic disorders. Illustrations and tables are added to emphasize important key points discussed in each chapter.

As said above, the knowledge about the science of sleep has grown exponentially in the last decade and research has shown the association between sleep disorders and many other medical conditions. Sleep disorders can contribute to comorbidities and, vice versa, medical/neurological conditions can present with specific sleep disorders. As knowledge of the importance of sleep grows, we see it imperative to bring forth this book as a reference source of useful information for both clinicians and researchers.

We want to thank all the authors for their expert knowledge and valuable contribution to this book. We also want to thank Mariah Gumpert and Gregory Sutorius of Springer for their administrative assistance.

San Francisco, CA, USA Lourdes M. DelRosso
Troina, Enna, Italy Raffaele Ferri

Contents

Part II Sleep in Neurologic Conditions

List of Contributors

Tommaso Baldini, MD Department of Biomedical and Neuromotor Sciences, University of Bologna, Bellaria Hospital, Bologna, Italy

Célyne H. Bastien, PhD School of Psychology, Laval University, Quebec, QC, Canada

Michael Blechner, MD Department of Pediatric Pulmonary and Sleep Medicine, St. Joseph's Children's Hospital, Paterson, NJ, USA

Roxanne Bolduc-Landry, BA School of Psychology, Laval University, Quebec, QC, Canada

Oliviero Bruni, MD Department of Developmental and Social Psychology, S. Andrea Hospital – Sapienza University, Rome, Italy

Daniel P. Cardinali, MD, PhD Faculty of Medical Sciences, University of Buenos Aires, CONICET, Buenos Aires, Argentina

Pontifical Catholic University of Argentina, Buenos Aires, Argentina

John C. Carter, MD Department of Medicine, Division of Pulmonary, Critical Care, and Sleep Medicin, Case Western Reserve University School of Medicine and MetroHealth Medical Center, Cleveland, OH, USA

Andrew L. Chesson Jr, MD Neurology and Division of Sleep Medicine, School of Medicine, LSU Health Sciences Center – Shreveport, Shreveport, LA, USA

Carlo Cipolli, PhD Department of Specialty, Diagnostic and Experimental Medicine, University of Bologna, Bologna, Italy

Stefan Clemens, PhD Department of Physiology, Brody School of Medicine, East Carolina University, Greenville, NC, USA

Patrizia Congiu, MD, PhD Neurophysiology Unit, Sleep Disorder Centre, Department of Medical Sciences and Public Health, University of Cagliari, Cagliari, Italy

Lourdes M. DelRosso, MD, MSEd Department of Pediatrics, University of Washington/Seattle Children's Hospital, Seattle, WA, USA

Swapan Dholakia, MD Department of Neurology, Atlanta VA Medical Center, Emory University School of Medicine, Decatur, GA, USA

Raffaele Ferri, MD Sleep Research Centre, Oasi Research Institute - IRCCS, Troina, Italy

Michela Figorilli, MD Neurophysiology Unit, Sleep Disorder Centre, Department of Medical Sciences and Public Health, University of Cagliari, Cagliari, Italy

Helmut Frohnhofen, MD Department of Internal Medicine and Geriatrics, Alfried Krupp Krankenhaus, Essen, Germany

Luigi De Gennaro, PhD Department of Psychology, University of Rome Sapienza, Rome, Italy

Karl Van Gundy, MD, FCCP Departments of Pulmonary, Critical Care, and Sleep Medicine, UCSF Fresno, Fresno, CA, USA

Division of Internal Medicine, Community Regional Medical Center, Fresno, CA, USA

Dirk M. Hermann, MD Department of Neurology, Universitätsklinikum Essen, Essen, Germany

Vishesh K. Kapur, MD, MPH Department of Medicine, Division of Pulmonary, Critical Care and Sleep Medicine, University of Washington, Harborview Medical Center, Seattle, WA, USA

Lynn Keenan, MD, FAASM Department of Internal Medicine, UCSF Fresno, Fresno, CA, USA

Jessica Lebel, MS School of Psychology, Laval University, Quebec, QC, Canada

Carey T. Lockhart, MD Department of Pediatric Sleep Medicine and Neurology, University of Washington, Seattle Children's Hospital, Seattle, WA, USA

Giuseppe Loddo, MD Department of Biomedical and Neuromotor Sciences, University of Bologna, Bellaria Hospital, Bologna, Italy

Silvia Miano, MD, PhD Neurocenter of Southern Switzerland, Civic Hospital, Lugano, Switzerland

Stefania Mondello, MD, MPH, PhD Department of Biomedical and Dental Sciences and Morphofunctional Imaging, University of Messina, Messina, Italy

Claude Nguyen, MD, MSEd Division of Stroke and Neurocritical Care, California Pacific Medical Center, Sutter Health, San Francisco, CA, USA

Marisa Pedemonte, MD, PhD Department of Physiology, School of Medicine, CLAEH University, Prado and Salt Lake, Punta del Este, Maldonado, Uruguay

Fabio Pizza, MD, PhD DIBINEM, Department of Biomedical and Neuromotor Sciences, University of Bologna, Bologna, Italy

IRCCS, Institute of Neurological Sciences of Bologna, UOC Neurological Clinic, Bellaria Hospital, Bologna, Italy

Giuseppe Plazzi, MD DIBINEM, Department of Biomedical and Neuromotor Sciences, University of Bologna, Bologna, Italy

IRCCS, Institute of Neurological Sciences of Bologna, UOC Neurological Clinic, Bellaria Hospital, Bologna, Italy

Thierry Provencher, BA School of Psychology, Laval University, Quebec, QC, Canada

Federica Provini, MD, PhD Department of Biomedical and Neuromotor Sciences, University of Bologna, Bologna, Italy

IRCCS Institute of Neurological Sciences, Bologna, Italy

Monica Puligheddu, MD, PhD Neurophysiology Unit, Sleep Disorder Centre, Department of Medical Sciences and Public Health, University of Cagliari, Caligari, Italy

Ibrahim J. Raphael, MD Department of Pulmonary, Sleep, and Critical Care Medicine, Oklahoma University Health Sciences Center, Oklahoma City, OK, USA

Michelle Sobremonte-King, MD Department of Pediatric Pulmonary and Sleep Medicine, The Breathing Institute, Children's Hospital Colorado, Aurora, CO, USA

Saiprakash B. Venkateshiah, MD Division of Pulmonary, Allergy, Critical Care and Sleep Medicine, Emory University School of Medicine, Atlanta VA Medical Center, Decatur, GA, USA

Joanna E. Wrede, MD Department of Pediatrics and Neurology, University of Washington, Seattle Children's Hospital, Seattle, WA, USA

Part I

Neurobiology of Sleep and Wakefulness

Neurobiology of Wakefulness

1

Lourdes M. DelRosso and Marisa Pedemonte

Introduction

The sleep-wake cycle is a biological process found in all animals. The identification of areas in the brain involved in sleep and wakefulness dates back to the works of Soca and Constantin von Economo in the early 1900s. Soca described a continuous and prolonged sleep, easily arousable at the beginning in a young patient with a tumor located over the *sella turcica* which compressed the anterior hypothalamic region [1]. Von Economo studied the brains of patients who died of encephalitis lethargica. This devastating infection manifested in two main forms. One group of patients suffered from insomnia and a second group fell into a "sleep coma." Von Economo identified different areas of the brain associated with sleep and wakefulness. He discovered that the brains of patients who suffered from profound sleepiness had significant loss of neurons in the posterior hypothalamus and mesencephalic reticular formation; hence these two areas were identified as crucial for wakefulness, while the brains of patients who suffered insomnia had loss of neurons in the anterior hypothalamus and preoptic forebrain, concluding that these areas were crucial for sleep [2].

Research in sleep neurophysiology has been tremendously influenced by the invention of electroencephalography by Hans Berger who recorded the first human EEG in 1924 [3]. In 1936, Frederick Bremer researched EEG patterns in two cat brain preparations. He transected the cat brain at two levels. The cut between the medulla and the spinal cord (encephale isole) demonstrated EEG patterns of sleep and wakefulness. The cut between the inferior and superior colliculi (cerveau isole) demonstrated persistent sleep waves. In 1949, Moruzzi and Magoun discovered that stimulation of the "reticular formation" in the brain stem would lead to EEG

L. M. DelRosso (✉)
Department of Pediatrics, University of Washington/Seattle Children's Hospital, Seattle, WA, USA

M. Pedemonte
Department of Physiology, School of Medicine, CLAEH University, Prado and Salt Lake, Punta del Este, Maldonado, Uruguay

L. M. DelRosso, R. Ferri (eds.), *Sleep Neurology*,
https://doi.org/10.1007/978-3-030-54359-4_1

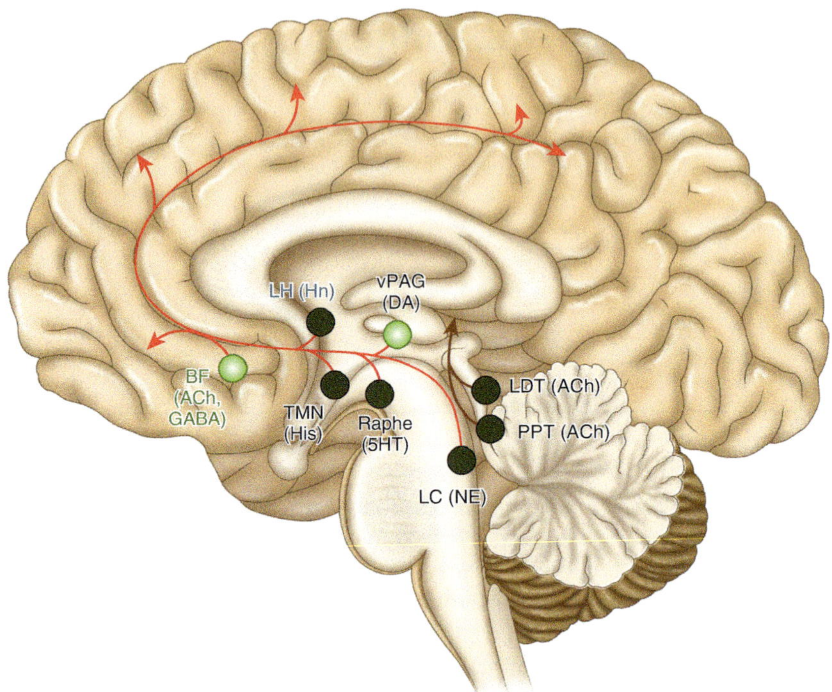

LC: Locus Ceruleus, NE: Norepinephrine, PPT: Pedunculopontine tegmentum, LDT: laterodorsal tegmentum, Ach: Acetylcholine, vPAG: ventral periaqueductal gray, DA: Dopamine, TMN: tuberomamilary nucleus, His: Histamine, BF Basal forebrain, LH lateral hypothalamus, Hn; hypocretin

Fig. 1.1 Nuclei and neurotransmitters involved in the arousal pathways

desynchronization and behavioral arousal. Lesions in the reticular formation of the brain stem produced persistent sleep [4]. These findings laid the foundations for what is currently known as the reticular activating system (Fig. 1.1).

Neural Circuits of Wakefulness

Moruzzi and Magoun discovered that wakefulness was achieved through the effects of the ascending reticular activating system (RAS) on the brain stem and cortex. RAS is conformed by several nuclei and neuronal projections that originate in the reticular formation of the brain stem. Brain stem nuclei include the locus ceruleus, dorsal raphe, median raphe, pedunculopontine nucleus, laterodorsal tegmentum, and parabrachial nucleus. Non-brain stem nuclei include the thalamic nuclei, hypothalamus, and basal forebrain. These nuclei cannot be clearly isolated from other structures by neuroimaging although some studies have attempted to demonstrate RAS fibers and connectivity using diffusion tensor imaging (Fig. 1.2) [5].

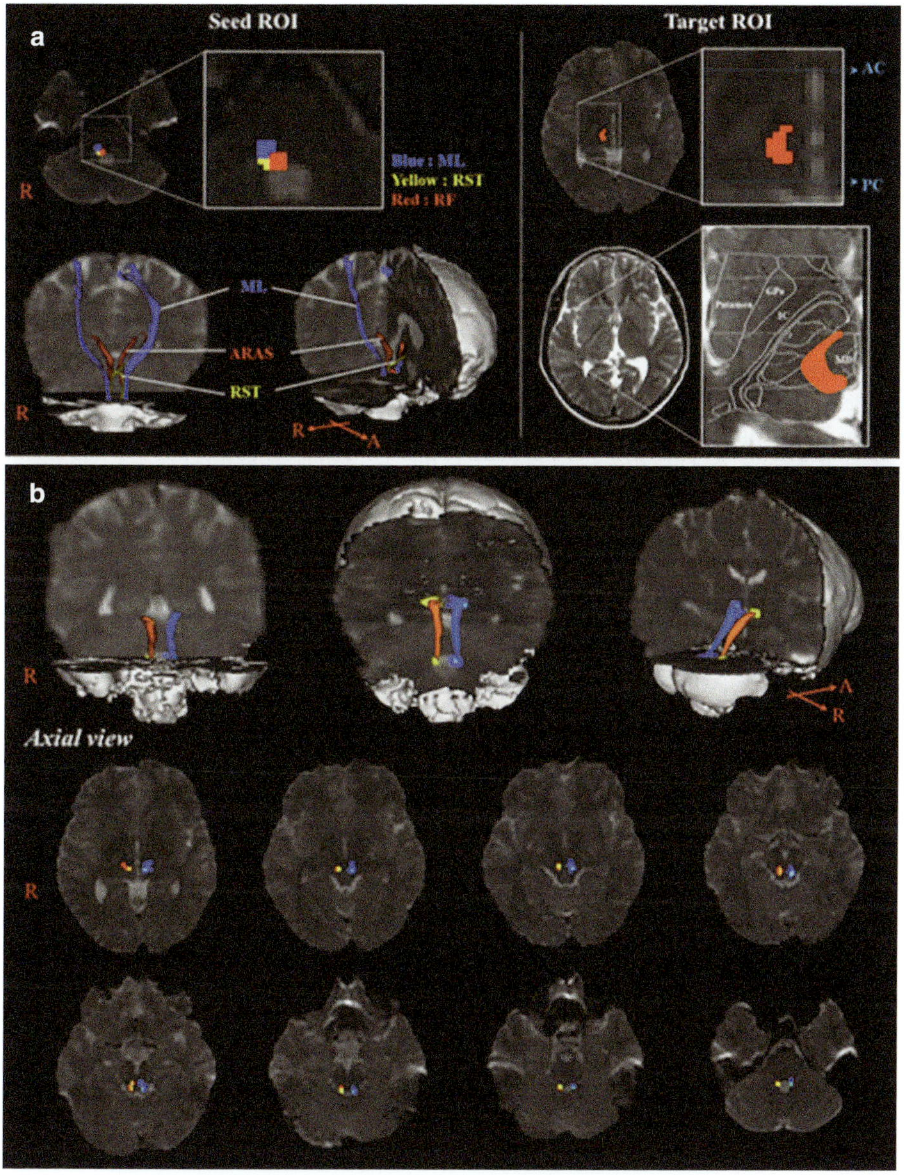

Fig. 1.2 (**a**) Seed regions of interest (ROI) are given on the pontine reticular formation (red color). The target ROI is given on the intralaminar nuclei of the thalamus at the level of the commissural plane. *ML* medial lemniscus, *RST* rubrospinal tract, *RF* reticular formation, *AC* anterior commissure, *PC* posterior commissure. (**b**) Pathways of the reconstructed ascending reticular activating system are shown at each level of the brain in a normal subject (26-year-old male). (From Yeo et al. [3])

Two nuclei are the major point of origin of the RAS: the pedunculopontine (PPN) and the laterodorsal tegmentum (LDT). These nuclei contain acetylcholine-synthesizing neurons. Every cell in the PPN generates and maintains beta and gamma activity during wakefulness via membrane oscillations that are mediated by voltage-dependent calcium channels and modulated by G proteins. These channels have separate intracellular pathways for wakefulness and for REM sleep (Fig. 1.3) [6].

From the PPT, signals travel via two pathways. The dorsal pathway, thought to play an important role in the thalamocortical transmission and on the EEG of sleep and wakefulness, projects to the thalamus (intralaminar, paraventricular, and reticular nuclei) and then to the cortex [7]. The ventral pathway projects to the hypothalamus and basal forebrain.

Arousal is therefore initiated and maintained by multiple brain regions, several nuclei in the brain stem, and their projections to the thalamus, hypothalamus, and forebrain. These nuclei synthesize neurotransmitters (acetylcholine, histamine, serotonin, norepinephrine, and hypocretin) (Table 1.1) that modulate wakefulness and sleep.

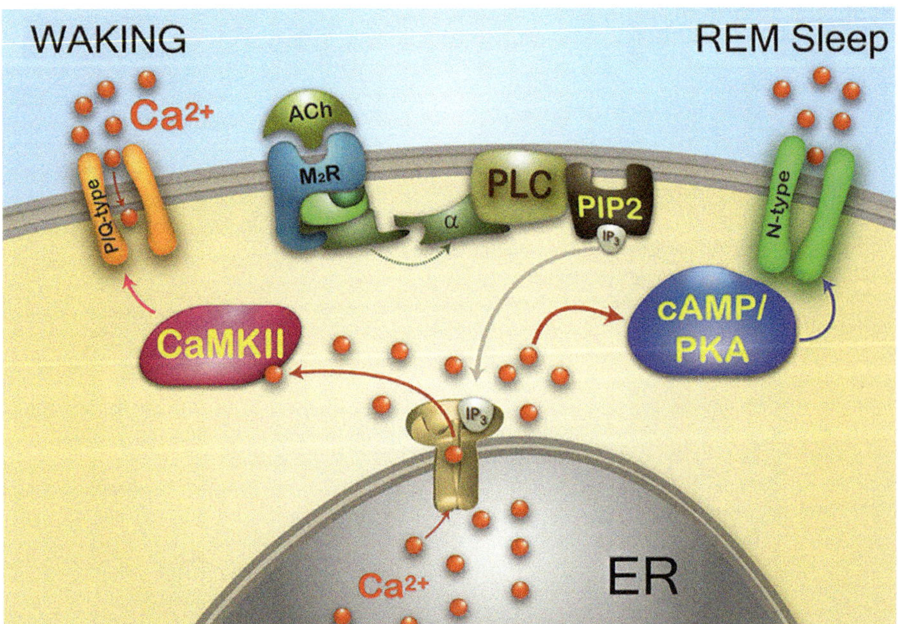

Fig. 1.3 Intracellular pathways and calcium channels differentially related to waking versus REM sleep. Representation of effects of acetylcholine (ACh) activation of a muscarinic 2 cholinergic receptor (M2R) acting through G protein coupling to phospholipase C (PLC) that in turn cleaves phospholipid phosphatidylinositol bisphosphate (PIP2) into inositol trisphosphate (IP3). IP3 is released and binds to IP3 receptors in the endoplasmic reticulum (ER) to release calcium (Ca^{2+}). One of the intracellular pathways activated involves CaMKII, which modulates P/Q-type calcium channels, and the other pathway involves cAMP/PKA, which modulates N-type calcium channels. The CaMKII/P/Q-type pathway mediates beta/gamma band activity during waking, while the cAMP/PKA/N-type pathway mediates beta/gamma band activity during REM sleep. (From Garcia-Rill E et al. [4])

Table 1.1 Neurotransmitters of wakefulness

Neurotransmitter	Location
Hypocretin	Lateral hypothalamus
Histamine	Tuberomammillary nucleus
Serotonin	Raphe nucleus
Acetylcholine	Pedunculopontine tegmentum Laterodorsal tegmentum
Dopamine	Ventral periaqueductal gray
Norepinephrine	Locus ceruleus

Anatomical Components of the Arousal System

Brain Stem Nuclei

The Locus Ceruleus (LC)

The LC demonstrates tonic activation during wakefulness. The activation of the LC increases norepinephrine (NE) throughout the brain. The LC receives non-neuropeptide and neuropeptide afferents [8].

1. Non-neuropeptide afferents
   (a) Glutamate from nucleus paragigantocellularis.
   (b) GABA from the nucleus prepositus hypoglossus.
   (c) Serotonin from the dorsal raphe.
2. Neuropeptide afferents
   (a) Corticotropin-releasing factor: usually released during stress to increase NE activity.
   (b) Hypocretin: local administration of hypocretin in the LC increases wakefulness and suppresses REM sleep [8].
   (c) Substance P: implicated in pain-related increase in activation of the LC.

Pedunculopontine Tegmentum (PPT) and Laterodorsal Tegmentum (LDT)

Neurons from the PPT/LDT are mainly cholinergic and glutamatergic and demonstrate activity during wakefulness and during REM sleep. The PPT/LDT initiates and maintains beta and gamma activity mediated by two types of voltage-dependent calcium channels: the P/Q-type channels regulated by CaMKII during wakefulness and the N-type channel regulated by cAMP/PK during REM sleep (Fig. 1.3) [6].

The PPT/LDT projects to the thalamus, cerebellum, cerebral cortex, and basal forebrain. Besides participating in wakefulness and REM sleep generation, the PPT is also important in learning and in locomotion.

Raphe Nuclei

The dorsal raphe nucleus in the middle of the brain stem contains serotonin-producing neurons. Firing rates of the dorsal raphe nucleus are higher during wakefulness.

Thalamus

The thalamus relays information from and to the cortex. Glutamatergic neurons in the thalamus relay sensory, motor, and limbic information to the cortex. Thalamic neurons participate in the formation of spindles during NREM sleep. During wakefulness, acetylcholine depolarizes thalamic neurons suppressing spindle production [9].

Hypothalamus

The hypothalamus is essential in the regulation of sleeping and waking, feeding, drinking, activity, and body temperature. The posterior hypothalamus is an important participant in RAS. Inactivation of the rostral and middle parts of the posterior hypothalamus has been associated with severe hypersomnia. During insomnia, posterior hypothalamic neurons are in an estate of hyperactivity. The sleep/wake components of the hypothalamus include the tuberomammillary nucleus (TMN*)*, the lateral hypothalamus (LH), and the ventrolateral preoptic nucleus (VLPO). The TMN is a discrete group of about 60,000 neurons located in the posterior hypothalamus that project mostly unmyelinated axons to the central nervous system [10]. It demonstrates significant increase in c-FOS expression (a marker of cell activation) during the wakefulness state, and it is the only source of neuronal histamine in the brain. Histaminergic neurons have the most wake-selective firing patterns and are inhibited by GABAergic projections form the VLPO [11]. Norepinephrine from the LC activates histaminergic neurons indirectly by inhibiting the GABAergic inputs. Dopamine from nearby hypothalamic areas excites histaminergic neurons. Hypocretin-producing neurons in the LH promote waking via excitatory effect primarily on the TMN, basal forebrain, and RAS neurons.

The Basal Forebrain (BF) The BF contains various neurotransmitters including GABAergic, glutamatergic, and cholinergic neurons (comprise 5%). Studies have shown that BF cholinergic neurons play an important role in wakefulness [12].

BF cholinergic neurons project to inhibit the VLPO decreasing c-FOS expression.

Neurotransmitters of Wakefulness

Neurotransmitters are chemical substances synthesized by neurons, stored in synaptic vesicles, and released into the synaptic cleft where they bind a postsynaptic receptor to elicit a specific response. A neurotransmitter can be excitatory or inhibitory depending on the type of receptor it binds to. In this section we will discuss the main neurotransmitters involved in wakefulness.

Acetylcholine

Acetylcholine-producing neurons, involved in wakefulness, are located in the brain stem (PPT and LDT) and in the basal forebrain. Cholinergic neurons can bind to muscarinic or nicotinic receptors. Muscarinic effects include depolarization of pyramidal neurons, subthreshold oscillations of the beta/gamma range, and blockade of slow afterhyperpolarization. Nicotinic effects include depolarization of interneurons and glutamate release [13]. Acetylcholine neurons exhibit higher activity during both wakefulness and REM sleep than during NREM sleep.

Norepinephrine

Norepinephrine neurons are predominantly located in the locus ceruleus of the brain stem. Norepinephrine neurons fire during wakefulness, diminish activity during NREM sleep, and are quiet during REM sleep. The mechanism by which norepinephrine promotes wakefulness is by excitation of various neuronal systems of the RAS via alpha-1 receptors. Beta receptors inhibit hyperpolarizations of cortical pyramidal neurons, allowing the faster firing during wakefulness.

Serotonin

Serotonin release is increased during wakefulness in the dorsal raphe nucleus, decreased during NREM sleep, and almost absent during REM sleep. The mechanism by which serotonin promotes wakefulness is complex, but it has been postulated to be via depolarization of histaminergic tuberomammillary neurons and basal forebrain GABA neurons projecting to the hippocampus and neocortex. There are various serotonin receptors, but agonists to 5HT1A, 5HT1B, 5HT2A/2C, or 5HT3 appear to increase wakefulness and decrease sleep [12].

Histamine

In the central nervous system, histamine is synthesized from L-histidine. Besides arousal, histamine is also involved in feeding, drinking, sexual behavior, locomotor activity, and analgesia. Histamine neurons are active during wakefulness and inactive during sleep. Various types of stress, hypoxia, and hypercapnia have been shown to activate histamine neurons. Histamine promotes wakefulness via excitatory effects on the nuclei of the RAS and inhibitory effects on the sleep-active projection neurons of the VLPO [13]. The effects of histamine are mediated by four histamine receptor types: H1R, H2R, H3R, and H4R. H1R, H2R, and H3R are found in the brain. H1R is responsible for the waking effect of histamine. H2R is implicated in attention and H3R in autoregulation [11].

Hypocretin (Orexin)

Hypocretin was first described in 1998 as a sleep- and eating-associated neurotransmitter. Neurons that produce hypocretin are located in the perifornical area of the dorsolateral hypothalamus and project to all the areas of the CNS and spinal cord. Hypocretin enhances synaptic transmissions and increases intracellular calcium [13]. The cell bodies of hypocretin-producing neurons are predominantly innervated by glutamatergic (excitatory) synapses compared to inhibitory (GABAergic) synapses. This particular array facilitates the excitation of the system in response to various stimuli [13]. Hypocretin receptors are found inside and outside the brain. Activation of hypocretin neurons during wakefulness is produced by intrinsic depolarization and by glutamatergic activation. Hypocretin activates thalamocortical neurons, serotonergic neurons in the raphe nuclei, dopaminergic neurons in the dorsal tegmental area, norepinephrine neurons in the locus ceruleus, and cholinergic neurons in the basal forebrain and the mesopontine tegmentum [9]. Two receptors have been identified, hypocretin 1 and hypocretin 2. Hypocretin neurons send dense projection to the TMN where they depolarize histaminergic neurons via hypocretin type 2 receptors. Hypocretin-containing neurons are active during wakefulness and inactive in both REM and NREM sleep.

Dopamine

Dopamine neurons in the ventrotegmental area (VTA) fire more during wakefulness and during REM sleep than during NREM sleep with subsequent increased dopamine release in the nucleus accumbens and prefrontal cortex. Dopamine neurons in the ventral periaqueductal gray (vPAG) fire during wakefulness and not during sleep. The vPAG projects to the RAS and receives input from the VLPO [12].

Glutamate

Glutamate is an excitatory neurotransmitter important in cognitive function. Glutamate-containing neurons are widely spread in the brain. Levels of glutamate in the hypothalamus and cortex are elevated during wakefulness and during REM sleep. Promotion of wakefulness is via the N-methyl-D-aspartate (NMDA) receptor.

Electroencephalogram (EEG) of Wakefulness

During wakefulness, the EEG represents synaptic potentials from pyramidal cells within the cortex and hippocampus. The EEG of wakefulness is characterized by low-voltage (5–10 microvolts), high-frequency (20–30 hertz) waves called beta waves and gamma waves (30–120 hertz). These frequencies are fired from the PPN during wakefulness and during REM sleep. Transection studies anterior to the PPN have shown to prevent these waves, while stimulation of the PPN produces gamma waves on EEG [6]. During relaxed wakefulness, the EEG frequency slows down to a frequency of 8–13 hertz called alpha rhythm (Fig. 1.4).

Physiology of Wakefulness

Considering the ontogenetic evolution of the human being, wakefulness emerges from sleep. The newborn baby sleeps almost all day, and the short periods of wakefulness to feed gradually, become longer. Wakefulness is "sculpted" or "modeled" with sensory inputs from the environment.

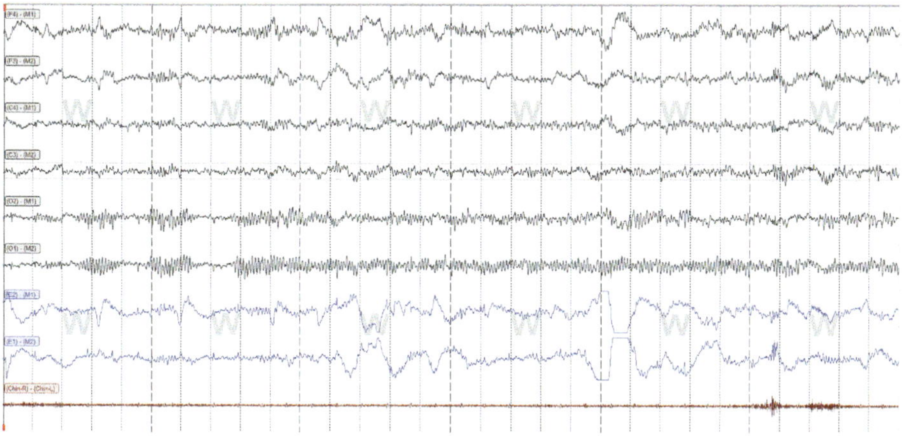

Fig. 1.4 30-second epoch during polysomnography. Relaxed wakefulness is characterized by alpha waves more predominantly seen in the occipital leads (O2-M1 and O1-M2). Also note frequent eye movements (E2-M1 and E1-M2) consistent with wakefulness

Development in humans is constantly influenced by a series of biological rhythms. In utero, the CNS is developing in an environment influenced by excitable tissues that are rhythmically organized. Moreover, all the sensory systems are formed under the influence of the biological rhythms of the mother. This process continues at birth, now also influenced by environmental cues, and persists throughout life in an uninterrupted manner, during the wakefulness-sleep cycle.

Day to day wakefulness begins in a circadian context given mainly by the fall of nocturnal melatonin and the increase of cortisol in the morning. However, there are many other rhythms that also modulate wakefulness, for example, the ultradian attentional rhythms, feeding, and cardiac, among others. Framed in these oscillations, during wakefulness, all homeostatic controls must be met, with precision and speed to maintain parameters within the normal ranges, although we are constantly changing the environment and the metabolic demands given by various activities. Deviations of the physiological ranges elicit error signals in thermo-, baro-, chemo-, mechano-, and nociceptive receptors sending information to the CNS, processing via feedback loops, and finally returning to normal values via autonomic, endocrine, or somatic responses. All these homeostatic controls [14] are made through two mechanisms that interact continuously: feedback or reactive homeostasis, where the responses are caused by changes in the environment or the organism itself, and predictive homeostasis (feedforward) that prevents future disturbances and generates preventive actions to mitigate the future energy cost [15, 16].

According to the Parmeggiani model [16, 17], the control commands, despite the immovable morphological hierarchy, vary the functional hierarchical array depending on the ultradian wakefulness-sleep cycle (quiet wakefulness, QW; NREM sleep and REM sleep).

Wakefulness constitutes two thirds of our lives. It is during wakefulness that we develop relationships through interactions, mental stimulation, and emotions. During wakefulness homeostatic controls maintain the vital parameters within physiological ranges through autonomic and hormonal controls. There is a global predominance of the sympathetic nervous system over the parasympathetic system that is exacerbated at times of maximum demand.

Throughout the day, functional and structural changes occur, Changes in cerebral blood flow, in metabolism, in the speed of reflex responses, alert levels, immunity response, among others. A "sleep debt" is generated that will be paid the following night for the good restoration of all the physiological functions. Therefore, to consider the physiology of a healthy organism, we must consider these two states acting synergistically and facing the sleep-wake cycle as a functional whole.

Conclusion

Wakefulness is a recurring state under circadian control, orchestrated by a complex interaction between neurotransmitters of the arousal system.

Signals arise from the brain stem and ascend through the thalamus, hypothalamus, and basal forebrain to promote and maintain wakefulness. Several nuclei with their respective neurotransmitters are involved in this system. Understanding the pathways involved in wakefulness will help understand the potential pathophysiologic basis of sleep-wake disorders.

References

1. Berger H. On the electroencephalogram of man. Electroencephalogr Clin Neurophysiol. 1969;28(Suppl):37.
2. Moruzzi G, Magoun HW. Brain stem reticular formation and activation of the EEG. Electroencephalogr Clin Neurophysiol. 1949;1(4):455–73.
3. Yeo SS, Chang PH, Jang SH. The ascending reticular activating system from pontine reticular formation to the thalamus in the human brain. Front Hum Neurosci. 2013;7:416. https://doi.org/10.3389/fnhum.2013.00416.
4. Garcia-Rill E, Luster B, D'Onofrio S, Mahaffey S, Bisagno V, Urbano FJ. Implications of gamma band activity in the pedunculopontine nucleus. J Neural Transm (Vienna). 2016;123(7):655–65.
5. Fuller PM, Sherman D, Pedersen NP, Saper CB, Lu J. Reassessment of the structural basis of the ascending arousal system. J Comp Neurol. 2011;519(5):933–56.
6. Zitnik GA. Control of arousal through neuropeptide afferents of the locus coeruleus. Brain Res. 2016;1641(Pt B):338–50.
7. Espana RA, Scammell TE. Sleep neurobiology from a clinical perspective. Sleep. 2011;34(7):845–58.
8. Blandina P, Munari L, Provensi G, Passani MB. Histamine neurons in the tuberomamillary nucleus: a whole center or distinct subpopulations? Front Syst Neurosci. 2012;6:33.
9. Haas HL, Lin JS. Waking with the hypothalamus. Pflugers Arch. 2012;463(1):31–42.

10. Chen L, Yin D, Wang TX, et al. Basal forebrain cholinergic neurons primarily contribute to inhibition of electroencephalogram delta activity, rather than inducing behavioral wakefulness in mice. Neuropsychopharmacology. 2016;41(8):2133–46.
11. Brown RE, Basheer R, McKenna JT, Strecker RE, McCarley RW. Control of sleep and wakefulness. Physiol Rev. 2012;92(3):1087–187.
12. Watson CJ, Baghdoyan HA, Lydic R. Neuropharmacology of sleep and wakefulness. Sleep Med Clin. 2010;5(4):513–28.
13. Gao XB, Hermes G. Neural plasticity in hypocretin neurons: the basis of hypocretinergic regulation of physiological and behavioral functions in animals. Front Syst Neurosci. 2015;9:142.
14. Cannon WB. Organization for physiological homeostasis. Physiol. Rev. 1929;9:399–31.
15. Parmeggiani PL. Systemic homeostasis and poikilostasis. Is REM sleep a physiological paradox? Imperial College Press, London; 2011.
16. Parmeggiani PL. Regulations of physiological functions during sleep in mammals. Experientia, 1982;38:1405–408.
17. Moore-Ede MC. Physiology of the circadian timing system: Predicting versus reactive homeostasis. Am. J. Physiol. 1986;250:R737–52.

Neurobiology of Sleep

2

Ibrahim J. Raphael and Marisa Pedemonte

Introduction

Sleep is a complex behavioral state that has perplexed the men and women of science for millennia. In Greek mythology, Hypnos the god of sleep was the son of Nyx the goddess of the night and Erebus god of the darkness. He lived with his twin brother Thanatos, the deity of death, in a cave in the underworld where no light or sound would enter. The entrance was full of poppies, a narcotic analgesic plant. Through the cave coursed Lethe, the river of oblivion whose consumption would induce forgetfulness and flowing murmurs and drowsiness. Lethe formed the border between night and day. Hypnos married Pasithea, daughter of Zeus and minor goddess of hallucination and relaxation. Pasithea gave birth to the Oneiroi, gods of dreams, Morpheus, Phobetor, and Phantasos personifying shape, fear, and imagination, respectively.

In the more modern days, sleep was long thought to result from an interruption in consciousness and was long considered akin to syncope, narcosis, or coma. Sleep was however noted to be distinct in its easy reversibility, i.e., waking up. In an attempt to better understand this physiologic phenomenon, several hypotheses and theories have been proposed throughout the years. Most of these were deemed obsolete, even at the time, motivating researchers and physicians alike to find more complete and satisfying explanations. One of the first major schools of thought supported the theory of lack of stimuli, in other words, implying that sleep is a consequence of the brain ceasing to receive external stimuli. For example, Exner and Rabl-Rükhard proposed that the ganglionic cells of the brain retract their dendrites during sleep.

I. J. Raphael (✉)
Department of Pulmonary, Sleep, and Critical Care Medicine, Oklahoma University Health Sciences Center, Oklahoma City, OK, USA

M. Pedemonte
Department of Physiology, School of Medicine, CLAEH University, Prado and Salt Lake, Punta del Este, Maldonado, Uruguay

L. M. DelRosso, R. Ferri (eds.), *Sleep Neurology*,
https://doi.org/10.1007/978-3-030-54359-4_2

Purkinje attributed sleep to the swelling of the gray matter exerting mass effect and strangulation of signaling pathways to and from the brain. Mauthner then added that sleep is due to a physiological recurrence of this swelling. Later came other concepts such as the anemia theory or the vasomotor theory of Mosso. In the early twentieth century, chemical theories became more popular in an attempt to involve not only the brain but also the human body as an entity. For example, Pflüger and Du Bois-Raymond suggested that carbon dioxide asphyxiation of the brain generated sleep. Piéron later reported that fatigue itself could be associated with the production of substances released in the blood stream and stimulating the onset of sleep. Weichhardt went a step further and hypothesized that the accumulation of those substances during wakefulness and their excretion during sleep explained the periodicity of sleep. This was later followed by the introduction of the hormonal theory of Mingazzini who thought that the periodicity of sleep could be explained by an alternating balance of two groups of endocrine glands. Decades of research and experiments have led us to much better understand the sleep state.

Parmeggiani's research since the 1960s has provided an important advance in the understanding of the changes in the physiological controls during sleep. All the changes demonstrated are based on two major concepts: NREM sleep maintains a parasympathetic predominance, while REM sleep is a poikilostatic state, with sympathetic bursts of activity [1, 2].

After centuries of relating sleep to death or diminishing brain functions, it is now recognized that the brain is actively working, on synaptic plasticity, recovery, and restoration of biochemical and physiological processes performed during wakefulness, as well as in cognitive, learning and memory processes [3].

Despite this marked activity, the central nervous system (CNS) continues to interact with the environment, although in a different way than during wakefulness. This is particularly appreciated in the auditory system [4, 5].

Thanks to advancements in technology, our understanding of sleep has increased exponentially in the past few decades. Sleep is necessary to maintain healthy cognitive and emotional function, performance, and safety as well as overall physical health. Several epidemiological, clinical, and basic science studies have elucidated without a reasonable doubt the importance of adequate sleep. Transitions between wakefulness and the various stages of sleep are gradual rather than discrete events. Sleep stages are broadly grouped as rapid eye movement (REM) and non-REM (NREM) stages depending on well-defined electrical patterns seen on polysomnography (PSG).

NREM Sleep

Neural Pathways

In 1916, von Economo described "lethargic encephalitis" in which patients suffered symptoms such as insomnia, reversal of sleep-wake periodicity, and somnambulism to name a few [6]. Later in the 1930s, he observed that this insomnia was commonly seen in patients with encephalitis affecting the preoptic area and basal forebrain [6]. Many studies have been able to confirm this by demonstrating that the destruction or inactivation of neurons in these areas resulted in a state of insomnia [7–9]. These findings suggested these areas of the brain play a key role in the regulation of sleep. Animal and immunocytochemistry experiments later demonstrated clusters of neurons in the ventrolateral preoptic (VLPO) and the median preoptic nucleus (MnPO) areas that were selectively activated during sleep [10–12]. These neurons have projections to the wake-promoting centers, namely, the histaminergic neurons of the ipsilateral tuberomammillary nucleus (TMN), the laterodorsal tegmental nuclei (LDT) and pedunculopontine tegmental nuclei (PPT), the locus coeruleus (LC), the serotoninergic dorsal raphe (DR), and the lateral hypothalamus [10, 11]. As mentioned in the previous chapter, these centers play an important role in the arousal system. Activation of the VLPO neurons resulted in an increase in GABA concentration in the wake-promoting centers, thus inhibiting their activity and thereby contributing to the onset and maintenance of sleep [6]. VLPO neuronal firing was markedly increased during NREM sleep, slowed during REM sleep, and absent in wakefulness [13–15]. In contrast, MnPO neurons were noted to become active immediately prior to NREM onset [16].

Neurotransmitters

Neurons in the VLPO and MnPO produce GABA and galanin. These inhibitory neurotransmitters act on the wake-promoting centers of the brain. Researchers have shown that drugs such as benzodiazepines, benzodiazepine-like drugs (Zolpidem), and barbiturates bind to GABA-A receptors whereas gamma hydroxybutyrate (sodium oxybate or Xyrem®) binds to GABA-B receptors [17–19]. It is likely that those drugs work by stimulating GABA receptors in the VLPO and MnPO to promote sleep maintenance.

Key Points

- MnPO neurons are most active prior to NREM, likely involved in sleep onset.
- VLPO neurons are most active during NREM, slower during REM, and inactive during wakefulness, likely involved in sleep maintenance.
- VLPO and MnPO neurons innervate wake-promoting centers, likely inhibit arousal from sleep.
- VLPO and MnPO neurons release the inhibitory neurotransmitters GABA and galanin.

Polysomnography Findings

NREM sleep is comprised of three substages labeled N1, N2, and N3 each defined by unique electroencephalography (EEG) and electrooculography (EOG) criteria. Electromyography (EMG) criteria are more important in scoring stage of REM sleep. During NREM, chin EMG can be of variable amplitude with a tendency to decrease in amplitude from stages N1 to N3. Stage-specific criteria are defined by the American Academy of Sleep Medicine (AASM) and are scored according to the stage occupying the majority of a 30-second epoch.

Stage N1 is characterized by a background of low-amplitude, mixed-frequency (LAMF) activity (4–7 Hz) for more than 50% of an epoch. Vertex (V) sharp waves are sharp waves with a maximal duration of 0.5 seconds. V waves can be observed in stages N1 and N2; however, their presence is not required for scoring. The EOG shows slow eye movements (SEM) that appear with closed eyes during wakefulness and during N1. They are conjugate, sinusoidal eye movements with an initial deflection lasting more than 0.5 seconds (Fig. 2.1).

Stage N2 is characterized by the appearance of K-complexes and sleep spindles. K-complexes are sharp waves with a negative followed by positive deflection with a minimum duration of 0.5 seconds, better seen on frontal leads. Sleep spindles represent inhibitory bursts originating from the reticulothalamic cells [20]. They are distinct segments of sinusoidal activity (11–16 Hz) lasting at least 0.5 seconds, best seen in central leads. EOG typically shows no eye movements or persistence of SEM (Fig. 2.2).

Stage N3 is a deeper stage of sleep characterized by slow wave activity seen in 20% or more of a 30-second epoch. Slow waves have a frequency of 0.5–2 Hz with a minimal amplitude of 75 microvolts better seen in the frontal leads. They originate from cortical neurons under dorsal reticulothalamic control and are

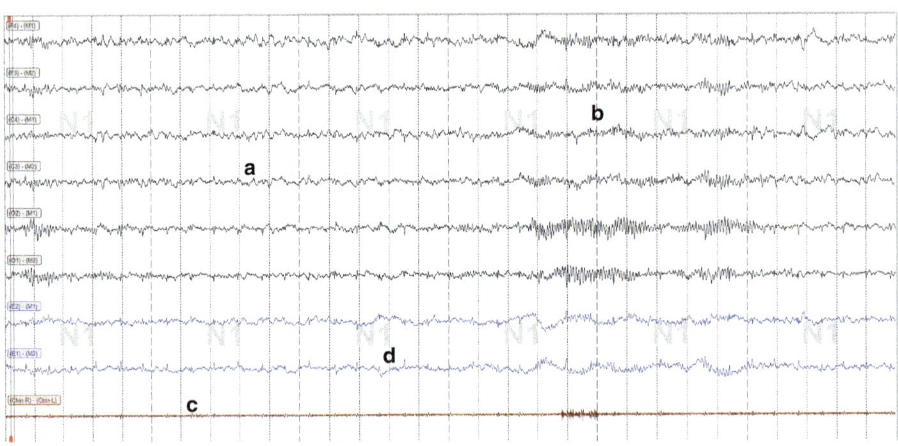

Fig. 2.1 This 30-second epoch has been labeled as N1. (**a**) A background of low-amplitude, mixed-frequency (LAMF) activity at a frequency of 4–7 Hz can be seen for the majority of this epoch. (**b**) An arousal is illustrated by an abrupt onset of alpha waves (8–13 Hz) best seen on the frontal leads. Notice the persistence of (**c**) chin tone and (**d**) slow eye movements

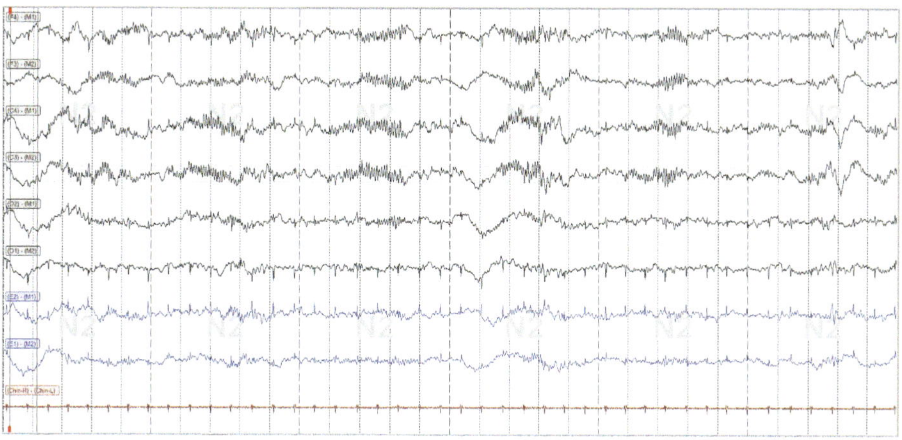

Fig. 2.2 This 30-second epoch has been labeled as N2. Several sleep spindles can be seen throughout this epoch. They are sinusoidal bursts better seen in the central leads, with a frequency of 11–16 Hz and last at least 0.5 seconds each

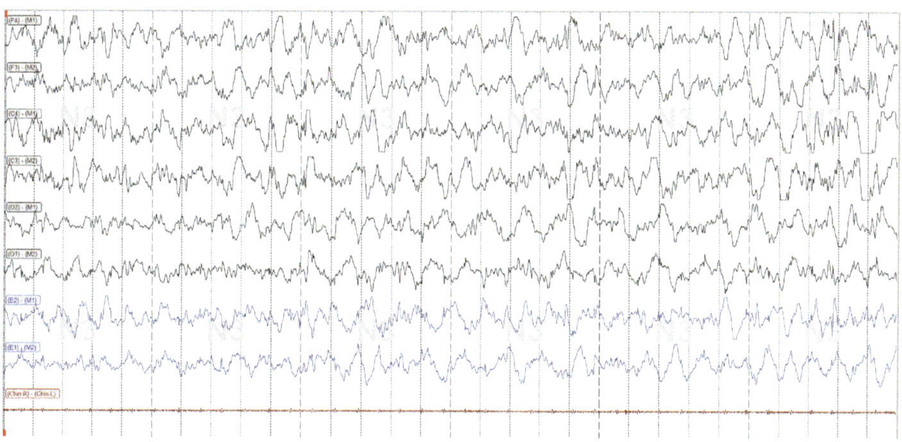

Fig. 2.3 This 30-second epoch has been labeled as N3. Slow wave activity with a frequency of 0.5–2 Hz with a minimal amplitude of 75 microvolts. Although occupying the totality of this epoch, only 20% or 6 seconds of slow wave activity is required for scoring

thought to reflect cortical synchronization [21]. Eye movements are not typically seen in this stage (Fig. 2.3).

Physiology

Sleep stages exhibit an ultradian rhythm that repeats four or five times each night. Typically sleep onset in adults is through N1, followed with N2, N3, and REM sleep. This sequence repeats itself approximately every 90 minutes. The physiology of the various organ systems is modulated and varies according to the sleep-wake cycle: motor activity, sensory information, cardiovascular and respiratory functions, endocrine, and temperature control.

Motor Activity

Motor activity is greatly diminished; movements are slight in the NREM stages or in the brief periods of awakening between the ultradian cycles of the night of sleep. In periods of stillness, muscle tone is diminished compared to wakefulness.

Sensory Information Processing

Despite the reduced motor activity and the loss of awareness, sensory systems continue to send information that is processed in the CNS during sleep; visceral and vascular receptors continue homeostatic control, and proprioceptive receptors control posture and tone. Receptors continue sending signals about the environment to warn about changes that may provoke arousals or awakenings. On the other hand, sensory networks and centers must process all the information entered during the day and generate new memories, decide forgetfulness, and participate in dreaming.

The evoked potentials and the unitary responses of certain neuronal groups are comparatively greater during NREM than during wakefulness. This fact was demonstrated for visual and auditory inputs. The thalamic and cortical auditory-evoked potentials show greater amplitude during NREM when compared to wakefulness and REM [22–25]. It has been shown that synchronous activity, both visual and auditory, at different levels of the pathways, is modulated by the sleep-wake cycle in very complex ways [4]. This neuronal activity is also related to other ultradian rhythms, such as the theta rhythm of the hippocampus whose temporal correlation would have the function of evaluating a novel stimulus, both in wakefulness and in sleep [26–28]. Inversely, the complete absence of auditory input produces modifications in sleep and wakefulness [29, 30].

Homeostatic Controls

NREM sleep can be considered to be a period in which homeostatic controls are conserved with the functional prevalence of parasympathetic influences associated with quiescence of sympathetic activity, energy conservation, predominance of anabolic metabolism, and general decrease in cardiovascular and respiratory functions [2].

Cardiovascular Functions

During NREM sleep the sensitivity of baroreceptor reflexes increases; the mean arterial pressure and heart rate decrease as a result of a drop in diastolic and systolic blood pressures. The lowest value is recorded during stage N3. The pulmonary artery blood pressure remains stable during all NREM sleep phases [31].

Respiratory Changes

During NREM sleep, metabolic respiratory control predominates over the neural, which ensures the homeostasis of arterial O_2 and CO_2 through information conveyed from central and peripheral chemoreceptors. Both N1 and N2 stages provoke an unstable respiratory rhythm with consecutive hypoventilations and hyperventilations called "periodic ventilation." This pattern alternates with periods of regular breathing. During N3 ventilation becomes regular, with higher amplitude and a lower respiratory rate. When NREM starts, automatic control mechanisms are released, with inactivation of telencephalic mechanisms that are active during wakefulness. The partial pressure of alveolar CO_2 increases because the chemoreceptor sensitivity to CO_2 is moderately reduced, whereas the partial pressure of alveolar and arterial O_2 decreases [32].

Endocrine Functions

Most hormones are produced or controlled in the hypothalamic-pituitary axis. In addition, the hypothalamus regulates autonomic actions to comply with homeostatic controls throughout the body and houses the control centers for hunger, satiety, and thirst. All these complex functions are modulated by circadian rhythms and the sleep-wake cycle. Melatonin, for example, secreted by the pineal gland, is considered a chemical code through which the brain "understands" that it is nighttime [33].

All hormones exhibit cyclical oscillations in their blood levels, which have been subdivided into three different categories: (a) hormones modulated by a particular stage of sleep (i.e., the peak in growth hormone levels during stage N3), (b) hormones highly influenced by the whole period of sleep (prolactin and thyrotrophin), and (c) hormones weakly modulated by sleep (adrenocorticotrophin and cortisol). There are hormones, such as gonadotropins, that are subject to the modulation of many rhythms, circadian, ultradian related to the sleep stages, and infradian taking part of the monthly cycle [32, 34].

Body Temperature

In homeothermic animals, during wakefulness, body temperature is controlled through the interaction between hypothalamic and cortical mechanisms. In NREM the automatic mechanisms are released from cortical control. In an environment of thermoneutrality, the homeostatic controls fix the body temperature to values lower than those of quiet wakefulness, due to the predominant conservative metabolism and muscle hypotonia.

Renal Function

Antidiuretic hormone, synthesized in the hypothalamus and secreted in the neurohypophysis, increases its secretion during sleep, which causes a marked increase in urinary concentration. Glomerular filtration, urine volume, and sodium, potassium, and calcium excretion decrease during sleep.

Digestive Functions

There is an overall decrease in gastric acid secretion during sleep in normal humans. Also the esophagus motility is consistently reduced.

REM Sleep

The term "REM sleep" was first coined by Aserinsky and Kleitman after noting that dreaming occurred in the presence of rapid eye movements while sleeping, along with a low-voltage EEG [35]. In 1959, Michel Jouvet, a French professor of experimental medicine, reported that an intact pontine tegmentum was necessary for the generation of REM sleep [36–38]. REM sleep was eliminated with transection in the pons; however, it was preserved after suprapontine transections [36–38]. Jouvet also observed that during REM sleep, cats entered a temporary vegetative-like state, with total absence of muscle tone, increased respiratory rate, decreased heart rate, and a fall in core body temperature by 2–3°C compared to the preceding sleep stage. These findings coupled with the wake-like brain activity on EEG and the presence of rapid eye movements led Jouvet to describe REM sleep as a "paradoxical sleep stage" due to its similarity with wakefulness [36].

REM is usually automatically associated with dreaming. Historically, dreams have been associated with prophecy telling, repressed primal desires, and unconscious mental reflection [39]. Yet dreams do not always occur in REM sleep. Siegel studied sleep in the platypus and reported that REM only occurs at the level of the brain stem without forebrain activation, a necessary component of higher cognitive function such as dreaming [40]. Moreover, not all subjects recall dreams when aroused from REM sleep [41], which brings up the question, do all people dream? And do dreams occur with every REM stage or only with specific REM periods? REM sleep was once thought to be an essential stage of sleep for physical and mental health, which has been challenged by the lack of impairment in health or cognition in patients with pontine lesions who are unable to reach REM sleep [42]. Furthermore, tricyclic antidepressants (TCA), monoamine oxidase inhibitors (MAOi), and selective serotonin and serotonin-norepinephrine reuptake inhibitors (SSRI and SNRI) are known REM suppressants [43]. Drug-induced REM suppression has not been correlated with a detrimental effect on overall health, behavior, or cognition [44].

To this day, the true role of REM sleep remains to be elucidated. Although neuronal pathways and neurotransmitters involved in REM sleep have been discovered, the full circuitry and coordination between these isolated discoveries is still unclear. In this section, we will explore the different neurotransmitters involved in REM sleep as well as their origin and target. In the end we will review the polysomnographic findings characteristic to REM sleep.

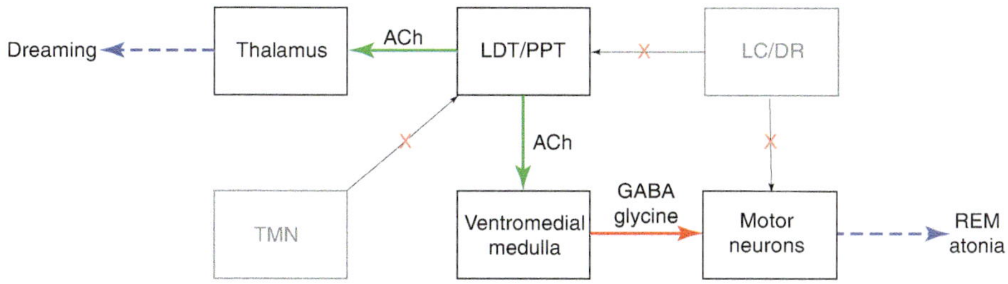

Fig. 2.4 Cholinergic REM-on pathway. The LDT/PPT sends excitatory cholinergic projections to the thalamus and ventromedial medulla. The ventromedial medulla sends inhibitory projections to the spinal and brain stem motor neurons thereby contributing to REM atonia. *REM* rapid eye movements, *LDT/PPT* laterodorsal and pedunculopontine tegmental nuclei

Acetylcholine

Acetylcholine (ACh), a neurotransmitter of wakefulness, is also active during REM sleep. Injection of a cholinergic agonist or acetylcholinesterase inhibitor directly into the pons produces long-lasting REM sleep and muscle atonia [45–48]. On the other hand, lesions in the LDT/PPT were associated with a marked reduction in REM sleep [49, 50]. Cholinergic neurons in the LDT/PPT (part of the ascending arousal system) have projections to the thalamus and the ventromedial medulla. Stimulation of the thalamocortical neurons by ACh is thought to play a key role in dreaming during REM sleep [16]. Stimulation of the ventromedial medullary neurons releases inhibitory neurotransmitters GABA and glycine, thereby inhibiting the activity of motor neurons in the brain stem resulting in muscle atonia [16, 51, 52] (Fig. 2.4).

Monoamines

Monoaminergic neurons are found in the locus coeruleus (LC), dorsal raphe (DR), and tuberomammillary nucleus (TMN). They release the monoamines norepinephrine (NE), serotonin (5-hydroxytryptamine or 5-HT), and histamine (HA), respectively. Monoamines increase muscle tone by direct stimulation of motor neurons and suppress REM sleep by inhibiting the REM-active cholinergic neurons [16, 53–55]. During REM sleep, monoaminergic neurons are suppressed, thereby disinhibiting the LDT/PPT and ceasing to stimulate the basal forebrain and spinal motor neurons. As mentioned above, medications such as SSRI, SNRI, MAOi, and TCA cause drug-induced REM suppression [43]. By increasing muscle tone in sleep, these drugs have also been shown to worsen or unmask underlying periodic leg movements (PLM) and REM behavior disorder (RBD) [56–63] (Fig. 2.5).

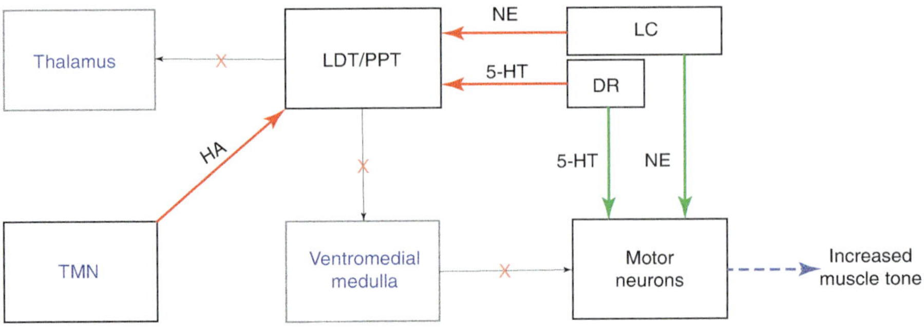

Fig. 2.5 Monoaminergic REM-off pathway. The LC, DR, and TMN inhibit the activity of the LDT/PPT preventing thalamic and ventromedial medulla stimulation. The LC and DR also stimulate the basal forebrain and spinal motor neurons, thereby increasing muscle tone. *REM* rapid eye movements, *LDT/PPT* laterodorsal and pedunculopontine tegmental nuclei, *TMN* tuberomammillary nucleus, *LC* locus coeruleus, *DR* dorsal raphe, *NE* norepinephrine, *5-HT* 5-hydroxytryptamine or serotonin, *HA* histamine

GABA

An alternate proposed pathway of REM sleep includes the sublaterodorsal nucleus or subcoeruleus (SLD). This collection of neurons ventral to the LC produces GABA and glutamate [64]. They have projections to the ventromedial medulla and the ventral horn of the spinal cord [65, 66]. When active, they promote REM sleep and atonia [67]. An insult to the SLD can reduce REM sleep, reduce atonia, and could potentially be an important factor in REM behavior disorder (RBD) [64, 68–70].

Melanin-Concentrating Hormone (MCH)

In addition to orexin, the lateral hypothalamus contains a cluster of neurons that produce the inhibitory neurotransmitters MCH and GABA [71]. These neurons innervate the LC and DR and are most active during REM, slow down during NREM, and are silent with wakefulness [72, 73]. Micro-infusions of MCH produce increased REM (and slow wave sleep), whereas micro-infusions of an MCH receptor antagonist produce decreased REM sleep [72, 74, 75].

Polysomnography Findings

REM sleep is portrayed by conjugate, sharp, and irregular derivations in EOG with an initial deflection lasting less than 0.5 seconds. Since rapid eye movements can also be seen during wakefulness with eyes open, the simultaneous presence of low/absent chin tone on EMG is required for scoring REM sleep. The background EEG is one of LAMF without K-complexes or sleep spindles (characteristic of stage N2). Sawtooth waves can also be present, better seen over the central leads with a frequency of 2–6 Hz.

Physiology

An important physiological characteristic of REM sleep is the loss of central homeostatic control. It has not yet been clarified why these brief periods of poikilostatic state exist in a homeostatic organism. Could it be an evolutionary advantage? Will it be essential to fulfill some special function still unknown? Velluti's hypothesis proposed the fragmentation of REM sleep throughout the night as an internal protection system to avoid long periods without homeostatic control [32, 76].

Metabolism and Blood Flow

REM sleep is a period of great activity. This has been demonstrated in human studies using positron emission tomography (PET) with *2-deoxy-D-glucose*. It has been demonstrated that during N3 the cerebral glucose metabolism decreases 12% compared to wakefulness, while during REM it increases 16% [77]. Previous research in animal models showed that cerebral oxygen exhibits a particular distribution of O2 availability during REM, which has been denominated "pO2 PS system" [78]. This system includes the reticular formation, the reticularis pontis oralis, the basal forebrain, the hypothalamus, the amygdala, and the cerebellum. This oscillating pattern, which was attributed to a decrease in the local homeostatic control, has also been interpreted as an increase in glucose degradation induced by anaerobic metabolism during neural activity increments [79–83].

Cardiovascular Functions

In humans, blood pressure and heart rate during REM become variable and exhibit transient increases of up to 40 mm Hg which overlap with a tonic hypotension, in periods with phasic events (rapid eye movements, muscle twitches). One of the factors that increase blood pressure is the increase in peripheral resistance caused by periods of vasoconstriction in skeletal muscles [31].

In animal models it has been shown that the rhythmicity of the heart rate is modulated by changing in autonomic tone during the sleep-wake cycle and by ultradian rhythms such as the hippocampal theta rhythm [84–86]. This interval variability was also demonstrated in humans [80].

Respiratory Changes

During REM sleep the respiratory rate is faster and irregular, with brief episodes of apnea and hypoventilation. Muscle atonia, characteristic of this sleep phase, contributes to these respiratory findings by decreasing the chest expansion and increasing upper airway resistance. The diaphragm maintains an irregular activity [87].

Body Temperature

During REM sleep there is absence of temperature regulation due to inhibition of muscle tone and absence of shivering which prevents heat generation. Body temperature falls along the night to reach the lowest levels by the last hours of sleep. However, just as sleep modulates thermoregulation, thermoregulation also modulates sleep; in situations that move away from thermoneutrality, the quality-quantity of sleep is compromised until it is abolished [88–90].

Sexual Functions

Penile erection or tumescence has been shown during REM sleep in human subjects between 3 and 79 years of age, although in adolescents this also happens in NREM. This characteristic is often used to differentiate between organic and psychogenic impotence. Women exhibit clitoral erections and increment of the vaginal blood flow during REM.

Conclusion

Sleep consists of NREM and REM sleep. NREM sleep has three substages N1, N2, and N3. Each state has different electroencephalographic and physiologic characteristics. Furthermore, various functions, hormones, and processes occur during each state of sleep. REM is characterized by muscle atonia and rapid eye movements. The similarity of the electroencephalogram of REM sleep and wakefulness made it to be called "paradoxical sleep stage." The understanding of the physiologic changes in each sleep stage is very important for the understanding of sleep, sleep disorders, and neurological conditions.

References

1. Parmeggiani PL. Sleep behaviour and temperature. In: Parmeggiani PL, Velluti RA, editors. The physiologic nature of sleep. London: Imperial College Press; 2005. p. 387–406.
2. Parmeggiani PL. Systemic homeostasis and poikilostasis. Is REM sleep a physiological paradox? London: Imperial College Press; 2011.
3. Tononi G, Cirelli C. Sleep and the price of plasticity: from synaptic and cellular homeostasis to memory consolidation and integration. Neuron. 2014;81:12–34.
4. Pedemonte M, Velluti RA. What individual neurons tell us about encoding and sensory processing in sleep. In: Parmeggiani PL, Velluti RA, editors. The physiologic nature of sleep. London: Imperial College Press; 2005. p. 489–508.
5. Velluti RA. Other sensory modalities in sleep. In: The auditory system in sleep. 2nd ed. London: Elsevier/Academic Press; 2018). Chapter 8. p. 185–95.
6. Von Economo C. Sleep as a problem of localization. J Nerv Ment Dis. 1930;71:249–59.
7. Lin JS, Sakai K, Jouvet M. Evidence for histaminergic arousal mechanisms in the hypothalamus of cat. Neuropharmacology. 1988;27:111–22.
8. McGinty DJ, Sterman MB. Sleep suppression after basal forebrain lesions in the cat. Science. 1968;160:1253–5.

9. Sakai K, el Mansari M, Lin JS, Zhang ZG, Vanni-Mercier G. The posterior hypothalamus in the regulation of wakefulness and paradoxical sleep. In: Mancia M, Marini M, editors. The diencephalon and sleep. New York: Raven Press; 1990. p. 171–98.
10. Sherin JE, Shiromani PJ, McCarley RW, Saper CB. Activation of ventro-lateral preoptic neurons during sleep. Science. 1996;271:216–9.
11. Gong H, McGinty D, Guzman-Marin R, Chew KT, Stewart D, Szymusiak R. Activation of c-fos in GABAergicneurones in the preoptic area during sleep and in response to sleep deprivation. J Physiol Lond. 2004;556:935–46.
12. Suntsova N, Guzman-Marin R, Kumar S, Alam MN, Szymusiak R, McGinty D. The median preoptic nucleus reciprocally modulates activity of arousal-related and sleep-related neurons in the Perifornical lateral hypothalamus. J Neurosci Off J Soc Neurosci. 2007;27(7):1616–30. https://doi.org/10.1523/JNEUROSCI.3498-06.2007.
13. Szymusiak R, Alamac N, Steiningera TL, McGinty D. Sleep-waking discharge patterns of ventrolateralpreoptic/anterior hypothalamic neurons in rats. Brain Res. 1998;803:178–88.
14. Takahashi K, Lin J, Saki K. Characterization and mapping of sleep-waking specific neurons in the basal forebrain and preoptic hypothalamus in mice. Neuroscience. 2009;161:269–92.
15. Suntsova N, Szymusiak R, Alam N, et al. Sleep-waking discharge patterns of median preoptic nucleus neurons in rats. J Physiol. 2002;543:665–77.
16. Espana R, Scammell T. Sleep neurobiology from a clinical perspective. Sleep. 2011;34:845–58.
17. Gottesmann C. GABA mechanisms and sleep. Neuroscience. 2002;111:231–9.
18. Sanger DJ. The pharmacology and mechanisms of action of new generation, non-benzodiazepine hypnotic agents. CNS Drugs. 2004;18(Suppl 1):9–15.
19. Vienne J, Bettler B, Franken P, Tafti M. Differential effects of GABAB receptor subtypes, {gamma}-hydroxybutyric acid, and baclofen on EEG activity and sleep regulation. J Neurosci. 2010;30:14194–204.
20. Weiner OM, Dang-Vu TT. Spindle oscillations in sleep disorders: a systematic review. Neural Plast. 2016. Article ID 7328725:30. https://doi.org/10.1155/2016/7328725.
21. Silber MH, Ancoli-Israel S, Bonnet MH, et al. The visual scoring of sleep in adults. J Clin Sleep Med. 2007;3(2):121–31.
22. Campbell K, Bell I, Bastien C. Evoked potential measures of information processing during natural sleep. In: Broughton RJ, Ogilvie RD, editors. Sleep, arousal, and performance. Boston/Basel/Berlin: Birkhauser; 1992. p. 89–116.
23. Coenen AM. Neuronal activities underlying the electroencephalogram and evoked potentials of sleeping and waking: implications for information processing. Neurosci Biobehav Rev. 1995;19:447–63.
24. Velluti RA. Interactions between sleep and sensory physiology. J Sleep Res. 1997;6:61–77.
25. Bastuji H, García-Larrea L. Evoked potentials as a tool for the investigation of human sleep. Sleep Med Rev. 1999;3:23–45.
26. Gambini JP, Velluti RA, Pedemonte M. Hippocampus theta rhythm synchronizes visual neurons in sleep and waking. Brain Res. 2002;926:37–141.
27. Pedemonte M, Gambini JP, Velluti RA. Novelty-induced correlation between visual neurons and the hippocampus theta rhythm in sleep and wakefulness. Brain Res. 2005;1062:9–15.
28. Liberman T, Velluti RA, Pedemonte M. Temporal correlation between auditory neurons and the hippocampal theta rhythm induced by novel stimulations in awake Guinea pigs. Brain Res. 2009;1298:70–7.
29. Pedemonte M, Peña JL, Torterolo P, Velluti RA. Auditory deprivation modifies sleep in the Guinea pig. Neurosci Lett. 1996;223:1–4.
30. Velluti RA, Pedemonte M, Suárez H, Bentancor C, Rodriguez-Servetti Z. Auditory input modulates sleep: an intra-cochlear implanted human model. J Sleep Res. 2010;19(4):585–90.
31. Silvani A, Lenzi P. Reflex cardiovascular control in sleep. In: Parmeggiani PL, Velluti RA, editors. The physiologic nature of sleep. London: Imperial College Press; 2005. p. 323–49.
32. Velluti RA. The physiological bases of sleep. In: The auditory system in sleep. 2nd ed. London: Elsevier/Academic Press; 2018). Chapter 2. p. 41–82.

33. Cardinali DP. The use of melatonin as a chronobiotic-cytoprotective agent in sleep disorders. In: Parmeggiani PL, Velluti RA, editors. The physiologic nature of sleep. London: Imperial College Press; 2005. p. 455–88.
34. Brandenberger G. Endocrine correlates of sleep in humans. In: Parmeggiani PL, Velluti RA, editors. The physiologic nature of sleep. London: Imperial College Press; 2005. p. 433–53.
35. Aserinsky E, Kleitman N. Regularly occurring periods of eye motility, and concomitant phenomena, during sleep. Science. 1953;118:273–4.
36. Jouvet M, Michel F, Courjon J. Sur un stade d'activité électrique cérébrale rapide au cours du sommeil physiologique. C R Séances Soc Biol Fil. 1959;153:1024–8.
37. Jouvet M, Michel F. Corrélations électromyographiques du sommeil chez le chat décortiqué et mésencéphalique chronique. C R Séances Soc Biol Fil. 1959;153:422–5.
38. Jouvet M, Mounier D. Effets des lésions de la formation réticulaire pontique sur le sommeil du chat. C R Séances Soc Biol Fil. 1960;154:2301–5.
39. Palagini L, Rosenlicht N. Sleep, dreaming, and mental health: a review of historical and neurobiological perspectives. Sleep Med Rev. 2011;15:179–86.
40. Siegel JM, Manger PR, Nienhuis R, Fahringer HM, Shalita T, Pettigrew JD. Sleep in the platypus. Neurosci. 1999;91:391–400.
41. Pagel JF. Non-dreamers. Sleep Med. 2003;4:235–41.
42. Lavie P, Pratt H, Scharf B, Peled R, Brown J. Localized pontine lesion: nearly total absence of REM sleep. Neurology. 1984;34:118–20.
43. Schweitzer PK. Drugs that disturb sleep and wakefulness. In: Kryger MH, Roth T, Dement WC, editors. Principles and practice of sleep medicine. Amsterdam: Elsevier Saunders; 2011. p. 542–60.
44. Rasch B, Pommer J, Diekelmann S, Born J. Pharmacological REM sleep suppression paradoxically improves rather than impairs skill memory. Nat Neurosci. 2009;12:396–7.
45. Velazquez-Moctezuma J, Gillin JC, Shiromani PJ. Effect of specific M1, M2 muscarinic receptor agonists on REM sleep generation. Brain Res. 1989;503:128–31.
46. Marks GA, Birabil CG. Comparison of three muscarinic agonists injected into the medial pontine reticular formation of rats to enhance REM sleep. Sleep Res Online. 2001;4:17–24.
47. George R, Haslett WL, Jenden DJ. A cholinergic mechanism in the brainstem reticular formation: induction of paradoxical sleep. Int J Neuropharmacol. 1964;3:541–52.
48. Shiromani P, Siegel JM, Tomaszewski KS, McGinty DJ. Alterations in blood pressure and REM sleep after pontine carbachol microinfusion. Exp Neurol. 1986;91:285–92.
49. Webster HH, Jones BE. Neurotoxic lesions of the dorsolateral pontomesencephalic tegmentum-cholinergic cell area in the cat. II. Effects upon sleep-waking states. Brain Res. 1988;458:285–302.
50. Shouse MN, Siegel JM. Pontine regulation of REM sleep components in cats: integrity of the pedunculopontine tegmentum (PPT) is important for phasic events but unnecessary for atonia during REM sleep. Brain Res. 1992;571:50–63.
51. Steriade M, Datta S, Pare D, Oakson G, Curro Dossi RC. Neuronal activities in brain-stem cholinergic nuclei related to tonic activation processes in thalamocortical systems. J Neurosci. 1990;10:2541–59.
52. Kayama Y, Ohta M, Jodo E. Firing of 'possibly' cholinergic neurons in the rat laterodorsal tegmental nucleus during sleep and wakefulness. Brain Res. 1992;569:210–20.
53. Fedirchuk B, Dai Y. Monoamines increase the excitability of spinal neurones in the neonatal rat by hyperpolarizing the threshold for action potential production. J Physiol. 2004;557:355–61.
54. Perrier JF, Delgado-Lezama R. Synaptic release of serotonin induced by stimulation of the raphe nucleus promotes plateau potentials in spinal motoneurons of the adult turtle. J Neurosci. 2005;25:7993–9.
55. Leonard CS, Llinas R. Serotonergic and cholinergic inhibition of mesopontine cholinergic neurons controlling REM sleep: an in vitro electrophysiological study. Neuroscience. 1994;59:309–30.
56. Husain MRG, Novak M, Jindal R, Shapiro CM. Periodic leg movements in patients on different antidepressant therapies. Sleep Res. 1997;26:380.

57. Yang C, White DP, Winkelman JW. Antidepressants and periodic leg movements of sleep. Biol Psychiatry. 2005;58:510–4.
58. Schenck CH, Mahowald MW, Kim SW, O'Connor KA, Hurwitz TD. Prominent eye movements during NREM sleep and REM sleep behavior disorder associated with fluoxetine treatment of depression and obsessive-compulsive disorder. Sleep. 1992;15:226–35.
59. Salin-Pascual RJ, Galicia-PoloL D-CR. Sleep changes after 4 consecutive days of venlafaxine administration in normal volunteers. J Clin Psychiatry. 1997;58:348–50.
60. Louden MB, Morehead MA, Schmidt HS. Activation by selegiline (Eldepryle) of REM sleep behavior disorder in parkinsonism. W V Med J. 1995;91:101.
61. Onofrj M, Luciano AL, Thomas A, Iacono D, D'Andreamatteo G. Mirtazapine induces REM sleep behavior disorder (RBD) in parkinsonism. Neurology. 2003;60:113–5.
62. Schenck CH, Mahowald MW. Motor dyscontrol in narcolepsy: rapid-eye-movement (REM) sleep without atonia and REM sleep behavior disorder. Ann Neurol. 1992;32:3–10.
63. Dorsey CM, Lukas SE, Cunningham SL. Fluoxetine-induced sleep disturbance in depressed patients. Neuropsychopharmacology. 1996;14:437–42.
64. Sakai K, Sastre JP, Salvert D, Touret M, Tohyama M, Jouvet M. Tegmentoreticular projections with special reference to the muscular atonia during paradoxical sleep in the cat: an HRP study. Brain Res. 1979;176:233–54.
65. Xi MC, Morales FR, Chase MH. Interactions between GABAergic and cholinergic processes in the nucleus pontisoralis: neuronal mechanisms controlling active (rapid eye movement) sleep and wakefulness. J Neuro Sci. 2004;24:10670–8.
66. Lu J, Sherman D, Devor M, Saper CB. A putative ip-op switch for control of REM sleep. Nature. 2006;441:589–94.
67. Boissard R, Gervasoni D, Schmidt MH, Barbagli B, Fort P, Luppi PH. The rat ponto-medullary network responsible for paradoxical sleep onset and maintenance: a combined microinjection and functional neuroanatomical study. Eur J Neurosci. 2002;16:1959–73.
68. Hendricks JC, Morrison AR, Mann GL. Different behaviors during paradoxical sleep without atoniadepend on pontine lesion site. Brain Res. 1982;239:81–105.
69. Boeve BF, Silber MH, Saper CB, et al. Pathophysiology of REM sleep behaviour disorder and relevance to neurodegenerative disease. Brain. 2007;130:2770–88.
70. Sastre JP, Jouvet M. Oneiric behavior in cats. Physiol Behav. 1979;22:979–89.
71. Bittencourt JC, Presse F, Arias C, Peto C, Vaughan J, Nahon JL, Vale W, Sawchenko PE. The melanin-concentrating hormone system of the rat brain: an immuno- and hybridization histochemical characterization. J Comp Neurol. 1992;319:218–45.
72. Verret L, Goutagny R, Fort P, et al. A role of melanin-concentrating hormone producing neurons in the central regulation of paradoxical sleep. BMC Neurosci. 2003;4:19.
73. Koyama Y, Takahashi K, Kodama T, Kayama Y. State-dependent activity of neurons in the perifornical hypothalamic area during sleep and waking. Neuroscience. 2003;119:1209–19.
74. Ahnaou A, Drinkenburg WH, Bouwknecht JA, Alcazar J, Steckler T, Dautzenberg FM. Blocking melanin-concentrating hormone MCH1 receptor affects rat sleep-wake architecture. Eur J Pharmacol. 2008;579:177–88.
75. Torterolo P, Sapogna S, Morales FR, Chase MH. MCH-containing neurons in the hypothalamus of the cat: searching for a role in the control of sleep and wakefulness. Brain Res. 2006;1119:101–14.
76. Velluti RA. A functional viewpoint on paradoxical sleep-related brain regions. Acta Physiol Pharmacol Latinoam. 1988;38:99–115.
77. Maquet P. Functional neuroimaging of normal sleep by positron emission tomography. J Sleep Res. 2000;9:207–31.
78. Velluti RA. An electrochemical approach to sleep metabolism: a pO_2 paradoxical sleep system. Physiol Behav. 1985;34:355–8.
79. Velluti R, Roig JA, Escarcena LA, Villar JI, García-Austt E. Changes of brain pO_2 during arousal and alertness in unrestrained cats. Acta Neurol Latinoam. 1965;11:368–82.
80. Garcia Austt E, Velluti RA, Villar JI. Changes in brain pO2 during paradoxical sleep in cats. Physiol Behav. 1968;3:477–85.

81. Velluti R, Monti JM. pO_2 recorded in the amygdaloid complex during the sleep-waking cycle in cats. Exp Neurol. 1976;50:798–805.
82. Velluti R, Velluti JC, García-Austt E. Cerebellum pO_2 and the sleep waking cycle in cats. Physiol Behav. 1977;18:19–23.
83. Franzini C. Brain metabolism and blood flow during sleep. J Sleep Res. 1992;1:3–16.
84. Pedemonte M, Rodriguez A, Velluti RA. Hippocampal theta waves as an electrocardiogram rhythm timer in paradoxical sleep. Neurosci Lett. 1999;276:5–8.
85. Pedemonte M, Goldstein-Daruech N, Velluti RA. Temporal correlation between heart rate, medullary units and hippocampal theta rhythm in anesthetized, sleeping and awake Guinea pigs. Autonomic Neurosci. 2003;107:99–104.
86. Pedemonte M, Rodriguez A, Velluti RA. Electroencephalographic frequencies associated with heart changes in RR interval variability during paradoxical sleep. Autonomic Neurosci. 2005b;123:82–6.
87. Parmeggiani PL, Sabattini L. Electromyographic aspects of postural, respiratory and thermoregulatory mechanisms in sleeping cats. Electroencephalogr Clin Neurophysiol. 1972;33:1–13.
88. Libert JP, Bach V. Thermoregulation and sleep in the human. In: Parmeggiani PL, Velluti RA, editors. The physiologic nature of sleep. London: Imperial College Press; 2005. p. 407–31.
89. Parmeggiani PL. Temperature regulation during sleep: a study in homeostasis. In: Orem H, Barnes C, editors. Physiology in sleep. New York: Academic Press; 1980. p. 98–136.
90. Parmeggiani PL. Physiologic regulation in sleep. In: Kryger MH, Roth T, Dement WC, editors. Principles and practice of sleep medicine. Philadelphia: Elsevier Saunders; 2005. p. 169–78.

Neurobiology of the Control of Sleep

3

Lourdes M. DelRosso, Joanna E. Wrede, Carey T. Lockhart, and Daniel P. Cardinali

Introduction

Two biological processes have been thought to independently regulate the timing and length of sleep: process C and process S. This two-process model of sleep regulation was first proposed in 1982 by a Swiss sleep researcher, Alexander Borbely [1]. Process C, the circadian process, refers to the roughly 24-hour sleep-wake cycle. Research in the 1970s determined that the suprachiasmatic nucleus (SCN) of the hypothalamus was capable of autonomous oscillation and was later found to be the primary controller of this circadian process [2]. Process S, or the homeostatic process, determines that the propensity to sleep is also based on the previous amount of wakefulness and sleep [3]. These two processes have different regulatory mechanisms. The main markers of process C are core body temperature (CBT) and melatonin [4]. The main markers of process S are

L. M. DelRosso (✉)
Department of Pediatrics, University of Washington/Seattle Children's Hospital, Seattle, WA, USA

J. E. Wrede
Department of Pediatrics and Neurology, University of Washington, Seattle Children's Hospital, Seattle, WA, USA
e-mail: Joanna.wrede@seattlechildrens.org

C. T. Lockhart
Department of Pediatric Sleep Medicine and Neurology, University of Washington, Seattle Children's Hospital, Seattle, WA, USA
e-mail: Carey.lockhart@seattlechildrens.org

D. P. Cardinali
Faculty of Medical Sciences, University of Buenos Aires, CONICET, Buenos Aires, Argentina

Pontifical Catholic University of Argentina, Buenos Aires, Argentina

L. M. DelRosso, R. Ferri (eds.), *Sleep Neurology*,
https://doi.org/10.1007/978-3-030-54359-4_3

non-rapid eye movement (NREM) sleep and slow wave activity (SWA) [5]. In the classic two-process model theory, when process S approaches the upper threshold of process C, sleep occurs. Conversely, wakefulness occurs when process S approaches the lower threshold of process C. These two processes are illustrated in Fig. 3.1 [1]. In the last two decades, new scientific findings have helped expand our knowledge of sleep regulation, recognizing that sleep homeostasis not only is a global brain process but also presents regional variations. Regional brain differences in the sleep electroencephalography (EEG), for example, can have implications on the markers of process S, as low-frequency power is predominant in the frontal derivations. The discovery of the circadian genes has moved research of sleep to the molecular/genetic level, showing that both processes have more complex interactions than what we initially thought. Sleep stages, on the other hand, are regulated by ultradian cycles. This refers to a particular order in which sleep stages occur and cycle through the night. In this chapter, we will discuss the two-process model, how process C and process S regulate sleep, and a brief introduction to the ultradian rhythm.

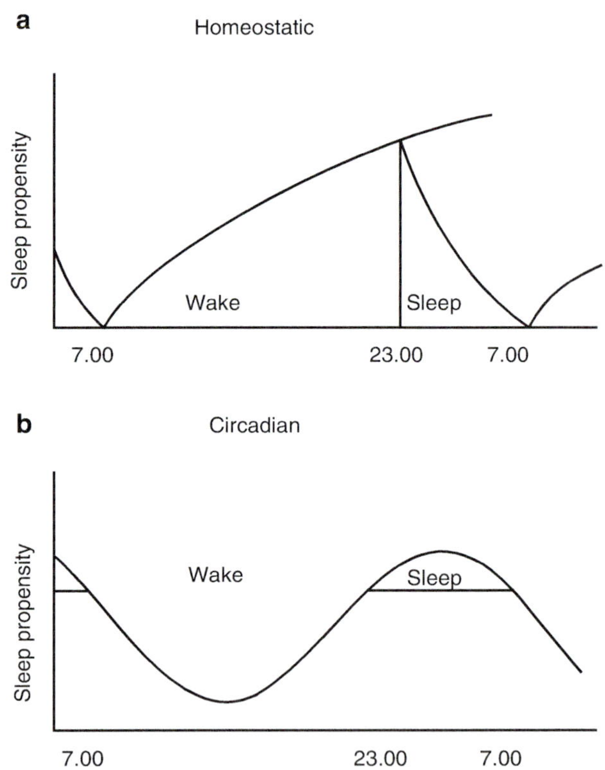

Fig. 3.1 (**a**) Homeostatic control of sleep. (**b**) Circadian control of sleep. (From Borbely [1])

Circadian Rhythm

Circadian Anatomy

The Pacemaker: The Suprachiasmatic Nucleus

The rest/activity cycle in most animals exhibits a 24-hour rhythm that is synchronized mainly to the light-dark cycle. In the absence of light or in the absence of any environmental cues, the circadian rhythms persist in free-running cycles that last a little longer than 24 hours in humans. Early studies have identified the suprachiasmatic nucleus (SCN) is the primary circadian oscillator [6]. The SCN is located in two small areas of the hypothalamus, located just superior to the optic chiasm, on either side of the third ventricle. Destruction of this area causes loss of circadian rhythmicity of the rest/activity cycle. In the majority of cases, the destruction is done experimentally in animals [7, 8]. In humans, accidental destruction of the SCN has been reported in a woman who suffered a gunshot wound that affected both optic nerves and suprachiasmatic nucleus resulting in irregular sleep-wake cycle as seen in Fig. 3.2 [9].

The paired right and left SCN each contains approximately 10,000 neurons that express circadian rhythmicity. They include neurons that synthesize γ-aminobutyric acid (GABA) among other peptides and neurotransmitters [10]. The SCN has been

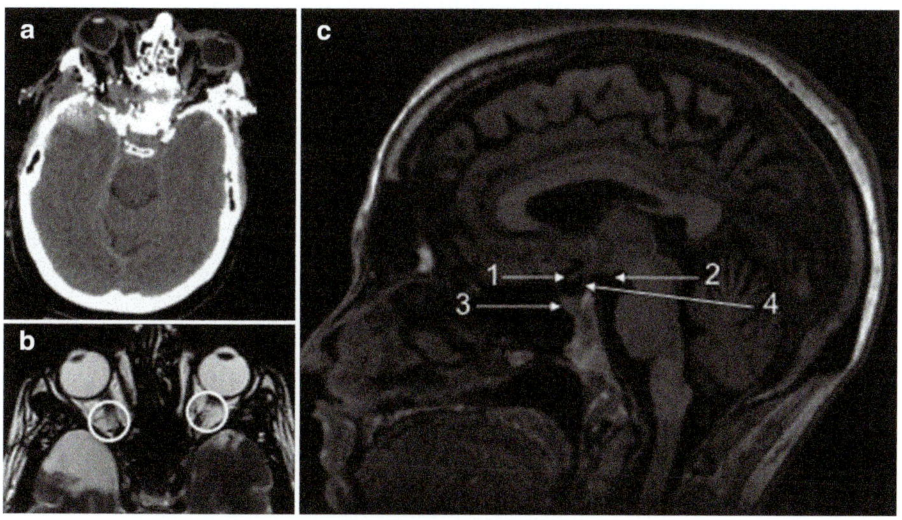

Fig. 3.2 (**a**) Computerized tomography of the head showing damage of the optic nerves bilaterally following a gunshot wound to the right temple. Note bone fragments along the bullet path across the suprachiasmatic area. (**b**) Axial T2-weighed magnetic resonance imaging (MRI) of the head showing bilateral optic nerve hyperintensities consistent with nerve damage (circles). (**c**) Sagittal T1-weighted MRI of the head. The optic chiasm (arrow 1) forms the anterior boundary of the hypothalamus. The infundibulum (arrow 4) to the pituitary gland (arrow 3) and the mammillary bodies (arrow 2) form the inferior boundary of the hypothalamus. The optic chiasm appears thinned along with the adjacent hypothalamus. (From DelRosso et al. [9])

traditionally divided into a ventrolateral area that synthesizes vasoactive intestinal peptide (VIP) and gastrin-releasing peptide (GRP) and a dorsomedial area that synthesizes vasopressin (VP), substance P, and somatostatin [11]. Another functional organization of the SCN refers to these areas as a shell (dorsomedial) and a core (ventrolateral) [12]. The core receives most of the light signal input from the retina.

Circadian Afferents

The major input to the SCN is from the retinohypothalamic tract (RHT). Photic information is received by ganglion cell receptors in the retina and transmitted to the SCN. These signals activate GRP and VIP cells in the core of the SCN. Research has shown that VIP release plays a pivotal role in circadian synchronization [13]. In the absence of VIP or the VIP receptor VIPC2R [14], the neurons in the SCN do not synchronize affecting many rest/activity cycles [15].

A secondary pathway by which light information reaches the SCN arises from the intergeniculate leaflet (IGL) and is called the geniculohypothalamic tract (GHT). The GHT neuron terminals release GABA and neuropeptide Y, and it is thought to play a role in transmitting the effect of more subtle photic stimuli and non-photic stimuli such as metabolic inputs to the SCN [16, 17].

Other afferent pathways to the SCN include serotonin-producing neurons projecting from the raphe nucleus, noradrenergic neurons from the locus coeruleus, histaminergic neurons from the posterior hypothalamus, and cholinergic projections from the pontine tegmentum and basal forebrain (BF). Recent evidence suggests that the SCN receives information directly or indirectly from about 80 different areas in the brain (Fig. 3.3) [18].

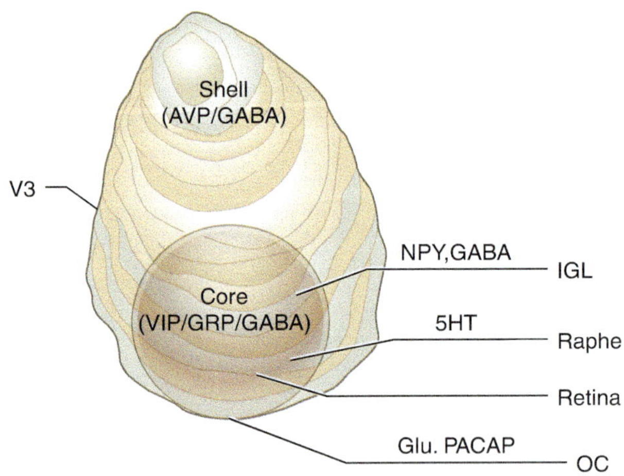

Fig. 3.3 Schematic of SCN with major afferents and associated neurotransmitters. *AVP* arginine vasopressin, *GABA* gamma-aminobutyric acid, *VIP* vasoactive intestinal peptide, *GRP* gastrin-releasing peptide, *NPY* neuropeptide Y, *5HT* serotonin, *Glu* glutamate, *OC* optic chiasm, *PACAP* pituitary adenylate cyclase-activating polypeptide, *IGL* intergeniculate leaflet, *Raphe* raphe nucleus. (Modified from Honma [46])

Circadian Efferents

The SCN has three major output pathways: (1) the rostral pathway to the preoptic area and paraventricular nucleus of the thalamus, (2) caudally to the retrochiasmatic area and ventromedial nucleus, and (3) to the ventral subparaventricular zone (SPZ) and to the hypothalamic paraventricular nucleus (PVN). From the PVN, signals synapse in the preganglionic sympathetic neurons of the spinal cord where melatonin secretion is regulated. The most dense pathway is to the SPZ. Lesions to the ventral part of the SPZ abolish circadian rhythms, and lesions in the dorsal SPZ affect body temperature. The ventral SPZ projects to the dorsomedial hypothalamus (DMH) that in turn connects to other areas involved in sleep and wakefulness. The largest projection is to the ventrolateral preoptic nucleus (VLPO), a group of sleep-active neurons that has inhibitory input to the major ascending arousal systems during sleep; lesions of the VLPO are associated with insomnia.

The function of the circadian clock can be simplistically summarized into three main areas: generation of the circadian rhythms, entrainment of the circadian rhythms, and coupling of the circadian rhythm with the social sleep-wake cycle. We will discuss these functions and how they are accomplished in the following sections.

Circadian Physiology

Generation of Circadian Rhythms

Over the past 20 years, the understanding of the molecular mechanism of circadian rhythmicity has increased. The SCN generates spontaneous oscillations by feedback loops of translation and transcription of circadian genes. This feedback cycle is cell-autonomous and allows for adaptability. Feedback manipulation of circadian genes can lead changes in timing, quality, and quantity of sleep. The firing frequency of the SCN is 6–10 Hz during the day and <1 Hz at night. The first circadian gene identified was called Clock (circadian locomotor output cycles kaput), which codes for a transcription factor CLOCK which plays a role in the persistence and period of circadian rhythms [19]. Other major circadian genes are Per, Cry, and Bmal1. While CLOCK is constitutively expressed with minimal circadian variability in protein levels, PER, CRY, and BMAL1 all have a diurnal pattern of expression [19].

There are multiple levels of feedback loops within this system (Fig. 3.4). The PER and CRY proteins form a complex that indirectly provide inhibition of the activation of the Per and Cry genes. Conversely, the complex formed by the activators BMAL1 and CLOCK enhances expression of Per and Cry. Binding of the PER-CRY complex to the BMAL1-CLOCK complex prevents the expression of Per and Cry. Likewise, Bmal1 expression undergoes feedback inhibition by its product, BMAL1, via the product of the Rev-Erba gene. When the PER-CRY complex binds to BMAL1-CLOCK, it reduces transcription of Rev-Erba, thus indirectly enhances Bmal1 expression. The net effect is that PER and CRY are expressed more in the daytime, and the formation of their complex inhibits activity of CLOCK-BMAL1. Then in the circadian night, PER and CRY are degraded, leading to a repeat of the

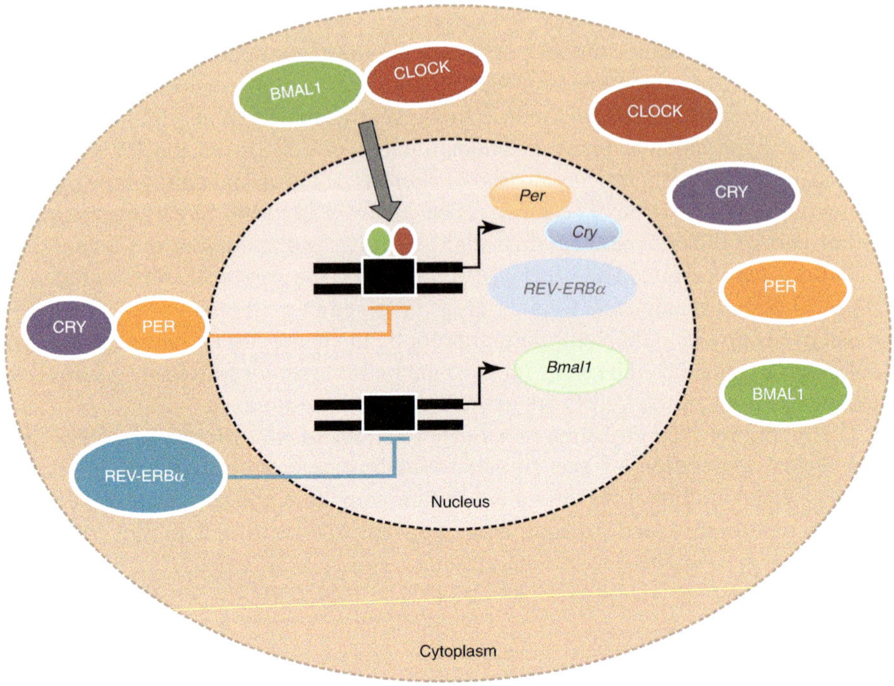

Fig. 3.4 Model of the circadian clock in mammals. CLOCK and BMAL1 act as master transcription factors to regulate (1) the Per and Cry genes in the core feedback loop of the clock, (2) the REV-ERB/ROR feedback loop regulating Bmal1 transcription, and (3) thousands of target genes that are clock outputs. The stability of the PER and CRY proteins is tightly regulated by E3 ubiquitin ligases in both the cytoplasm and nucleus that determine circadian period. (From Krakowiak and Durrington [47])

cycle with next turn of the circadian clock [20]. There are additional molecular cascades downstream that further sculpt and refine circadian output.

Mutations or manipulation of circadian genes can lead changes in timing, quality, and quantity of sleep; for example, mutations of Clock can alter circadian period, and mutations of Bmal1 can abolish circadian rhythmicity.

Circadian rhythms are not isolated to the control of sleep. It is estimated that 20–40% of genes have some circadian periodicity to them, although most oscillate in only one organ. While about 3% of genes in the hypothalamus have a circadian pattern of expression, other organs, including the liver, kidney, and lung, have more than 10% of genes expressed in a circadian pattern [21]. Circadian regulation of gene expression is not limited to the transcriptional processes but includes every regulatory step in gene expression, including splicing, transcriptional termination, polyadenylation, nuclear/cytoplasmic transport, miRNA regulation, translation, protein phosphorylation, and RNA degradation [22, 23]. A substantial fraction of genes in any cell or tissue undergo circadian oscillations at the mRNA level. In a recent publication on a diurnal transcriptome atlas of a primate across major neural and peripheral tissues, 82% of genes were found rhythmic in at least one tissue. Hence, regulation of circadian gene expression includes both transcriptional and posttranscriptional mechanisms, which in turn can be modulated by various factors (including metabolic ones) interacting with these processes [24]. Misalignment of circadian patterns across organ systems can contribute to human disease and disability. For example, the role of circadian rhythms in metabolism and metabolic disorders has been extensively studied. This is reflected in a known connection between shift work and metabolic diseases. One related finding is that 10 days of circadian asynchrony can lead to insulin resistance and increased fat deposition in humans.

Circadian Entrainment

Circadian rhythms are a part of organisms. In fact they exist naturally in the absence of any external stimuli but can be entrained to adapt to changing photoperiods or changes in social, ecologic, or behavioral conditions. Phase control of the circadian clock is achieved by "zeitgebers." Zeitgebers (literally "time givers" in German) are defined as environmental cues that synchronize or reset the circadian clock. Light is the most powerful zeitgeber. Others include food, sound, temperature, and even stress or social interactions. Entrainment of circadian rhythms starts occurring in the first few months of life as infants. The SCN is cycling at birth, but entrainment to the external environment occurs in the first few weeks of life. By 3 weeks there is a difference between day and nighttime sleep durations and associated diurnal changes in body temperature [25].

The physiologic process of entrainment to light cues begins when melanopsin-containing photosensitive ganglion cells on the retina receive light signals and transmit them via the RHT to the SCN where activation of nerve endings releases glutamate. Glutamate acts on N-methyl-D-aspartic acid (NMDA) and non-NMDA receptors to increase Per gene expression. Experimental exposure of the SCN to glutamate mimics the effect of light. Light activation of the SCN produces an influx of calcium and activation of Per genes.

The RHT also projects to the anterolateral hypothalamus, subparaventricular zone, supraoptic region, and intergeniculate leaflet of the thalamus. RHT also releases substance P and pituitary adenylate cyclase-activating peptide (PACAP), both of which play a role in circadian rhythm shifts. Damage to rods and cones in mice does not affect the circadian cycle, but removal of the melanopsin-containing retinal ganglion cells abolishes circadian entrainment.

Destruction of the SCN abolishes rhythmicity, resulting in an "irregular sleep-wake rhythm disorder." Patients with neurodegenerative disorders like Alzheimer's disease are at increased risk. The author reported a unique case of a

38-year-old woman with a gunshot wound to the head that severed both optic nerves (Fig. 3.2). The woman exhibited an irregular pattern of sleep and wakefulness with an estimated average total sleep time of 8.8 hours per 24-hour period. Polysomnography hypnogram corresponding to day 1 of actigraphy demonstrated fragmented sleep-wakefulness pattern that was not completely identified by actigraphy alone.

Circadian Coupling Disorders

Circadian dysregulation occurs when there is an abnormal phase in the sleep-wake cycle. The *International Classification of Sleep Disorders Third Edition* [26] criteria for circadian rhythm sleep-wake disorder include a disruption in the sleep-wake period either due to a change in the endogenous circadian timing or due to a misalignment between the circadian clock and the socially desired sleep-wake cycle. This misalignment in circadian timing must cause sleep-related symptoms such as insomnia, excessive sleepiness, or both. Circadian rhythm disorders are divided into advanced sleep-wake-phase disorder (ASWPD), delayed sleep-wake-phase disorder (DSWPD), irregular sleep-wake disorder, non-24-hour sleep-wake disorder, shift work disorder, jet lag disorder, and circadian sleep-wake disorder not otherwise specified [26].

ASWPD can be seen when sleep occurs earlier than desired sleep times as seen with familial advanced sleep-phase syndrome (FASPS). This is also seen with advanced age.

In DSWPD sleep periods occur later than desired as seen with the delayed sleep-phase syndrome (DSPS). When the sleep-wake cycle is quasiperiodic, following a nearly 24-hour cycle, but free running around the clock with lack of entrainment to light-dark cycles, this can be seen in non-24-hour sleep-wake syndrome, as seen in those with total blindness, with inability for light signal to reach the SCN (not seen in cortical blindness where the retinal ganglion pathway to the SCN remains intact).

Markers of Process C

Dim Light Melatonin Onset

Melatonin secretion from the pineal gland peaks in the evening, signaling readiness for sleep onset. Melatonin induces vasodilation and is associated with the subjective feeling of drowsiness.

Body Temperature

Body temperature declines, and core body temperature (CBT) reaches its nadir during night sleep. CBT is the temperature of the brain and internal organs and is subject to circadian oscillations [27, 28]. The explanation for the relationship between the thermoregulatory systems and sleep regulatory system is likely that of sleep's purpose of energy conservation. The overall regulation of CBT is affected by both the circadian and homeostatic processes. A rostral projection from the SCN to the preoptic anterior hypothalamus conveys the circadian signal to the thermoregulatory system as melatonin endogenously downregulates CBT as it rises in the evening [29, 30]. Heat distribution from the core to the periphery is completed during the third sleep cycle resulting in the nadir CBT. Of note, this thermoregulatory system is not active during REM sleep, and body temperature during REM sleep actually rises due to increased metabolic activity.

Body temperature, like melatonin, is also controlled by environmental cues, one of the major cues being body position. In the supine body position, blood is redistributed together with heat, from the core to the periphery resulting in increased skin temperature and decreased CBT, thus increasing sleepiness (e).

Cortisol

Sleep onset is associated with short-term inhibition of cortisol secretion, particularly during slow wave sleep. Cortisol then rises dramatically before the waking period and is regulated by the circadian process. SCN neurons project both directly and indirectly to the dorsomedial hypothalamic nucleus (DMH), which then sends efferent neurons to the PVH. There are corticotropin-releasing hormone-containing neurons in the PVH that project to the median eminence where corticotropin-releasing hormone is released to circulation activating the release of cortisol from the anterior pituitary. The adrenocorticotropic hormone in the blood then results in rhythmic induction of corticosteroid secretion from the adrenal gland [31].

Homeostatic Control of Sleep

Process S refers to sleep homeostasis, wherein accumulated time spent awake leads to increasing sleep pressure. The pressure of this sleep debt then decreases during sleep. Under typical diurnal human conditions, this sleep pressure rises and falls entrained to day and night. When S approaches a lower boundary, it triggers wake. As it approaches an upper boundary, sleep is triggered.

Anatomy

Adenosine is the major neuromodulator felt to be involved in the homeostatic process of sleep. Adenosine inhibits the central nervous system. In cats, the basal forebrain and, to a lesser extent, the neocortex, adenosine increases during sustained waking [32]. Brain adenosine rises when ATP production is reduced. One theory is that increased adenosine is a signal of reduced brain energy reserves that develop during waking and that sleep is induced as an energy-restorative state [33]. Another supporting discovery implicating adenosine's role in sleep homeostasis is that caffeine, a potent alerting agent, is an adenosine antagonist.

Physiology

In 2008, Landolt postulated the role of adenosine in the homeostatic control of sleep. Adenosine has been shown to promote slow wave sleep and rapid eye movement (REM) sleep in animal models. Adenosine accumulates in the extracellular space during wakefulness and acts to inhibit wake-promoting neurons primarily in the basal forebrain via A1 receptors. During sleep, extracellular adenosine levels decline.

Accumulation of adenosine in the basal forebrain has correlated with both increased SWA and sleep duration. Its accumulation therefore corresponds to the degree of sleep pressure [34].

Interestingly, caffeine blocks adenosine receptors, which is proposed as a primary mechanism of its alerting action.

Markers

Markers which correspond to this sleep pressure include non-rapid eye movement (NREM) sleep electroencephalography (EEG) slow wave activity (SWA), which increases under conditions of increasing sleep pressure. Theta activity in waking is a marker of the rising limb of S as one moves toward wakefulness [35].

Process S is impacted by lack of sleep regardless of time of day. For example, NREM sleep SWA during daytime naps increase with duration of prior wakefulness. Following naps, the SWA in that night's sleep period decreases predictably. Suppression of SWA without disturbing sleep has been shown to result in a rebound increase in SWA in the next sleep period. Studies have also shown that increased mental or physical work results in increase of SWA and shorter sleep onset latency, which is also considered a marker of sleep pressure.

Ultradian Rhythm

As we have seen in previous chapters, sleep is not a homogeneous state, but it is divided into different stages with various electroencephalographic and physiologic characteristics. In general terms sleep is divided into non-rapid eye movement (NREM) sleep stage and rapid eye movement (REM) sleep. NREM is further divided into three stages: N1,N2, and N3. As humans fall asleep, they usually enter sleep into N1 and spend few minutes in this stage, as N2 ensures forming a more consolidated sleep, after which N3 appears. After approximately an hour of NREM sleep, the brain usually enters REM sleep. This cycle repeats itself four to five times in a night. REM sleep deprivation studies have also demonstrated that REM sleep is regulated by homeostatic control; however, the mechanism of NREM-REM cycling during sleep remains unknown [36].

Circadian Draining of Waste During Sleep

Clearance of metabolic waste from the brain is essential for CNS homeostasis. The circadian control of this process is indicated by the occurrence of a daily rhythm in the permeability of the blood-cerebrospinal fluid (CSF) barrier, as shown by measuring relative metabolite levels between CSF and blood. Indeed, the choroid plexus, an important component of the blood-CSF barrier, displays circadian clock activity [37], regulating timed fluid exchange between the blood and CSF. In the fruit fly, a circadian clock in glial cells of the hemolymph-brain barrier regulating xenobiotic efflux was demonstrated [38] giving rise to the hypothesis that glial cells including astrocytes regulate blood-brain barrier function and modulate metabolic waste clearance in mammals.

The conventional view of solute movements in the extracellular space (ECS) of the CNS has been based on diffusion, and the mechanisms modulating the variations in the abundance of products to be eliminated have been sought in the coupling of blood flow and metabolism, processes which are influenced by regulators such as biogenic amines, adenosine, NO, H+, and K+. In the last years, a new interpretation has been put forward, i.e., the glymphatic hypothesis [39–41]. The term has been coined to describe the active, lymphatic-like movements in the ECS which occur in the brain. It has been suggested that this process, which is normally dependent on the activity of lymphatic vessels, can nevertheless take place in the brain, which is devoid of such vessels, because of water exchange. More specifically, it has been hypothesized that solute movement is driven by perivascular astrocytes, which are strongly enriched in aquaporin-4 (AQP4), and by changes in the vascular lumen. AQP4 is predominantly expressed in the end feet of these astrocytes. Water release via AQP4 may be responsible for an actively driven fluid exchange between para-arterial and paravenous spaces, connected by a convective flow through the other parts of the ECS, especially concerning the interstitial fluid. Additionally, it has been assumed that arteriolar pulsations as well as respiration-dependent venular collapse and reinflation may further enhance the flow through occurring in the ECS [42].

Sleep-related variations in cerebral waste products were described [38–40]. The elimination of amyloid β peptide (Aβ) has been reported to be considerably enhanced during sleep [43]. The concept of glymphatic Aβ clearance has received support from the observation that elimination of injected radiolabeled Aβ peptide was strongly reduced in AQP4−/− mice [39].

The exchange of solutes between the CSF and the interstitial fluid occurs mostly during NREM sleep when the cortical interstitial space increases by more than 60% and provides a low resistance path for the movement of CSF and interstitial fluid in the brain parenchyma. The contribution of the glymphatic system to the repair function of sleep has been discussed in relation to age-associated deviations from the normal sleep profile, including decreased and delayed CSF penetration of fluid along perivascular pathways and pial surface [44]. Various neurological disease states such as stroke, traumatic brain injury, and AD have been interpreted in terms of the contribution by glymphatic dysfunction [42]. Altogether, the data indicate that the circadian clock regulates a variety of crucial processes in the brain, including sleep, brain metabolism, and maintenance of flow balance of compounds into and out of the brain. Disruption of the circadian clock, and/or any of the signaling pathways regulated by it, can lead to inefficient anabolic and catabolic biochemical processes [45].

Conclusion

In this chapter we have discussed the processes that control and regulate sleep. Far from a quiet state, sleep is orchestrated by circadian, ultradian, and homeostatic regulation. We discussed the role of the suprachiasmatic nucleus as the pacemaker of the circadian rhythm and the entrainment with light and melatonin. Circadian sleep disorders occur when there is a misalignment between the internal pacemaker and the socially established sleep time, causing significant nighttime or daytime dysfunction. Prolonged awakening increases sleep pressure during homeostatic control. REM and NREM occur during cycles in sleep, and finally, clearance of metabolic waste from the brain is essential to keep a healthy homeostatic process.

References

1. Borbely AA. A two process model of sleep regulation. Hum Neurobiol. 1982;1(3):195–204.
2. Ibuka N, Kawamura H. Loss of circadian rhythm in sleep-wakefulness cycle in the rat by suprachiasmatic nucleus lesions. Brain Res. 1975;96(1):76–81.
3. Knowles JB, Coulter M, Wahnon S, Reitz W, MacLean AW. Variation in process S: effects on sleep continuity and architecture. Sleep. 1990;13(2):97–107.
4. Lewy AJ, Sack RL. The dim light melatonin onset as a marker for circadian phase position. Chronobiol Int. 1989;6(1):93–102.
5. Finelli LA, Baumann H, Borbely AA, Achermann P. Dual electroencephalogram markers of human sleep homeostasis: correlation between theta activity in waking and slow-wave activity in sleep. Neuroscience. 2000;101(3):523–9.
6. Moore RY, Eichler VB. Loss of a circadian adrenal corticosterone rhythm following suprachiasmatic lesions in the rat. Brain Res. 1972;42(1):201–6.
7. Ibuka N, Nihonmatsu I, Sekiguchi S. Sleep-wakefulness rhythms in mice after suprachiasmatic nucleus lesions. Waking Sleeping. 1980;4(2):167–73.
8. Stephan FK, Zucker I. Circadian rhythms in drinking behavior and locomotor activity of rats are eliminated by hypothalamic lesions. Proc Natl Acad Sci U S A. 1972;69(6):1583–6.
9. DelRosso LM, Hoque R, James S, Gonzalez-Toledo E, Chesson AL Jr. Sleep-wake pattern following gunshot suprachiasmatic damage. J Clin Sleep Med. 2014;10(4):443–5.
10. Antle MC, Silver R. Orchestrating time: arrangements of the brain circadian clock. Trends Neurosci. 2005;28(3):145–51.
11. Morin LP, Allen CN. The circadian visual system, 2005. Brain Res Rev. 2006;51(1):1–60.

12. Moore RY. The suprachiasmatic nucleus and the circadian timing system. Prog Mol Biol Transl Sci. 2013;119:1–28.
13. An S, Harang R, Meeker K, Granados-Fuentes D, Tsai CA, Mazuski C, et al. A neuropeptide speeds circadian entrainment by reducing intercellular synchrony. Proc Natl Acad Sci U S A. 2013;110(46):E4355–61.
14. Harmar AJ, Marston HM, Shen S, Spratt C, West KM, Sheward WJ, et al. The VPAC (2) receptor is essential for circadian function in the mouse suprachiasmatic nuclei. Cell. 2002;109(4):497–508.
15. Colwell CS, Michel S, Itri J, Rodriguez W, Tam J, Lelievre V, et al. Disrupted circadian rhythms in VIP- and PHI-deficient mice. Am J Physiol Regul Integr Comp Physiol. 2003;285(5):R939–49.
16. Rosenwasser AM, Turek FW. Neurobiology of circadian rhythm regulation. Sleep Med Clin. 2015;10(4):403–12.
17. Saderi N, Cazarez-Marquez F, Buijs FN, Salgado-Delgado RC, Guzman-Ruiz MA, del Carmen BM, et al. The NPY intergeniculate leaflet projections to the suprachiasmatic nucleus transmit metabolic conditions. Neuroscience. 2013;246:291–300.
18. Morin LP. Neuroanatomy of the extended circadian rhythm system. Exp Neurol. 2013;243:4–20.
19. Vitaterna MH, King DP, Chang AM, Kornhauser JM, Lowrey PL, McDonald JD, et al. Mutagenesis and mapping of a mouse gene, clock, essential for circadian behavior. Science. 1994;264(5159):719–25.
20. Hastings MH, Maywood ES, Brancaccio M. Generation of circadian rhythms in the suprachiasmatic nucleus. Nat Rev Neurosci. 2018;19(8):453–69.
21. Zhang R, Lahens NF, Ballance HI, Hughes ME, Hogenesch JB. A circadian gene expression atlas in mammals: implications for biology and medicine. Proc Natl Acad Sci U S A. 2014;111(45):16219–24.
22. Takahashi JS. Transcriptional architecture of the mammalian circadian clock. Nat Rev Genet. 2017;18(3):164–79.
23. Mure LS, Le HD, Benegiamo G, Chang MW, Rios L, Jillani N, et al. Diurnal transcriptome atlas of a primate across major neural and peripheral tissues. Science. 2018;359(6381) https://doi.org/10.1126/science.aao0318.
24. Mermet J, Yeung J, Naef F. Systems chronobiology: global analysis of gene regulation in a 24-hour periodic world. Cold Spring Harb Perspect Biol. 2017;9(3):a028720.
25. Hellbruegge T, Lange JE, Stehr K, Rutenfranz J. Circadian periodicity of physiological functions in different stages of infancy and childhood. Ann N Y Acad Sci. 1964;117:361–73.
26. American Academy of Sleep Medicine. International classification of sleep disorders 3rd ed. Darien: American Academy of Sleep Medicine; 2014.
27. Krauchi K. The thermophysiological cascade leading to sleep initiation in relation to phase of entrainment. Sleep Med Rev. 2007;11(6):439–51.
28. Tayefeh F, Plattner O, Sessler DI, Ikeda T, Marder D. Circadian changes in the sweating-to-vasoconstriction interthreshold range. Pflugers Arch. 1998;435(3):402–6.
29. Moore RY, Danchenko RL. Paraventricular-subparaventricular hypothalamic lesions selectively affect circadian function. Chronobiol Int. 2002;19(2):345–60.
30. Krauchi K, Cajochen C, Pache M, Flammer J, Wirz-Justice A. Thermoregulatory effects of melatonin in relation to sleepiness. Chronobiol Int. 2006;23(1–2):475–84.
31. Chou TC, Scammell TE, Gooley JJ, Gaus SE, Saper CB, Lu J. Critical role of dorsomedial hypothalamic nucleus in a wide range of behavioral circadian rhythms. J Neurosci. 2003;23(33):10691–702.
32. Thakkar MM, Winston S, McCarley RW. A1 receptor and adenosinergic homeostatic regulation of sleep-wakefulness: effects of antisense to the A1 receptor in the cholinergic basal forebrain. J Neurosci. 2003;23(10):4278–87.
33. Benington JH, Heller HC. Restoration of brain energy metabolism as the function of sleep. Prog Neurobiol. 1995;45(4):347–60.
34. Mabuchi H, Nakahashi H. A major inhibitor of phenytoin binding to serum protein in uremia. Nephron. 1988;48(4):310–4.

35. Borbely AA, Daan S, Wirz-Justice A, Deboer T. The two-process model of sleep regulation: a reappraisal. J Sleep Res. 2016;25(2):131–43.
36. Weber F. Modeling the mammalian sleep cycle. Curr Opin Neurobiol. 2017;46:68–75.
37. Myung J, Schmal C, Hong S, Tsukizawa Y, Rose P, Zhang Y, et al. The choroid plexus is an important circadian clock component. Nat Commun. 2018;9(1):1062.
38. Zhang SL, Yue Z, Arnold DM, Artiushin G, Sehgal A. A circadian clock in the blood-brain barrier regulates xenobiotic efflux. Cell. 2018;173(1):130–9. e10
39. Iliff JJ, Wang M, Liao Y, Plogg BA, Peng W, Gundersen GA, et al. A paravascular pathway facilitates CSF flow through the brain parenchyma and the clearance of interstitial solutes, including amyloid beta. Sci Transl Med. 2012;4(147):147ra11.
40. Iliff JJ, Lee H, Yu M, Feng T, Logan J, Nedergaard M, et al. Brain-wide pathway for waste clearance captured by contrast-enhanced MRI. J Clin Invest. 2013;123(3):1299–309.
41. Jessen NA, Munk AS, Lundgaard I, Nedergaard M. The Glymphatic system: a Beginner's guide. Neurochem Res. 2015;40(12):2583–99.
42. Nedergaard M. Neuroscience. Garbage truck of the brain. Science. 2013;340(6140):1529–30.
43. Boespflug EL, Iliff JJ. The emerging relationship between interstitial fluid-cerebrospinal fluid exchange, amyloid-beta, and sleep. Biol Psychiatry. 2018;83(4):328–36.
44. Zeppenfeld DM, Simon M, Haswell JD, D'Abreo D, Murchison C, Quinn JF, et al. Association of Perivascular Localization of Aquaporin-4 with cognition and Alzheimer disease in aging brains. JAMA Neurol. 2017;74(1):91–9.
45. Albrecht U, Ripperger JA. Circadian clocks and sleep: impact of rhythmic metabolism and waste clearance on the brain. Trends Neurosci. 2018;41(10):677–88.
46. Honma S. The mammalian circadian system: a hierarchical multi-oscillator structure for generating circadian rhythm. J Physiol Sci. 2018;68(3):212.
47. Krakowiak K, Durrington H. The role of the body clock in asthma and COPD: implication for treatment. Palm Ther. 2018;4:29–43. Received May 1, 2018. Published online June 21, 2018. https://doi.org/10.1007/s41030-018-0058-6.

The Neurological Consequences of Sleep Deprivation

4

Lynn Keenan and Karl Van Gundy

Introduction

Intuitively, we all know that sleep is important to our health through our personal experience of a night of diminished sleep, whether imposed by work or pleasure, family obligations, or sickness. Sleep deprivation has been blamed for several disasters affecting public safety and the environment. Closer to our day to day activities is the sleep loss of those in medical training, with increased risk for motor vehicle accidents, percutaneous sticks, and medical errors and decreased ability to learn.

In this chapter we will review the current work on the effects of sleep deprivation, or the extension of wakefulness beyond the typical 16–18 hours, and sleep restriction – not getting enough sleep per 24 hours for multiple nights.

Incidence of Sleep Deprivation

The American Academy of Sleep Medicine and the Sleep Research Society have recommended adults to sleep more than 7 hours a night for optimal health. Sleeping less than 7 hours on a regular basis has been associated with higher mortality, with more weight gain, depression, diabetes, hypertension, heart disease, and stroke [1]. In a review of the 2005–2008 data of the National Health and Nutrition Examination

L. Keenan (✉)
Department of Internal Medicine, UCSF Fresno, Fresno, CA, USA
e-mail: lkeenan@fresno.ucsf.edu

K. Van Gundy
Departments of Pulmonary, Critical Care, and Sleep Medicine, UCSF Fresno, Fresno, CA, USA

Division of Internal Medicine, Community Regional Medical Center, Fresno, CA, USA
e-mail: kvangundy@fresno.ucsf.edu

L. M. DelRosso, R. Ferri (eds.), *Sleep Neurology*,
https://doi.org/10.1007/978-3-030-54359-4_4

Survey (NHANES), over one third of American adults report chronically getting less than 7 hours of sleep, with the greatest percentage in the age group 40–59 at 40.3% [2]. According to the National Sleep Foundation poll, 15% of subjects sleep less than 6 hours per night on weekdays. Children 6–12 are recommended to sleep 9–12 hours a night, and teens from 13 to 18 years old are recommended to sleep 8–10 hours, yet a review of youth risk behaviors shows that 57.8% of middle school students and 72.7% of high school students are sleep restricted [3]. Although acute sleep deprivation may happen to most people at some point in life, shift workers generally have less sleep overall and more incidences of acute sleep deprivation. The Bureau of Labor Statistics estimates 15 million Americans do shift work.

Medical conditions, sleep disorders, extended work hours, shift work, family care needs, and lifestyle choices can all impose restriction to sleep opportunities. Although there appears to be some trend in recent years to improving sleep duration in America, the impact of sleep restriction on cognitive function, work performance, accidents, and relationships can be considerable [4].

Effects of Acute Sleep Deprivation

Social consequences. Sleep deprivation has been blamed for some of the most glowing examples of catastrophes, such as the Exon Valdez or Three Mile Island accidents. Closer to home is the risk of impaired driving related to lack of sleep. In a group of males, ages 19–35, performance was measured in a driving simulator after alcohol use or sleep deprivation. Performance after 19 or 22 hours of wakefulness was comparable to blood alcohol levels of 0.05% and 0.08%, respectively [5]. In a survey of house staff in 1996, 49% reported falling asleep driving, mostly after call. Multiple studies have also been done on the adverse effects of prolonged shifts on doctors in training. The 2004 study by Landrigan and Lockley, comparing intern schedules including 30-hour shifts vs schedules with 8–16-hour shifts, found the prolonged shift schedule had 35.9% more serious medical errors, with 20.8% increase in serious medication errors and 5.6 times as many diagnostic errors [6].

Individual consequences. On the multiple sleep latency test (MSLT), which tests the propensity to fall asleep in a dark, quiet room over a series of four or five naps, the average healthy adult will fall asleep in less than 2 minutes after a night without sleep, and the time from sleep onset to slow wave sleep is halved [7]. Working memory capacity (WMC), or the ability to work with information, attention, and concentration, is adversely affected by sleep deprivation. In a study evaluating internal medicine residents on a rotation with call every fourth night compared to a non-call rotation, a standardized WMC test showed a decrement in speed, accuracy, and math performance with the prolonged shifts. Some performances did not fully recover until the 4th day after call [8]. If tasks are not time dependent, cognitive slowing occurs, but if time pressured, accuracy declines [9].

Neurobiology of Sleep Deprivation

With sleep deprivation, there is a breakdown of the brain's ability to recognize outside cues and stimulation, and the quality of information stored in memory is degraded [10]. Sleep deprivation has shown effects similar to aging. Comparing the fMRI of a group of young subjects, rested and sleep deprived, to an older group, brain networks' functional connectivity became impaired in the young with sleep deprivation to be comparable to the older brain connectivity [11]. Pasula looked at how sleep deprivation affected young (19–38) and old (59–82) in three aspects of working memory – encoding new information, displacement of old information by new information, and episodic recall of stored memory. In a rested condition, young have better encoding and displacement, but following sleep deprivation of 32 hours, younger adult performance declined to a point similar to that observed in normally rested older adults in each parameter. Although both groups declined with sleep deprivation, the young had the greatest decline [12].

Divergent thinking is affected by sleep loss, which includes being able to assimilate information as it changes, updating strategies based on the new information, being able to come up with innovative solutions, assessing risk, maintaining interest in outcomes, and communication skills [13]. Those sleep deprived tend to perseverate on ineffective solutions rather than being able to come up with a new strategy. It also takes more effort to behave appropriately. Although tasks may begin well, performance can deteriorate with greater time on task. Situational awareness can decline with growing neglect of information viewed as nonessential [9, 14, 15]. Cognitive performance errors can involve both errors of omission, such as not responding when appropriate, and errors of commission, such as responding when no stimulus was present [15].

Acute sleep deprivation can also affect the emotional tone of memories. After a night of sleep deprivation, recall of encoded memories is associated with a more negative affective tone [16]. False memories also increase with sleep deprivation [17]. An area that can particularly affect relationships is the inability to recognize human facial emotions accurately when sleep deprived, an effect that was most significant in women [18]. In one study using fMRI to evaluate brain responses to emotional recall, there was a larger hippocampal and medial prefrontal cortex response in the rested group compared to sleep deprived, but in the sleep deprived, there was a stronger response to negative emotions in the amygdala compared to the rested group [19]. Hence negative emotions are preferentially remembered in sleep-deprived states. Judgment of reward was also skewed in sleep, with higher expectation of a win and attenuated response to gambling losses [20].

Sleep deprivation can affect pain perception as well. Studies in rats have shown pain hypersensitivity and diminished effect of opioids after sleep deprivation [21]. This would suggest that our response to medications is dependent not only on the dose, absorption, and physiological effect but also on the amount of sleep (Table 4.1).

Table 4.1 Effects of sleep deprivation on cognitive function

Effects of sleep deprivation on cognitive function
Attention lapses with delayed or inappropriate responses
Errors of commission – over responding to wrong stimulus
Slowed cognitive throughput delays task completion
Time-pressured tasks have increased cognitive errors
Reaction time slows
Performance deteriorates as task duration increases
Reduced learning of new cognitive tasks
Decreased ability to change approach if solution is ineffective
Mood appropriate behavior is more difficult
Decreased assimilation of changing information
More difficult to update strategies based on new information
Less innovation
Impaired ability for accurate risk assessment
More difficulty maintaining interest in outcomes
Less ability to read human expressions accurately
Poor insight into performance deficits

Effects of Chronic Sleep Restriction

With over a third of Americans chronically sleep restricted, it is a much more prevalent problem than acute sleep deprivation, though may be less studied. One of the largest controlled studies of sleep restriction was performed by Van Dongen and involved 48 healthy adults randomized to 2 weeks of sleep limited to 4, 6, or 8 hours of sleep per night and compared to no sleep for three nights [22]. Subjects had neurobehavioral testing every 2 hours, including tests of behavioral alertness with the psychomotor vigilance task (PVT), measures of working memory with a digit symbol substitution task, and a serial addition/subtraction task to measure cognitive throughput. Subjects were also evaluated with the Stanford Sleepiness Scale for self-ratings of subjective sleepiness. In addition, they had polysomnography 10 of the 14 nights. Those subjects in the 4 and 6 hour per night sleep category showed significant cumulative performance deficits relative to those in the 8-hour sleep condition. The 4-hour and 6-hour groups showed progressive decline in reaction speed, alertness, and cognitive accuracy compared to the 8-hour group. The 4-hour group reached levels of impairment in cognitive throughput performance (serial addition/subtraction) equal to one night of complete sleep deprivation, and tests in alertness and working memory declined to levels of two nights of complete sleep deprivation. The 6-hour group reached levels of one night of sleep deprivation in working memory and behavioral alertness. The rate of decline in performance was near linear over the days of sleep restriction. Of particular interest was the change in subjective sleepiness. In the 3 days without sleep group, the subjective sleepiness worsened in a near linear fashion, but in the 4- and 6-hour groups, subjective sleepiness worsened over the initial few days and then became fairly steady, with no significant difference between the two groups. At the end of the 2 weeks, when objective performance was at its worst, the subjects reported feeling only slightly sleepy. This lack of subjective sleepiness despite objective impairment may be the reason many feel they can adapt to sleep restriction with few consequences.

In Van Dongen's study, stage N1, N2, and REM sleep all showed reductions varying with time allowed to sleep, but there was little variation in amounts of slow wave sleep among the groups. Other studies have shown that with sleep restriction to 4 hours, the total amount of slow wave sleep is conserved by visual scoring, but slow wave activity derived from power spectral analysis shows dynamic increases in sleep restricted for more than 1 day to 4 hours, suggesting slow waves are essential and protected in limited sleep [23, 24].

Another laboratory-based study of sleep restriction was performed in truck drivers allowed 3, 5, 7, and 9 hours in bed for 1 week. Performance on the psychomotor vigilance test (PVT) showed a progressive worsening of reaction time in the 3-hour group along with increased lapses. The 5- and 7-hour group showed initial decline in reaction speed, which then stabilized by the end of the week. The 9-hour group performance remained stable. A decreased mean sleep latency, in a dose-response effect, was found in the 3- or 5-hour group, but not the 7–9-hour group [25].

Sleep restriction has also been correlated with oculomotor responses, with eyelid closure and slow rolling eye movements, which are part of the initial transition from wake to stage 1 sleep. Slower eye movements have been shown to correlate with lapses on the PVT testing as well as simulated driving [26]. Driving simulation tests as well as epidemiologic studies have shown increased driving accidents as well as near misses with sleep restriction to less than 7 hours a night [24].

Surveys of medical residents have indicated between 41% and 57% reporting fatigue as the cause of their most serious errors, and those reporting 5 or fewer hours of sleep per night were twice as likely to be named in a malpractice suit [6, 27].

Finally, mortality studies have shown worsened mortality for people getting chronically less than 6.5 hours of sleep [28].

Pathophysiology of Sleep Loss

The cognitive and behavioral changes due to sleep loss are felt to be related to microsleeps from "sleep-wake instability" which reflect shifts in the balance between the maintenance of alertness and the sleep-initiating systems [15, 24]. The microsleeps may be half a second to several seconds and may not be a readily observed change in behavior but can be detected by testing, such as the PVT [29]. The wake-state instability theory posits two competing systems exerting influence on the performance of a sleep-deprived individual, with motivated behavior driving sustained alertness from rostral areas and the more caudal and central areas pushing the homeostatic drive to sleep [30]. The cognitive variability reflects the lapses, which can lead to errors of omission, and the times of overcompensation, with errors of commission [15]. Even between the lapses, there are effects from the burden of the sleep debt. The wake-state instability suggests multiple mechanisms interacting, and there are a growing number of molecules and pathways that may be involved. Overlaying the homeostatic drive for sleep and motivation to be awake are the circadian rhythm forces for optimal times of day for alertness and sleep. When chronically sleep deprived, the circadian influences make the neurobehavioral deficits greatest around 8 AM and relative protected time between 4:00 and 8:00 PM [31]. In addition, there are genetic factors that impact an individual's response to sleep loss [32]. Repeated exposure to sleep loss by an individual tends to give a consistent, trait-like response [31].

PET scans and functional MRI scans have also shown decreased metabolic rates in the frontal and temporal lobes, thalamus, basal ganglia, and cerebellum following sleep deprivation, but total brain metabolism remained the same [15]. After 24 hours of sleep deprivation, one night of recovery sleep only partially reversed these metabolic changes [33]. fMRI studies have also compared brain activation in rested and sleep-deprived states in those more resistant to sleep deprivation and those more vulnerable, based on testing performance in working memory tasks [34, 35]. The frontoparietal activation is reduced in sleep-deprived states but less so in those less vulnerable to sleep deprivation. The "cognitive reserve hypothesis" suggests that some may have more preexisting cognitive resources or a greater ability to recruit alternative neural resources when sleep deprived [36].

Using molecular tracers in PET studies, sleep deprivation has shown an increase in cerebral A1 adenosine receptor binding, particularly the orbitofrontal cortex [37]. Dopamine cell firing may increase as a response to the homeostatic drive in the sleep-deprived state [38].

In a recent study in sleep-deprived rats, inflammatory cytokines (TNFα, IL-1β) increased in the hippocampus and piriform cortex. IL-6 and CRP have also been shown to increase with sleep deprivation [39, 40]. Sleep deprivation has also been tied to increased reactive oxygen species (ROS) production [41, 42], whereas sleep has been associated with clearing of oxidative stress [42].

An interesting brain washing in sleep has been found with the glymphatic system. Cerebral spinal fluid circulates through the brain, exchanges with interstitial fluid, and exits along paravenous pathways, clearing interstitial waste [43, 44]. This washing increases during sleep [45]. One of the proteins removed is amyloid-β, which increases during wakefulness. Chronic sleep restriction accelerates amyloid-β burden, suggesting sleep loss may play a role in Alzheimer's disease [46]. Looking at healthy men after unrestricted sleep showed a 6% drop in amyloid-β, whereas a night of sleep deprivation negated this drop [47].

Mouse studies of sleep deprivation have shown even a few hours of sleep loss had increased phagocytic activity of the cerebral cortex, especially the presynaptic components of large synapses. This may be clearing worn aspects of heavily used synapses [48]. Sleep restriction greater than 5 days showed microglial activation and enhanced phagocytosis of synaptic elements [48]. Sleep-deprived rats have shown increased oxidative stress through glial activation, as well as altered fragile X mental retardation protein expression. Interestingly, melatonin was able to reverse these changes to control values [49]. Sleep-deprived rats have also shown increased catalase, glutathione peroxidase, and superoxide dismutase, along with memory

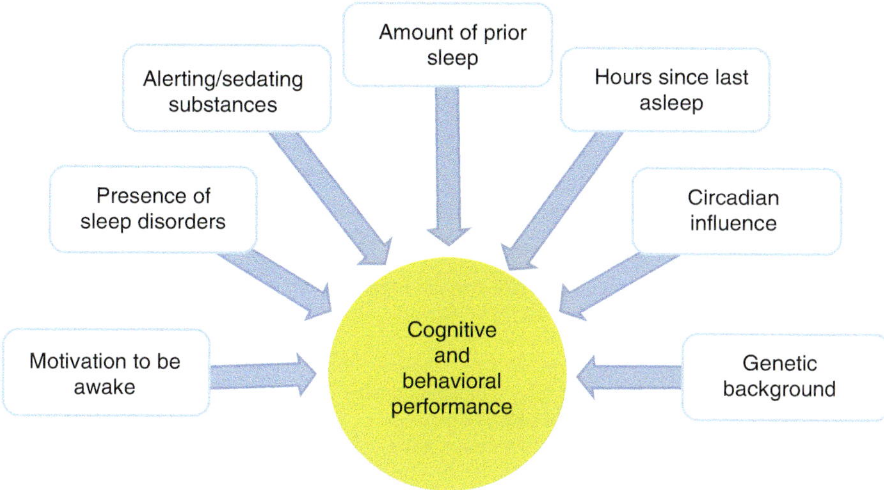

Fig. 4.1 Factors affecting cognitive and behavioral performance

deficits. Melatonin was able to prevent the memory impairment induced by the sleep deprivation, along with normalization of the antioxidant mechanisms [50].

Curcumin has also shown some protective effects against sleep deprivation. Sleep-deprived rats have shown working memory deficits and dendritic changes in the prefrontal cortex. However, rats given curcumin during the sleep deprivation had fewer errors on memory tasks and less structural changes [51]. Rats deprived of REM sleep show object recognition impairment and loss of cells in the hippocampus, both of which recovered with curcumin compared to water [52] (Fig. 4.1).

Recovery from Sleep Loss

Recovery from sleep loss is impacted by the type of sleep loss, acute or chronic, as well as the amount and number of nights of recovery. Some neurobehavioral performances recover slower than others, though the individual may not be aware of ongoing impairment.

In a study of acute sleep deprivation, subjects were awake for either one or two nights, followed by recovery sleep of either 6 or 9 hours per night for five nights. With one night of sleep loss, subjective sleepiness and response speed on PVT recovered after one night of 9-hour sleep, but mean sleep latency didn't recover until after the second night of extended sleep. Those given only 6 hours per night for recovery continued to show subjective sleepiness and reduced response times even after five nights of 6-hour sleep. With severe sleep deprivation of two nights, subjective sleepiness recovered after a night of 9 hours, but reaction time remained significantly below baseline even after five nights of recovery sleep [53].

Table 4.2 Minimizing neurological effects of sleep deprivation

Prior to anticipated sleep deprivation
Evaluate and treat for possible sleep disorders
Have prolonged sleep on one or two nights before the anticipated sleep loss
Nap prophylactically before a night shift
During prolonged shift
Use strategic napping if able
Taking breaks, especially with exercise, will reduce attentional lapses
Consider bright light at the beginning of shift and short exposures during shift to maintain alertness
Space caffeine intake – effect will last 4–6 hours for alertness
Modafinil or armodafinil can be used for shift work
Recovery sleep
Extend sleep to at least 8–10 hours for two to three nights
Consider adding melatonin, especially if sleeping during the day
Make room dark, cool, and quiet, using eye mask or ear plugs if needed
Remember reactive times and cognitive functioning may be impaired until adequate recovery time

In studying sleep restriction, Belenky looked at groups getting 3-, 5-, 7-, and 9-hour time in bed for a week, followed by three nights of 8-hour time in bed. Although subjective sleepiness and reaction time by PVT showed improvement with recovery sleep, reaction time did not return to baseline by the end of the study for the 3-, 5-, and 7-hour groups. Another study looking at sleep restricted to two third the habitual sleep for a week (average less than 5 hours) required two nights of 10-hour sleep to recover [54]. Prophylactic sleep can be beneficial, as a study giving a group 10 hours of sleep before sleep restriction showed performance declined at a slower rate and recovered quicker, even though subjective sleepiness was the same in both groups [55] (Table 4.2).

Summary

Sleep loss is highly prevalent in our society, but the effects on cognitive performance and reaction speed are often underestimated by the sleep-deprived individual. Ensuring adequate sleep prior to extended shifts and several nights of extended sleep to recover should be a priority to maintain optimal function.

References

1. Watson N. Recommended amount of sleep for a healthy adult: a joint consensus statement of the American Academy of Sleep Medicine and the Sleep Research Society. Sleep. 2015;38(6):843–4.
2. Centers for Disease Control and Prevention. Effect of short sleep duration on daily activities – United States, 2005–2008. MMWR Morb Mortal Wkly Rep. 2011;60(8):239–42.
3. Wheaton AG, Jones SE, Cooper AC, Croft JB. Short sleep duration among middle school and high school students – United States, 2015. MMWR Morb Mortal Wkly Rep. 2018;67(3):85–90.

4. Basner M, Dinges DF. Sleep duration in the United States 2003–2016: first signs of success in the fight against sleep deficiency? Sleep. 2018;41(4) https://doi.org/10.1093/sleep/zsy012.
5. Arnedt JT, Wilde GJ, Munt PW, MacLean AW. How do prolonged wakefulness and alcohol compare in the decrements they produce on a simulated driving task? Accid Anal Prev. 2001;33(3):337–44.
6. Landrigan CP, Rothschild JM, Cronin JW, Kaushal R, Burdick E, Katz JT, et al. Effect of reducing interns' work hours on serious medical errors in intensive care units. N Engl J Med. 2004;351(18):1838–48.
7. Dinges DF. Differential effects of prior wakefulness and circadian phase on nap sleep. Electroencephalogr Clin Neurophysiol. 1986;64(3):224–7.
8. Gohar A, Adams A, Gertner E, Sackett-Lundeen L, Heitz R, Engle R, et al. Working memory capacity is decreased in sleep-deprived internal medicine residents. J Clin Sleep Med. 2009;5(3):191–7.
9. Durmer JS, Dinges DF. Neurocognitive consequences of sleep deprivation. Semin Neurol. 2005;25(1):117–29.
10. Ratcliff R, Van Dongen HPA. The effects of sleep deprivation on item and associative recognition memory. J Exp Psychol Learn Mem Cogn. 2018;44(2):193–208.
11. Zhou X, Wu T, Yu J, Lei X. Sleep deprivation makes the young brain resemble the elderly brain: a large-scale brain networks study. Brain Connect. 2017;7(1):58–68.
12. Pasula EY, Brown GG, McKenna BS, Mellor A, Turner T, Anderson C, et al. Effects of sleep deprivation on component processes of working memory in younger and older adults. Sleep. 2018;41(3) https://doi.org/10.1093/sleep/zsx213.
13. Harrison Y, Horne JA. The impact of sleep deprivation on decision making: a review. J Exp Psychol Appl. 2000;6(3):236–49.
14. Cai DJ, Shuman T, Harrison EM, Sage JR, Anagnostaras SG. Sleep deprivation and Pavlovian fear conditioning. Learn Mem. 2009;16(10):595–9.
15. Goel N, Rao H, Durmer JS, Dinges DF. Neurocognitive consequences of sleep deprivation. Semin Neurol. 2009;29(4):320–39.
16. Tempesta D, De Gennaro L, Natale V, Ferrara M. Emotional memory processing is influenced by sleep quality. Sleep Med. 2015;16(7):862–70.
17. Chatburn A, Kohler MJ, Payne JD, Drummond SPA. The effects of sleep restriction and sleep deprivation in producing false memories. Neurobiol Learn Mem. 2017;137:107–13.
18. van der Helm E, Gujar N, Walker MP. Sleep deprivation impairs the accurate recognition of human emotions. Sleep. 2010;33(3):335–42.
19. Sterpenich V, Albouy G, Boly M, Vandewalle G, Darsaud A, Balteau E, et al. Sleep-related hippocampo-cortical interplay during emotional memory recollection. PLoS Biol. 2007;5(11):e282.
20. Venkatraman V, Chuah YM, Huettel SA, Chee MW. Sleep deprivation elevates expectation of gains and attenuates response to losses following risky decisions. Sleep. 2007;30(5):603–9.
21. Nascimento DC, Andersen ML, Hipolide DC, Nobrega JN, Tufik S. Pain hypersensitivity induced by paradoxical sleep deprivation is not due to altered binding to brain mu-opioid receptors. Behav Brain Res. 2007;178(2):216–20.
22. Van Dongen HP. The cumulative cost of additional wakefulness: dose-response effects on neurobehavioral functions and sleep physiology from chronic sleep restriction and total sleep deprivation. Sleep. 2003;2:117–26.
23. Guilleminault C, Powell NB, et al. Preliminary observations on the effects of sleep time in a sleep restriction paradigm. Sleep Med. 2003;4(3):177–84.
24. Banks S, Dinges DF. Behavioral and physiological consequences of sleep restriction. J Clin Sleep Med. 2007;3(5):519–28.
25. Belenky G, Wesensten NJ, Thorne DR, Thomas ML, Sing HC, Redmond DP, et al. Patterns of performance degradation and restoration during sleep restriction and subsequent recovery: a sleep dose-response study. J Sleep Res. 2003;12(1):1–12.
26. Russo M, Thomas M, Thorne D, Sing H, Redmond D, Rowland L, et al. Oculomotor impairment during chronic partial sleep deprivation. Clin Neurophysiol. 2003;114(4):723–36.

27. Raphael I, Almodovar K, Bedrosian A, Brandow C, Choi G, Saini R, et al. 1061 the FRESNO project: Fresno residents' experience in sleep and nighttime optimization. Sleep. 41:A394–5. Abstract Supplement
28. Cappuccio FP, D'Elia L, Strazzullo P, Miller MA. Sleep duration and all-cause mortality: a systematic review and meta-analysis of prospective studies. Sleep. 2010;33(5):585–92.
29. Lim J, Dinges DF. Sleep deprivation and vigilant attention. Ann N Y Acad Sci. 2008;1129:305–22.
30. Doran SM, Van Dongen HP, Dinges DF. Sustained attention performance during sleep deprivation: evidence of state instability. Arch Ital Biol. 2001;139(3):253–67.
31. Goel N, Basner M, Rao H, Dinges DF. Circadian rhythms, sleep deprivation, and human performance. Prog Mol Biol Transl Sci. 2013;119:155–90.
32. Dennis LE, Wohl RJ, Selame LA, Goel N. Healthy adults display long-term trait-like neurobehavioral resilience and vulnerability to sleep loss. Sci Rep. 2017;7(1):14889.
33. Wu JC, Gillin JC, Buchsbaum MS, Chen P, Keator DB, Khosla Wu N, et al. Frontal lobe metabolic decreases with sleep deprivation not totally reversed by recovery sleep. Neuropsychopharmacology. 2006;31(12):2783–92.
34. Mu Q, Mishory A, Johnson KA, Nahas Z, Kozel FA, Yamanaka K, et al. Decreased brain activation during a working memory task at rested baseline is associated with vulnerability to sleep deprivation. Sleep. 2005;28(4):433–46.
35. Chee MW, Choo WC. Functional imaging of working memory after 24 hr of total sleep deprivation. J Neurosci. 2004;24(19):4560–7.
36. Stern Y. What is cognitive reserve? Theory and research application of the reserve concept. J Int Neuropsych Soc. 2002;8(3):448–60.
37. Elmenhorst D, Meyer PT, Winz OH, Matusch A, Ermert J, Coenen HH, et al. Sleep deprivation increases A1 adenosine receptor binding in the human brain: a positron emission tomography study. J Neurosci. 2007;27(9):2410–5.
38. Volkow ND, Wang GJ, Telang F, Fowler JS, Logan J, Wong C, et al. Sleep deprivation decreases binding of [11C]raclopride to dopamine D2/D3 receptors in the human brain. J Neurosci. 2008;28(34):8454–61.
39. Opp MR. Cytokines and sleep. Sleep Med Rev. 2005;9(5):355–64.
40. Alkadhi K, Zagaar M, Alhaider I, Salim S, Aleisa A. Neurobiological consequences of sleep deprivation. Curr Neuropharmacol. 2013;11(3):231–49.
41. Silva RH, Kameda SR, Carvalho RC, Takatsu-Coleman AL, Niigaki ST, Abilio VC, et al. Anxiogenic effect of sleep deprivation in the elevated plus-maze test in mice. Psychopharmacology. 2004;176(2):115–22.
42. Ikeda M, Ikeda-Sagara M, Okada T, Clement P, Urade Y, Nagai T, et al. Brain oxidation is an initial process in sleep induction. Neuroscience. 2005;130(4):1029–40.
43. Iliff JJ, Wang M, Liao Y, Plogg BA, Peng W, Gundersen GA, et al. A paravascular pathway facilitates CSF flow through the brain parenchyma and the clearance of interstitial solutes, including amyloid beta. Sci Trans Med. 2012;4(147):147ra11.
44. Nedergaard M. Neuroscience. Garbage truck of the brain. Science. 2013;340(6140):1529–30.
45. Xie L, Kang H, Xu Q, Chen MJ, Liao Y, Thiyagarajan M, et al. Sleep drives metabolite clearance from the adult brain. Science. 2013;342(6156):373–7.
46. Aguirre CC. Sleep deprivation: a mind-body approach. Curr Opin Pulm Med. 2016;22(6):583–8.
47. Ooms S, Overeem S, Besse K, Rikkert MO, Verbeek M, Claassen JA. Effect of 1 night of total sleep deprivation on cerebrospinal fluid beta-amyloid 42 in healthy middle-aged men: a randomized clinical trial. JAMA Neurol. 2014;71(8):971–7.
48. Bellesi M, de Vivo L, Chini M, Gilli F, Tononi G, Cirelli C. Sleep loss promotes astrocytic phagocytosis and microglial activation in mouse cerebral cortex. J Neurosci. 2017;37(21):5263–73.
49. Kwon KJ, Lee EJ, Kim MK, Jeon SJ, Choi YY, Shin CY, et al. The potential role of melatonin on sleep deprivation-induced cognitive impairments: implication of FMRP on cognitive function. Neuroscience. 2015;301:403–14.

50. Alzoubi KH, Mayyas FA, Khabour OF, Bani Salama FM, Alhashimi FH, Mhaidat NM. Chronic melatonin treatment prevents memory impairment induced by chronic sleep deprivation. Mol Neurobiol. 2016;53(5):3439–47.
51. Noorafshan A, Karimi F, Karbalay-Doust S, Kamali AM. Using curcumin to prevent structural and behavioral changes of medial prefrontal cortex induced by sleep deprivation in rats. EXCLI J. 2017;16:510–20.
52. Noorafshan A, Karimi F, Kamali AM, Karbalay-Doust S, Nami M. Restorative effects of curcumin on sleep-deprivation induced memory impairments and structural changes of the hippocampus in a rat model. Life Sci. 2017;189:63–70.
53. Lamond N, Jay SM, Dorrian J, Ferguson SA, Jones C, Dawson D. The dynamics of neurobehavioural recovery following sleep loss. J Sleep Res. 2007;16(1):33–41.
54. Dinges DF, Pack F, Williams K, Gillen KA, Powell JW, Ott GE, et al. Cumulative sleepiness, mood disturbance, and psychomotor vigilance performance decrements during a week of sleep restricted to 4–5 hours per night. Sleep. 1997;20(4):267–77.
55. Rupp TL, Wesensten NJ, Bliese PD, Balkin TJ. Banking sleep: realization of benefits during subsequent sleep restriction and recovery. Sleep. 2009;32(3):311–21.

Neurobiology of Dreams

5

Carlo Cipolli and Luigi De Gennaro

Introduction: Terminology and History of Psychophysiological Research on Dreaming

Dreaming (also termed "dream experience," "sleep mentation," or "mental activity during sleep") is a state of consciousness universally experienced by humans during their sleep. Notwithstanding conspicuous interindividual differences in the frequency of dream recall in the everyday life (between the so-called good, poor, and non-recallers: see Cohen, 1979 [1]) and contrary to what individuals usually believe, dreaming occurs several times per night.

The experimental investigation on the psychophysiology of dreaming is conventionally dated in 1953, when Aserinsky and Kleitman [2] described the cyclic architecture of sleep (approximately every 90 minutes throughout the nocturnal sleep) and reported a frequent occurrence of dream experience in association with rapid eye movements (REMs) and desynchronization of cortical low-voltage fast EEG activity. After a few years, also the brain circuitry controlling the muscle atonia during REM (or "paradoxical") sleep was identified in the dorsal pontine brain stem [3]. From then on, the combination of REMs, activated EEG, and muscle atonia was considered the most reliable set of neurophysiological markers of dreaming [4].

The seemingly exclusive association between REM sleep and dreaming (the so-called "REM sleep = dreaming" equation), suggested by the early studies [5, 6], was soon challenged by clinical and experimental evidence of a possible bidirectional dissociation between REM sleep and dreaming. In particular, a more accurate

---

C. Cipolli (✉)
Department of Specialty, Diagnostic and Experimental Medicine, University of Bologna, Bologna, Italy
e-mail: carlo.cipolli@unibo.it

L. De Gennaro
Department of Psychology, University of Rome Sapienza, Rome, Italy
e-mail: luigi.degennaro@uniroma1.it

L. M. DelRosso, R. Ferri (eds.), *Sleep Neurology*,
https://doi.org/10.1007/978-3-030-54359-4_5

interview of subjects after awakening in laboratory context disclosed that perceptually vivid and bizarre ("dreamlike") mental experiences occur often at early onset of sleep (stage 1 and 2 of NREM sleep [7]) and sometimes during slow wave sleep (stage 3 and 4 of NREM sleep [8]), in some cases even indistinguishable from those developed during REM sleep [9].

From the late 1960s to the early 1990s, the theoretical debate was focused on the controversy over the quantitative (as postulated by the so-called "one dream-generator model") or qualitative differences ("two dream-generators model") between the mental activities developed during REM and NREM sleep. The first model assumed that there is "one generator" to produce sleep mentation in all sleep stages [10]: the different qualitative features of REM and NREM sleep mentation should depend on some variations in both the activation and subsequent organization of amnestic materials in a fairly coherent narrative structure, because the cortical synchronization of NREM sleep does not allow a complex processing and consolidation of contents of the ongoing dream experience [11]. Conversely, Hobson and McCarley's activation-synthesis model [12] posits two distinct "generators" for REM and NREM sleep mentation because of the different neurophysiological and neurochemical machinery of the two main types of sleep. Dream experience developed during REM sleep (where cholinergic "REM-on" neurons and monoaminergic "REM-off" neurons interact in the pontine tegmentum [13]) should be the result of a random and nonspecific activation provided by the brain stem and of a more or less coherent "interpretation" (i.e., synthesis in partially coherent and bizarre visual story) of the oculomotor information transmitted via pontogeniculooccipital (PGO) pathways to the cortex [12]. The "synthesis" portion of the model relies on the isomorphic assumption that the forebrain interprets subcortical processes similarly during waking and sleep [14].

The above controversy limited the attention of researchers for the transition from the perceptually grounded consciousness of waking to the hallucinatory simulation of dreaming: notwithstanding Foulkes' [7] findings and the Hori et al. [15] identification of visual imagery in all substages of sleep onset, the distinctive features of mental experiences occurring at these substages remain largely unexplored [16].

In the last two decades, the indications provided by studies using neuroimaging and multichannel EEG techniques have led to overcome the rigid distinction between sleep stages and the pseudo-dichotomy between the qualitative characteristics of mental activities developed during REM and NREM sleep (Table 5.1). The interest of dream researchers is now oriented toward the differences and similarities in the effectiveness of specific neural networks supposed to be responsible of the cognitive and/or emotional functions (as identified by reliable neuropsychological indicators) involved in dreaming during sleep as well as in mental activities during other behavioral states (active wake, mind wandering, resting). In particular, neuroimaging and neurophysiological studies on resting state and mind wandering have shown a wide overlap of their phenomenological features and brain mechanisms with those of dreaming. Complementarily, refined investigations of dream features using accurate procedures of dream collection (using articulated interview and trained subjects [17]) have shown important variations along a continuum from

Table 5.1 Methodological approaches to dreaming

Phenomenological or naturalistic approach
Dream data are collected at home from subjects as spontaneous (oral or written) daily reports (dream diary or log) provided by subjects after awakening at home as dream reports may be collected through diary (or log) compiled by subjects in an oral or written way for prospective measurement of dream frequency and features or through questionnaire or interviews (e.g., telephonic) for retrospective measurement of dream frequency and features
Psychophysiological approach
Dream data are collected as spontaneous or guided reports after one or more awakenings provoked at particular sleep stages and cycle during polygraphically monitored (a) undisturbed sleep, (b) previous partial or total sleep deprivation, or (c) after experimental manipulation of dream experience through stimuli delivered before or during sleep
Neuropsychobiological approach
Dream data are collected as spontaneous reports after one or more awakenings provoked at planned stages and cycle during sleep, monitored using both electropolygraphic and imaging techniques, of normal subjects and patients suffering from acute or chronic brain pathologies, sleep disorders, or psychiatric diseases

thought-like to dreamlike activity not only between but also within sleep stages [18]. Taken together, the new items of evidence provided by recent advances in electrophysiological and neuroimaging methods of sleep recording and analysis have led to both overcome several methodological limitations due to the asynchrony between dream generation during sleep and its recall after awakening and prompt to study dreaming through the conceptual apparatus now applied to investigate the consciousness in wake. In many studies, dreaming is now conceptualized as a "natural" extension of waking consciousness, namely, as a state characterized by internally generated multisensorial (overall visual and auditory), motor (sometimes dramatic), cognitive, and emotional experiences, whose contents are often inserted into a storylike plot [19] but without voluntary control of their flow [20].

Before examining the currently available indications on the neural bases of dreaming, it seems useful to summarize the frequency of recall and the phenomenological characteristics of dream experiences, as a result from over 60 years of laboratory and clinical studies.

Dream Phenomenology, Recall Frequency, and Functions of Dreaming

Psychophysiological studies carried out after the milestone of Foulkes' [7] investigation on NREM mentation have consistently shown that the amount of dream experience, as conservatively estimated through its verbal report after provoked awakening in laboratory, fluctuates between about 84% in REM sleep and about 50% in NREM sleep (roughly considered as a unique sleep type: for review, see Nielsen [11]). Moreover, the length of dream reports (as measured by counting content words [21]) is substantially higher after REM compared to NREM sleep awakening (with a ratio varying from 2 to 5: 1) but fluctuates differently according to whether the time is spent in NREM or REM sleep: reports are longer from 0 to

15 minutes (and even more when awakening takes place close to a prior REM sleep episode) and from 45 to 60 minutes of NREM sleep and shorter between, while the pattern is opposite for REM sleep [11].

Dream experiences reported after awakening from REM sleep usually appear more perceptually vivid, bizarre, emotional (often negatively toned, with prominence of anxiety and fear), dramatic, physically involving, and rich with characters and visual scenes than those reported after NREM sleep, whose contents are more thought-like (i.e., conceptual, logical, and mundane [14]). These main qualitative differences (as well as some minor ones in the emotional intensity, insight or dream awareness, and verbal and motor imagery) appear reliable, as largely independent of report length [11] and remarkably consistent across studies: in REM dreams visual events are reported almost always, auditory elements in about 60%, and movement and tactile sensations in about 15%, smell and taste rarely [22]. The amount and intensity of dream features, however, are not uniform over the nocturnal sleep (Table 5.2): the contents with such dreamlike features as perceptual vividness and bizarreness of contents are more frequent (a) at sleep onset (stages 1 and 2 of NREM sleep), when reports at times are as long as those from REM sleep awakening and contain dramatic and complex episodes and not simply isolated images [23], and (b)

Table 5.2 Examples of dream reports after awakenings from different stages of nighttime sleep

Stage 2 – NREM sleep of first cycle
Question: What was going through your mind before awakening?
"I was with my roommate and we were talking about restaurants... and there was other people there and we were talking about the best kinds of food... and we were going through places to go... Something was happening before we were hungry... I guess we were complaining about Davis food and then somebody made a joke about how there's got to be someplace good"
Stage 3 – NREM sleep of second cycle
Question: What was going...?
"I was thinking... I was with somebody, but I can't remember... We were driving, may be, like a plane... it was some kind of wide open space, with no car..."
Stage REM of first cycle
Question: What was going...?
"I was dreaming that there was this TV show and they were trying to figure out if this girl had been taken advantage of, cause they were trying to sue her and this guy and my friend had a hole in her foot and it was burning... And we didn't know how to get the fire out and somebody made some lewd comment to me and my dad tried to kill him..."
Stage REM of fourth cycle
Question: What was going...?
"I was in my grandma's kitchen. We were watching on TV ... and this man won a lottery game... he happened to break off the little chute that picks the names and... they advertised that little chute had been broken... and everyone in the town was like this guy who had won a large amount of money... I remember the scene switched to the lotto place and people were in this room trying to fix this lotto machine by putting super glue on the parts and trying to push them back together and it didn't work. They tried like the times, but it never worked. So my parents and family were going to this office to enter their entries... and it didn't have a machine to enter it into and a gay made all these excuses because they had never put the entries in. People started getting really upset and saying "well, you could put it in a box and so get the machine fixed." And so he took the entries and put them in a box... My mom said that she would only wait 2 days for them to call her up for the lottery thing..."

in both NREM [24] and REM sleep of late night [25], when also the length and complexity of storylike organization of dream experience increase along with REM sleep duration (5 vs. 10 minutes [26]).

The frequency and characteristics of dream reports (and their within-subject variability) are influenced, besides by sleep stages and cycles, by several situational, methodological, and individual factors [1, 27]. Situational factors concern the context (laboratory/home), the schedule of awakening (at prefixed stage and duration or at given interval from the onset of experimental session), the modality of awakening (abrupt/gradual: see Goodenough et al., 1965 [28]), and the possible presence of interfering tasks upon awakening [29]. Methodological factors concern the procedure of dream reporting (free recall/questionnaire/affirmative probes/guided recall [17]) and the measurement (using Likert-like scales and/or content analysis scales) of the qualitative (perceptual, emotional, and organizational) features of dream experience [30]. The Likert-like scales allow the subjects themselves and/or external raters to graduate specific features or the whole quality of dream experience (e.g., the continuity, vividness, and bizarreness of a dream experience may be measured using the dream fantasy scale [31]; see Table 5.3), while the content analysis scales allow to identify the occurrences of specific themes, characters, situations, and types of activity in dream reports [32, 33] (Table 5.4). The two kinds of measures can also be combined to investigate accurately how dream qualities vary in groups of individuals differing for age and sex or exposure to stressful or traumatic events (as reflected in threatening dreams and nightmares [34]).

Individual factors concern gender, age, personality traits (such as openness to experience, psychological boundaries, and absorption [35]), the tendency to suppress negative thoughts and emotions [36], and the attitude toward dreams, motivation, emotional reactivity, and cognitive styles [37].

The striking contrast between the experimental evidence of abundant production of dream experiences over all sleep stages and the small number of dreams (rarely more than one per night) recalled at home even by individuals classified as good recaller has raised the entwined issues of the fate and function of the wealth of nonrecalled dream experiences. Partial but plausible indications have been provided by

Table 5.3 The dreamlike fantasy scale

0. No content reported, feels mind was blank
1. No content reported, feels something was going through his mind, but forgets what
2. Conceptual content, everydayish
3. Conceptual content, bizarre or unusual topics
4. Perceptual content, nonhallucinatory, everydayish, undramatic
5. Perceptual content, nonhallucinatory, bizarre or unusual, dramatic
6. Perceptual content, hallucinatory, everydayish, undramatic
7. Perceptual content, hallucinatory, bizarre or unusual, dramatic

Adapted from Foulkes D et al. [31]

Table 5.4 Content analysis of dream reports: categories of Hall and Van de Castle's system

The most frequently used coding system of content analysis is that developed by Calvin S. Hall and Robert Van de Castle, primarily published in a book [32] and now available on the web, together with several examples
Since its publication, this system of dream content analysis has been used by many investigators (psychologists, psychiatrists, and anthropologists) in several countries. In its basic form, it treats dream reports as stories or plays in which there may be one or more: Characters (animals, humans, friends, strangers) Social interactions (aggression, friendliness, sexuality) Activities (thinking, talking, running) Success and failure Misfortune and good fortune Emotions (happiness, sadness, surprise, fear) One or more settings (indoors vs. outdoors, familiar vs. unfamiliar) Objects (e.g., chairs, cars, streets, body parts) Descriptive modifiers of humans, animals, and objects (e.g., tall, fast, crooked) Temporal references Elements from the past Food and eating references
Obviously, some dream contents may fit into more than one category (e.g., punching someone is both an aggressive interaction and a physical activity)
The categories of Hall and Van de Castle's system may encompass those used in other systems of dream content analysis with the advantage of a great reliability resulting from population studies which were replicated three times (for detail, see Domhoff GW [33])

studies on the so-called memory sources of dreaming. First, dream contents result from a deep elaboration and rebinding of several sources, in particular episodic memories (i.e., recent and remote events [38]). Second, such sources as recent events or concerns are processed not only over the following night (the so-called day-residue effect [39]) but also up to 5–7 nights later ("dream-lag effect" [40, 41]). Third, dreams reported after multiple awakenings over the same night show both repeated incorporations of the same sources (such as presleep stimuli [42]) as contents and high frequencies of semantically equivalent or similar (the so-called "interrelated" contents [43]), regardless of the number of awakenings, sleep stage, or delay between awakenings [29].

Taken together, these findings indicate that the activation of memory sources during sleep is not random but oriented by some current concerns and/or relationships with previously accessed items of information [44]. Consistently, neuroimaging and neurophysiological studies on the activation during sleep of the systems involved in the (re)processing of information which is individually relevant ("salient") at emotional or motivational level have shown opposite effects: the negative emotional valence of events is de-potentiated in normal subjects [45] and further consolidated in patients with post-traumatic stress disorder (PTSD: for a discussion, see Nielsen and Levin [34]). Also the faster fade over time of the negative than positive emotionally toned autobiographical memories in both experimental tasks [46] and dream experiences [47] suggests that sleep concurs to reduce the negative emotional charge of conflicting waking events. All these findings appear compatible with the general process (postulated by Stickgold and Walker [48] in their triage model of memory consolidation during sleep) of "tagging" of salient information for consolidation by means of emotional arousal, future relevance, and/or deliberate intention.

On the contrary, the possibility that also recently acquired information and individually salient or task-related experiences (either replayed or associated with older memories during dreaming) are further consolidated via their incorporation into dream content appears more controversial (as pointed out by Schredl [49]). For example, visuomotor skills improve more after nighttime sleep in low dream recallers (LDR) compared with high dream recallers (HDR), although the latter has a higher rate of incorporation of a new (mirror tracing) task into dream content and a better initial performance [50]. It seems thus more cautious to argue that (a) dreaming is not simply a function of the consolidation process of recent information, either replayed [51] or associated with older memories during dreaming [52], and (b) only a part of the neural activity underlying dreaming, which is distributed across several cortical and subcortical structures, is influenced by the brain structures subserving different memory systems [53].

Neuropsychology of Dreaming

The early indications on the neural bases of dreaming were provided by studies on the clinical-anatomical correlations in patients with acute brain lesions. This approach, which is usually called "neuropsychology of dreaming," was applied long before the discovery of REM sleep: the anoneiria (or Charcot-Wilbrand syndrome) was first described in the nineteenth century [54]. After the discovery of REM sleep and under the assumption of a close correlation between REM sleep and dream recall rate, the attention in clinical context was initially addressed to the subcortical structures as putatively responsible for dream generation. In detail, since pontine brain stem lesions suppress or reduce human REM sleep [55], a concomitant suppression or reduction of dream experience was expected. However, a review of all the published cases of brain stem lesions reporting also information on dreaming documented [56] that a drastic reduction of REM sleep was accompanied by a cessation of dreaming in only 1 of the 26 patients examined (4 of whom with extensive pontine lesions by Solms himself [57]). The lack of stable associations between pontine lesions and changes of dream recall has been confirmed also in a recent review [58]. Coherently, also patients with an autoactivation deficit following a bilateral damage to the basal ganglia have spontaneous dreams during REM sleep in the absence of thoughts during wakefulness [59], on the one hand, and dream recall rate is not affected in depressed patients by the pharmacological suppression of REM sleep induced by the phenelzine therapy for depressed patients, which eliminates REM sleep without altering slow wave sleep [60]. Hence, pontine brain stem (or, more in general, subcortical) lesions in humans do not abolish dreaming, and REM sleep is not a necessary prerequisite for the occurrence of dream experience. Moving from subcortical to cortical regions, two reviews [61, 62] suggested that a complete loss of dreaming is associated to unilateral or bilateral temporo-parieto-occipital lesions. A systematic review of 361 patients with acute cortical lesions [57] has confirmed that the cessation of dreaming is caused by a damage of either (1) a posterior system, mostly unilateral, in correspondence of or near the temporo-parieto-occipital (TPO) junction, or (2) an anterior system, mostly bilateral, underlying the ventromedial prefrontal cortex (vmPFC) and including the white matter tracts surrounding the anterior horns of the lateral ventricles.

Notably, lesions of the posterior system affect dreaming during sleep as well as visual imagery in waking [63]. These complementary items of evidence clearly corroborate the basic idea that dream experience, mental imagery, and late stages of visual perception share some neural mechanisms (in particular, visual associative cortices [64]).

The anterior system is less frequently associated to the global cessation of dreaming, caused by bilateral lesions of the white matter tracts, which surround the anterior horns of the lateral ventricles, underlying vmPFC. Since the ventromedial white matter contains dopaminergic projections from the limbic system to the frontal lobe, most patients with prefrontal leucotomy complain of global cessation of dreaming [56].

A unifying theoretical framework is still lacking. However, the fact that brain-damaged patients do not show significant changes in dreaming when lesions, independently of severity, are located in brain areas different from those of the two above systems is suggestive of high specificity of the neural networks involved in dreaming.

Further indications of the role played by specific structures in dream generation may be gathered by means of (a) longitudinal neuroimaging and EEG studies [65, 66] on brain-damaged patients (the recovery of dreaming after the acute phase remains almost completely unexplored at neurophysiological level; notwithstanding it occurs in 2/3 of anoneiric patients over the following few years [67]) and (b) phenomenological studies focused on the temporary or persistent loss of dream features similar to those impaired in waking experiences of patients with specific neurological syndromes (involving visual distortions such as Fregoli's delusional hyperidentification: for a discussion, see Schwartz and Maquet [68]). Combining the indications provided by the two types of studies would allow to better understand the functioning during sleep of the brain structures putatively responsible of specific features of dreaming as well as of everyday waking experience (for a discussion, see Dang-Vu et al. [69]).

Electrophysiology of Dreaming

Several advances in the electrophysiological study of dreaming, regarding the relationship between the physiological (electrical activity of cortical and, more recently, subcortical structures) and psychological levels of dreaming (as shown by quantitative and qualitative aspects of dream reports), have been made possible by the new techniques both of sleep recordings with high spatial resolution (i.e., high-density EEG and intracerebral stereo-EEG) and of transcranial stimulation. The former techniques allow assessing the spatiotemporal dynamics of peculiar EEG features, the latter to detect possible causal links between targeted brain areas and specific cognitive processes involved in dreaming (see next section).

In keeping with the "local sleep theory" [70], the regional differences in EEG activity at the cortical and subcortical level may be indicative of specific functions in dream generation and recall. Accordingly, the association of mediotemporal lobe activity and dream recall (considered as a peculiar form of episodic memory during sleep) has been investigated in a stereo-EEG study on patients with drug-resistant epilepsy [71], on whom a phase synchronization of rhinal-hippocampal EEG activity in the gamma range and an increased EEG coherence in the theta range had proved to be predictive of successful memory formation [72, 73]. The comparison of the rhinal-hippocampal and intrahippocampal EEG connectivity during waking, NREM, and REM sleep in good and poor dream recallers showed a close relationship between this connectivity and the capability to recall dreams as well as waking memories: patients who recalled dreams showed a significantly higher EEG coherence for all the frequency bands investigated (from 1 to 44 Hz, in particular, in the low-frequency theta range) and for all states (waking, NREM, and REM sleep) than patients who did not recall [71].

Other studies have provided indirect but important suggestions of possible relationships between the activity of subcortical structures and dream features. For example, the deactivation of human thalamic medial pulvinar nucleus (PuM) during REM sleep (as indicated by the great amount of its EEG delta activity in epileptic patients [74]), which is at odds with the general thalamocortical cholinergic activation, suggests the presence of some deactivated cortical areas (directly connected with PuM) among other activated areas (connected with other thalamic nuclei). If this dissociation in the level of cortical activation during REM sleep would be proved responsible of the reduced spatiotemporal cortical connectivity of REM sleep compared to waking (as postulated by Massimini et al. [75]), the cognitively bizarre features of dreaming could be explained as a result of the inactivation of cortical areas (such as posterior cingulate and dorsolateral prefrontal and parietal cortices) connected with the PuM [74]. Pertinent evidence may be gathered comparing mediotemporal and hippocampal EEG indicators and dream reports upon awakening from NREM and REM sleep in epileptic patients.

More conventionally, some studies have evaluated on normal subjects the association between dream recall and EEG activity in the sleep periods preceding the awakening from nocturnal sleep or daytime nap using multielectrode recordings and quantitative analysis of EEG, which are characterized by a good temporal resolution and an acceptable spatial resolution.

The first investigation of the EEG activity during sleep onset REM periods (SOREMPs) disclosed a complex relation between alpha activity and, respectively, lack of SOREMP dreams and appearance of NREMP dreams [76]. This relation has been partly confirmed for nocturnal sleep, where dream recall results are associated with lower alpha power (mainly in correspondence of temporoparietal areas) overall in stage 2 NREM sleep [77]. The involvement of alpha activity in dreaming during stage 2 NREM sleep (Fig. 5.1) has been found in another study [78], together with

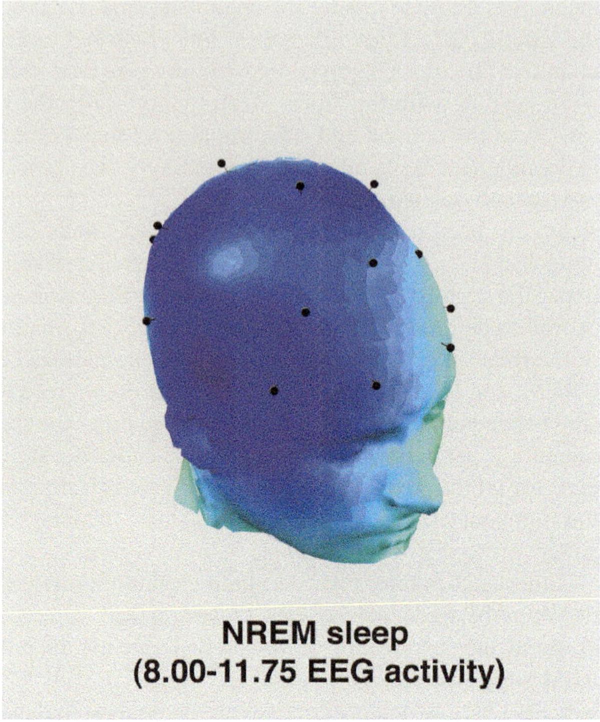

Fig. 5.1 A three-dimensional topographic distribution of correlation (rho) values between the number of dreams recalled upon morning awakenings and the amount of EEG alpha activity in NREM sleep. (Redrawn with data published by Marzano et al. [78])

a positive relation between the theta cortical brain oscillations (i.e., the genuine oscillatory activity) and dream recall after REM sleep (Fig. 5.2). Hence, stage 2-specific cortical brain oscillations in the sleep preceding awakening appear predictive of dream recall, namely, a higher frontal theta activity for REM sleep and a lower alpha oscillatory activity of the right temporal region for stage 2 NREM sleep, as also confirmed by a subsequent within-subject study [79].

The findings that the same regions which control successful memory encoding in waking are also involved in dream recall strengthen both the hypothesis that the neurophysiological mechanisms of encoding and recall of episodic memories are similar across different states of consciousness and the general notion of continuity between waking and sleep mentation. This notion is also supported by the available items of evidence that (a) theta and alpha oscillations are associated with memory encoding and retrieval during wakefulness; (b) a more efficient encoding of words and faces is associated to a lower alpha activity (10–12 Hz) in the right temporoparietal cortical region [80]; (c) an enhancement in alpha power is correlated with decreased memory performance [81]; and (d) successful encoding and retrieval of declarative information are associated with an increase in the theta power in scalp recordings [82]. Complementarily, intracranial recordings have confirmed that (a) an increase in frontal theta oscillations during the encoding phase is predictive of subsequent recall [83] and (b) theta oscillations in wakefulness mediate the prefrontal-mediotemporal dialogue in memory encoding [84].

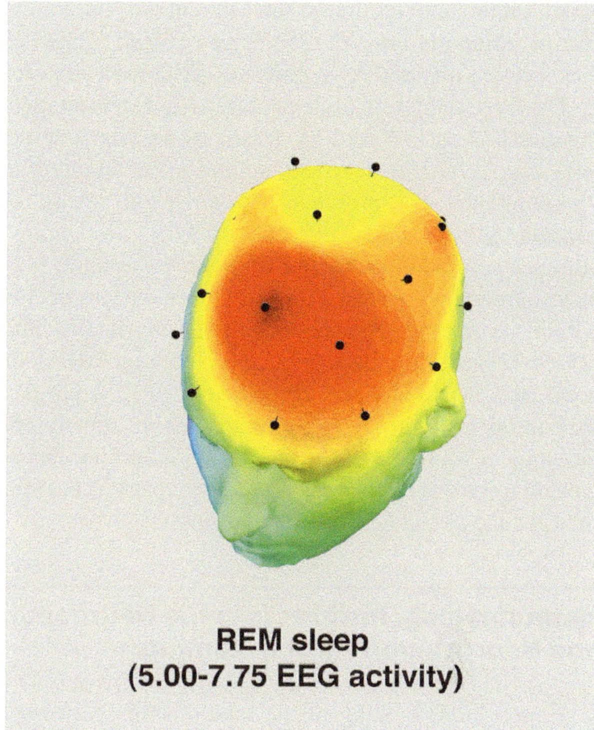

Fig. 5.2 A three-dimensional topographic distribution of correlation values (rho) between the number of dreams recalled upon morning awakenings and the amount of EEG theta activity in REM sleep EEG. (Redrawn with data published by Marzano et al. [78])

Important insights into the issue of whether the success/failure in dream recall is explained mainly by individually stable (i.e., trait-like) EEG or state-specific EEG correlates of dream recallers may be expected from within- rather than between-subject studies, carried out with multiple awakenings per session or (preferably) one awakening per multiple sessions. The former experimental condition has been applied in two studies on the cortical EEG correlates of dreaming during a 40-h multiple nap protocol under constant routine condition. These studies have shown that REM dream recall is preceded by higher alpha and beta activity in the occipital cortex, while NREM dream recall (a) is preceded in young subjects by lower delta and sigma EEG activity over frontal and centro-parietal derivations, respectively [85], and (b) is associated to more frontal delta and centro-parietal sigma EEG activity in older than young subjects, while the lower frontal-central alpha and beta activity in REM sleep of older compared with young subjects has no clear relation with the presence/absence of dream recall [86]. These age-related differences raise the question of the possible parallelism between specific patterns of frequency and topography EEG activity and the age-related variations in the effectiveness of cognitive processes [87].

The discrepancy between the results of studies using protocols with multiple naps or one awakening per session may be explained by the alterations of the chronobiological modulation of dreaming induced by multiple awakenings per session, which ineluctably confound the influence of ultradian (NREM-REM sleep stages) and circadian rhythms (the sequence of sleep cycles) on dream generation [88] and thus weaken the predictive power of EEG indicators for dream recall.

The large reduction of delta activity in correspondence of left frontal and temporoparietal areas followed by dream recall after one-per-day awakening from daytime stage 2 NREM sleep [89] suggests that an increased cortical EEG activation is a prerequisite for the presence of dream recall, at least for NREM sleep. Consistently, a recent within-subject study using a serial awakening protocol has shown that the presence compared to absence of dream experience is associated with (a) a decrease in low-frequency power (1–4 Hz) over parieto-occipital area, both in REM and NREM sleep, and (b) a less rapid activity in the range of high-frequency power (25–50 Hz) over the parieto-occipital area in NREM sleep and the frontal and temporal area in REM sleep, respectively [90]. Additionally, the similarly increased high-frequency activity in a cortical region mostly corresponding to the fusiform face area in both REM sleep dreams with familiar faces as contents and their reports in waking [90] is further supportive for a sort of anatomical-functional "continuity" between the two states of consciousness.

Neuroimaging; Insights into the Neuroanatomy and Neurochemistry of Dreaming

A deeper understanding of the relationships between specific dream features and brain activity during sleep was prompted in the 1990s by the advent of brain imaging techniques (in particular, positron emission tomography, PET; single-photon emission computed tomography, SPECT; structural and functional magnetic resonance imaging, s/fMRI). The first (and only) study combining PET and EEG recordings and the control of dream recall showed an activation during REM sleep in the pontine tegmentum, left thalamus, both amygdaloid complexes, anterior cingulate cortex, and right parietal operculum and a (bilateral) deactivation in a vast area of the dorsolateral prefrontal cortex (dlPFC), parietal cortex (supramarginal gyrus), posterior cingulate cortex, and precuneus [91]. Two contemporary studies (but without collection of dream reports [92, 93]) disclosed also a strong activation of the high-order occipitotemporal visual cortex (consistent with both the vividness of many dreams and clinical observations on brain-damaged patients) and a concomitant decreased activation of the primary visual cortex during REM sleep. Substantial differences in brain activity during REM compared with NREM sleep, in which activation diminishes from waking levels [92], were confirmed by most neuroimaging studies, regardless of the technique applied [94, 95].

The subsequent observation of a greater activation of the dorsomedial prefrontal cortex (dmPFC) in waking mind wandering (i.e., a self-generated thought process that proceeds independently of an external task) as compared with goal-directed intentional thinking [96] disclosed the importance of the presence of some dream-like features (such as visual imagery and bizarreness) in mental activity not only during stage 2 NREM but also in relaxed wakefulness (as first shown by Foulkes and Fleisher, 1975 [97]). The comparison of neuroimaging data concomitant with different waking states (daydreaming, mind wandering, task-oriented thinking) makes it possible (a) to measure the functional connectivity of specific neural networks identified using diffusion tensor imaging (DTI) and (b) to assess the relationships between the variations in the functioning of these networks during waking and sleep stages and the variations in the effectiveness of specific cognitive processes.

In this way, the (often implicitly) postulated isomorphism between the metabolism of the neuroanatomy of REM sleep and the functional neurophysiology of dreaming is overcome, and neuroimaging data are interpreted more cautiously, namely, in heuristic rather than (and nearly always post-factum) explanatory terms. For example, the discovery that a default network of brain regions active during resting state is in part reactivated during REM sleep but deactivated during NREM sleep has suggested that its two subsystems have distinct functions in dream generation. The "simulation subsystem," as centered on the hippocampal formation (the medial temporal lobe subsystem is more active in REM than NREM [98]), may explain the facilitated access during REM sleep to memories [99] and, thus, their conversion as dream contents. The "self-referential subsystem," as centered on the dmPFC, which shows incomplete connectivity during REM sleep [100], may be responsible for the lack of insight (i.e., dream awareness) into the current mental state [95].

Several convergent indications from lesional and neuroimaging studies on specific brain networks supposed to modulate the most distinctive characteristics of dreams and dreamlike mental activities [94, 95] appear integrated into the most recent models of the neurobiological bases of dreaming. The main indications regard (1) the activation of the amygdala complex, anterior cingulate cortex, and orbitofrontal cortex could be related to the emotional features of dreams; (2) the increased activation of the occipitotemporal visual cortex could be associated with visual dream imagery; (3) the relative hypoactivation of the dlPFC could be associated with alterations in logical reasoning, working memory, episodic memory, and executive functions, which frequently occur in REM sleep dreams; and (4) the activation of mesial temporal areas, and specifically the hippocampus, could account for the access and processing of episodic (recent and remote) memories commonly recognizable in specific dream contents.

The involvement of the first two networks is consistent with the results of brain lesion studies, while the role of mesial temporal areas is supported by recent electrophysiological [101, 102] and morpho-anatomical findings [103, 104]. Conversely, the neural bases of mind wandering seem to involve more than just default network activity [98, 105], and also several perceptual and emotional features of daydreaming seem to differ from those of REM and NREM daytime naps [106]. The availability of partially discrepant findings, however, does not invalidate the models but, rather, prompts to elaborate more specific and testable hypotheses. Appropriate testing may be provided by combining complementary bottom-up (from lesional and neuroimaging data to cognitive features of mental experiences) and top-down research strategies (from dream phenomenology to brain maps, as suggested by Schwartz and Maquet [68]) and by using refined psychometric measures of the functioning of specific cognitive processes involved in dreaming (e.g., working memory [107]) and waking experiences. Pertinent examples of this integrated neurobiological and neuropsychological approach have been provided by some investigations on how the variations in the activity of given cortical and subcortical areas are associated to the trait- and state-like individual differences in specific cognitive (such as recall frequency and types of content) and emotional (such as bad dreams and nightmares) features of dreaming.

The items of evidence gathered by comparing appear crucial to explain the predictive power of neurobiological data of sleep for the varying effectiveness of the cognitive processes involved in dream generation and recall. The comparison of subjects with high and low dream recall rates (HRs and LRs) during REM and stage 3 NREM sleep [101] has shown that the regional cerebral blood flow (rCBF), measured by PET with $[^{15}O]H_2O$, is higher in the temporoparietal junction (TPJ) in both REM and NREM sleep, while in the medial prefrontal cortex (mPFC) is higher in HRs than LRs during only REM sleep. Notably, these neurofunctional correlates of dream recall in healthy subjects are fairly consistent with the evidence of a global anoneiria in patients with lesions involving the TPO junction and the vmPFC [57]. It seems therefore plausible that the differences in spontaneous brain activity in TPJ and vmPFC during REM sleep are the neurophysiological substrate of stable ("trait-like") differences, at least in subjects belonging to the distribution ties of dream recall rate (HRs and LRs). Moreover, some morpho-anatomical findings suggest a role also of mesial temporal areas in dreaming (see below).

A comprehensive neurobiological model (affect network dysfunction, AND) has been proposed by Nielsen and Levin [34] to account for nightmares. This parasomnia is characterized by awakening primarily from REM sleep and vivid recall of disturbing dream, with intense (prevalently fear-related) emotions. Both idiopathic and post-traumatic (associated with post-traumatic stress disorder, PTSD) nightmares are distinct from sleep terrors, which arise from NREM sleep and involve fear-related arousals but not vivid dreams. The AND model stems on both neuroimaging data and evidence of emotional regulation during nightmares at physiological (alignment of fearful dream imagery and increased activation of heart and respiration rate in the few minutes preceding awakening from REM sleep [108]) and cognitive level (progressively decreased intensity and variability of emotions in REM periods [109]). The model posits that nightmares result from a dysfunction in the neural network subserving the adaptive function of fear memories in normal dreaming (usually leading to the so-called bad dreams, namely dreams with intense negative emotions), through the processes of acquisition and extinction of fear memories, which are activated during (overall REM) sleep as well as in waking by the network of the amygdala (for emotional activation and control of fear memory acquisition), hippocampus (for control of memory context), mPFC (for storage and control of extinction memories), and anterior cingulated cortex (for regulation of affect distress). The affect load should determine in normal dreaming a recombination (rebinding) of attributes of existing fear memories in new potentially fear-extinguishing contexts, whereas in nightmare a distress, largely regulated by the anterior cingulated cortex, in response to intense negatively toned emotional stimuli (such as memories of prior abuse, neglect, and trauma). The AND model is fairly predictive, given that using both structural and functional neuroimaging techniques it has been shown that in PTSD patients compared to healthy controls sleep disturbances and nightmares are associated with (a) loss of gray matter volume (GMV) in the amygdala, hippocampus, anterior cingulate, and insula; (b) increased rCBF in the midbrain, precuneus, and insula; and (c) decreased rCBF in the anterior cingulated cortex [110, 111].

Moreover, this model is compatible also with data of one type of "dreaming excess" observed in patients with lesions in the medial prefrontal cortex, the anterior cingulate cortex, and the basal forebrain [57]. An increased frequency and vividness of dreams may be present in patients both with frontal limbic lesions, with intrusion of dream content into waking life (the so-called anoneirognosis, namely a difficulty to distinguish between internally generated dream experiences and externally driven perceptions), and with temporal limbic seizure activity, with "recurring nightmares" (i.e., recurrent and stereotypical nightmares accompanying complex partial seizures).

Notably, the insights provided by neuroimaging data into the neurobiological bases of dreaming are not limited to cortical regions but also extend to subcortical nuclei. In particular, the metabolic decline of the brain stem and thalamus during NREM sleep (both involved in generating cortical slow waves [112]) and the increased activation in limbic and paralimbic structures during REM sleep [91] suggest important and complementary roles of the hippocampus and amygdala in dreaming. Indeed, the retrieval of cues of fear-conditioning stimuli delivered before sleep affects the emotional tonality of dreams in both REM and NREM sleep [113]; in waking, gamma-band coherence between the rhinal cortex and hippocampus increases when a stimulus is successfully encoded [72], while the gamma activity increases in the entorhinal cortex layers projecting to the hippocampus when the stimulus is retrieved [114]. Further items of evidence that the emotional load in dreams is related to amygdala activation have been provided by two studies which evaluated the differences in the brain tissue of the hippocampus-amygdala complex and measured the volume of gray matter and its alterations by means of microstructural analyses [i.e., volume and diffusivity (magnitude of neuronal water diffusion)] and diffusion tensor imaging (DTI) analysis of MRI scans. In normal subjects the interindividual differences in the length of reports, emotional intensity, and bizarreness of dream contents reported in the next morning are closely related to some differences in the brain tissue of the hippocampus-amygdala complex: indeed, a decreased microstructural integrity of the left amygdala is associated with shorter dream reports, lower emotional load, and lower bizarreness, while that of the right amygdala only with lower bizarreness. Moreover, in Parkinson's disease patients, the relationships between the visual vividness of dream reports and the volume of the amygdala, thickness of mPFC, and hypodopaminergic activity show a specific role for the mesolimbic dopaminergic system in qualitative (but not quantitative) aspects of dream recall [104].

Interestingly, the structural MRI methods can shed light also on the association of dream recall and content with the neurochemical mechanisms of specific brain circuits: the visual vividness of dream reports is positively associated to the amygdala volume and mPFC thickness, while the richness of dream report is negatively related to the dopamine agonist dosage (the higher the dosage, the poorer the report).

With the caution due to taking the dopamine agonist dosage as a measure of the hypodopaminergic state in PD patients, it seems worth stressing that the suggested role of the amygdala-mPFC dopamine network in dream recall is coherent with the indications provided by neuropsychological studies on brain-damaged patients [56, 57] and PET measurements on healthy subjects [102]. These findings converge to indicate that an anterior system mediates basic aspects of dream generation and recall within a general and pivotal role of the posterior system (i.e., TPJ) for sensorial aspects of dreams. Although an integrated model of the specific role of the TPJ and the mesolimbic system in dream experience is still lacking, the above neurochemical findings appear compatible with the reward activation model proposed by Perogamvros and Schwartz [115], according to which the activation of the mesolimbic-dopaminergic reward system contributes to the emotional and motivational components of dreaming.

Recent Advances in Dream Research Through New Methods of Sleep Recording and Analysis

The knowledge of the neurobiological bases of dreaming has been undoubtedly extended by the recent advances in electrophysiological and neuroimaging techniques of sleep recording/analysis. Indeed, these techniques allow now to investigate in healthy individuals (as well as in patients with brain damage, neurodegenerative diseases, sleep disorders, or parasomnias) the neural bases also of previously inaccessible aspects of dreaming, such as the presence of short periods of dream lucidity during REM sleep and the reliability of the dream contents reported after awakening.

The lack of awareness of the dreaming state and the entwined absence of voluntary control of the flow of dream contents have long puzzled dream researchers, given that it was well known that some individuals (so-called lucid dreamers) could experience a short period of lucidity while dreaming.

Some important insights into the neural correlates of the periods of dream lucidity have been provided by transcranial stimulation techniques. In particular, transcranial direct current stimulation (tDCS), which can modulate cortical excitability over PFC areas, has shown that (a) cathodal-frontal and anodal-parietal stimulation increases reports of visual dream imagery during stage 2 NREM [116]; (b) interhemispheric paired pulse transcranial magnetic stimulation (TMS) has shown a lower interhemispheric connectivity during REM sleep, which may account for such dream features as the lack of insight, time distortion, and amnesia [117]; and (c) the stimulation of PFC can induce dream lucidity during REM sleep in some subjects [118].

The recent evidence that "lucidity" is a dimension of dreaming that can be learnt by training subjects [119] but also predicted by a shift in EEG power, especially in the 40 Hz range of frontal brain regions (when the dream becomes "lucid" compared with when it is "nonlucid"), is of great theoretical interest [120].

Consistently, it has been shown that transcranial alternating current stimulation (tACS) in the lower gamma band during REM sleep can induce self-reflective awareness in dreams [121] and synchronous oscillations around 40 Hz in frontal regions are in correspondence of periods of lucid SOREMP dreaming in narcoleptic patients. These patients often report spontaneously their capability to maintain or recover lucidity while dreaming and modify unpleasant or frightening dreams [122].

Complementarily, a greater level of reactivation during REM sleep of the cerebral regions involved in learning a new task has shown to be accompanied in healthy subjects by the reactivation of the task-related information [123, 124]. The neuronal activation of the sensorimotor cortex (measured by fMRI and near-infrared spectroscopy, NIRS) during lucid REM dreaming (i.e., in correspondence of increased gamma band activity over frontal and frontolateral areas) appears comparable to that of a given movement (hand clenching) both imagined and performed in waking by the same subjects [123].

Finally, tDCS can influence the neural substrates also of mind wandering in waking: the propensity to mind wandering of subjects performing a monotonous task with a periodical sampling of their thoughts is enhanced by stimulation through an anode electrode on the left dorsolateral PFC and a cathode electrode on the right supraorbital area [125]. These findings strengthen the general hypothesis that the same cerebral areas are activated during specific sleep stages as well as in waking, as suggested by the greater level of reactivation of the cerebral regions involved in learning a new task in trained subjects than in controls during REM sleep [126] and SWS [127].

Also, the issue of the reliability of dream contents reported after awakening has been successfully approached using two distinct combinations of the new techniques of electrophysiological and neuroimaging recordings and analysis.

First, the possibility of identifying when specific visual contents are experienced during sleep has been addressed using noninvasive techniques (e.g., multivariate pattern analysis) [128]: capable of decoding perceptual or mental states with high accuracy and reasonable temporal resolution. By provoking repeated awakenings just after sleep onset and asking subjects to recall what they had seen, it has been found that fMRI patterns can be reliably linked, on a probabilistic basis, to the same visual information (i.e., categories of objects such as cars, men, or women) in dream reports [129]. Much more complex pattern recognition algorithms are obviously required to decode information from more simultaneous modalities (dream contents being visual, auditory, and tactile) before attempting to extend this form of "brain reading" to all dream features. However, this possibility is supported, in principle, by data collected with intracranial EEG recordings across the medial temporal lobe and neocortex during sleep and wakefulness and during visual stimulation with fixation [130]. Indeed, single neurons exhibit reduced firing rates before REMs and increasing rates immediately after REMs, similarly to the activity pattern upon image presentation during fixation in wakefulness.

Second, fMRI measurements have disclosed important differences in the activation of specific brain areas (right inferior frontal gyrus, right superior and middle temporal gyrus) when a dream which has been actually experienced by a subject is recalled compared with both a dream simply listened to or read by the subject in wakefulness and a daytime mental activity of the subject [131]. These findings corroborate the reliability of dream reports, besides suggesting a new strategy for investigating both the interindividual (HDR vs. LDR) and intraindividual (success vs. failure) differences in dream recall.

These insights into how the brain generates dreams during sleep, as well as dreamlike experiences in waking, converge to indicate (a) a substantial similarity in the functioning of the cognitive processes involved in their generation and recall; (b) the usefulness of neuroimaging and multichannel sleep recordings in the process of identifying the neural basis of dreaming as well as the factors responsible for its intra- and interindividual variability; and (c) the importance of general models of mental experience during sleep and waking to formulate testable hypotheses on the neural basis of specific features and functions of dreaming.

References

1. Cohen DB. Sleep and dreaming: origins, nature and functions. Oxford: Pergamon Press; 1979.
2. Aserinsky E, Kleitman N. Regularly occurring periods of eye motility, and concomitant phenomena, during sleep. Science. 1953;118:273–4.
3. Jouvet M, Michel F. Corrélations éléctromyographiques du sommeil chez le chat decortiqué et mésencéphalique chronique. C R Séances Soc Biol Fil. 1959;153(3):422–5.
4. Rechtschaffen A, Kales A. A manual of standardized terminology, techniques and scoring system for sleep stages of human subjects. Los Angeles: Brain Research Institute; 1968.
5. Dement W, Kleitman N. The relation of eye movements during sleep to dream activity: an objective method for the study of dreaming. J Exp Psychol. 1957;53:339–46.
6. Dement W, Kleitman N. Cyclic variations in EEG during sleep and their relation to eye movements, body motility, and dreaming. Electroencephalogr Clin Neurophysiol. 1957;9:673–90.
7. Foulkes D. Dream reports from different stages of sleep. J Abnorm Soc Psychol. 1962;65:14–25.
8. Monroe LJ, Rechtschaffen A, Foulkes D, Jensen J. Discriminability of REM and NREM reports. J Pers Soc Psychol. 1965;2:456–60.
9. Cavallero C, Cicogna P, Natale V, Occhionero M, Zito A. Slow wave sleep dreaming. Sleep. 1992;15:562–6.
10. Foulkes D. The psychology of sleep. New York: Scribner; 1966.
11. Nielsen TA. A review of mentation in REM and NREM sleep: "covert" REM sleep as a possible reconciliation of two opposing models. Behav Brain Sci. 2000;23:851–66.
12. Hobson JA, McCarley RW. The brain as a dream-state generator: an activation-synthesis hypothesis of the dream process. Am J Psychiatr. 1977;134:1335–48.
13. Hobson JA, McCarley RW, Wyzinnski PW. Sleep cycle oscillations: reciprocal discharge by two brainstem neuronal groups. Science. 1975;189:55–8.
14. Hobson JA, Pace-Schott EF, Stickgold R. Dreaming and the brain: toward a cognitive neuroscience of conscious states. Behav Brain Sci. 2000;23:793–842.
15. Hori T, Hayashi M, Morikawa T. Topographical EEG changes and hypnagogic experience. In: Ogilvie RD, Harsh JR, editors. Sleep onset: normal and abnormal processes. Washington, DC: American Psychological Association; 1994. p. 237–53.
16. Nielsen TA. Microdream neurophenomenology. Neurosci Conscious. 2017;1:1–17. https://doi.org/10.1093/nc/nix001.

17. Hurlburt RT, Heavey CL. Telling what we know: describing inner experience. Trends Cogn Sci. 2001;5:400–3.
18. Windt JM. Reporting dream experience: why (not) to be skeptical about dream reports. Front Hum Neurosci. 2013;7:708.
19. Cipolli C, Bolzani R, Tuozzi G. Storylike organization of dream experience in different periods of REM sleep. J Sleep Res. 1998;7:13–9.
20. Pace-Schott EF. Dreaming as a story-telling instinct. Front Psychol. 2013;4:159.
21. Antrobus J. REM and NREM sleep report: comparison of word frequencies by cognitive classes. Psychophysiology. 1983;20:562–8.
22. Strauch I, Meier B. In search of dreams: results of experimental dream research. New York: State University of New York Press; 1996.
23. Vogel GW. Sleep-onset mentation. In: Arkin AM, Antrobus JS, Ellman SJ, editors. The mind in sleep: psychology and physiology. New York: Lawrence Erlbaum Associates; 1978. p. 97–108.
24. Pivik T, Foulkes D. NREM mentation: relation to personality, orientation time, and time of night. J Consult Clin Psychol. 1968;32:144–51.
25. Snyder F. The phenomenology of dreaming. In: Meadow L, Snow L, editors. The psychodynamic implications of the physiological studies on dreams. Springfield: Thomas; 1970. p. 124–51.
26. Cipolli C, Guazzelli M, Bellucci C, Mazzetti M, Palagini L, Rosenlicht N, et al. Time-of-night variations in the story-like organization of dream experience developed during rapid eye movement sleep. J Sleep Res. 2015;24:234–40.
27. Schredl M, Wittmann L, Ciric P, Götz S. Factors of home dream recall: a structural equation model. J Sleep Res. 2003;12:133–41.
28. Goodenough DR, Lewis HB, Shapiro A, Jaret L, Sleser I. Dream reporting following abrupt and gradual awakenings from different types of sleep. J Pers Soc Psychol. 1965;2:170–9.
29. Cipolli C, Ferrara M, De Gennaro L, Plazzi G. Beyond the neuropsychology of dreaming: Insights into the neural basis of dreaming with new techniques of sleep recording and analysis. Sleep Med Rev. 2017;35:8–20.
30. Winget CN, Kramer M. Dimensions of dreams. Gainesville: University of Florida Press; 1979.
31. Foulkes D, Spear PS, Symonds JD. Individual differences in mental activity at sleep onset. J Abnorm Psychol. 1966;71:280–6.
32. Hall CS, Van de Castle RL. The content analysis of dreams. New York: Appleton-Century-Crofts; 1966.
33. Domhoff GW. Finding meaning in dreams: a quantitative approach. New York: Plenum Publishing; 1996.
34. Nielsen TA, Levin R. Nightmares: a new neurocognitive model. Sleep Med Rev. 2007;11:295–310.
35. Beaulieu-Prévost D, Zadra A. Absorption, psychological boundaries and attitude towards dreams as correlates of dream recall: two decades of research seen through a meta-analysis. J Sleep Res. 2007;16:51–9.
36. Malinowski JE. Dreaming and personality: wake-dream continuity, thought suppression, and the big five inventory. Conscious Cogn. 2015;38:9–15.
37. Zadra A, Robert G. Dream recall frequency: impact of prospective measures and motivational factors. Conscious Cogn. 2012;21:1695–702.
38. Fosse MJ, Fosse R, Hobson JA, Stickgold RJ. Dreaming and episodic memory: a functional dissociation? J Cogn Neurosci. 2003;15:1–9.
39. Nielsen TA, Stenstrom P. What are the memory sources of dreaming? Nature. 2005;437:1286–9.
40. Nielsen TA, Kuiken D, Alain G, Stenstrom P, Powell RA. Immediate and delayed incorporations of events into dreams: further replication and implications for dream function. J Sleep Res. 2004;13:327–36.
41. Blagrove M, Fouquet NC, Henley-Einion JA, Pace-Schott EF, Davies AC, Neuschaffer JL, et al. Assessing the dream-lag effect for REM and NREM stage 2 dreams. PLoS One. 2011;6:e26708.

42. Cipolli C, Fagioli I, Mazzetti M, Tuozzi G. Incorporation of presleep stimuli into dream contents: evidence for a consolidation effect on declarative knowledge during REM sleep? J Sleep Res. 2004;13:317–26.
43. Rechtschaffen A, Vogel G, Shaikun G. Interrelatedness of mental activity during sleep. Arch Gen Psychiatry. 1963;9:536–47.
44. Horton CL, Malinowski JE. Autobiographical memory and hyperassociativity in the dreaming brain: implications for memory consolidation in sleep. Front Psychol. 2015;6:874.
45. van Rijn E, Eichenlaub JB, Lewis PA, Walker MP, Gaskell MG, Malinowski JE, et al. The dream-lag effect: selective processing of personally significant events during rapid eye movement sleep, but not during slow wave sleep. Neurobiol Learn Mem. 2015;122:98–109.
46. van der Helm E, Walker MP. Overnight therapy? The role of sleep in emotional brain processing. Psychol Bull. 2009;135:731–48.
47. Vallat R, Chatard B, Blagrove M, Ruby P. Characteristics of the memory sources of dreams: a new version of the content-matching paradigm to take mundane and remote memories into account. PLoS One. 2017;12:e0185262.
48. Stickgold R, Walker MP. Sleep-dependent memory triage: evolving generalization through selective processing. Nat Neurosci. 2013;16:139–45.
49. Schredl M. Is dreaming related to sleep-dependent memory consolidation? In: Axmacher N, Rasch B, editors. Cognitive neuroscience of memory consolidation. Cham: Springer International Publishing Switzerland; 2017. p. 173–82.
50. Dumel G, Carr M, Marquis L-P, Blanchette-Carrière C, Paquette T, Nielsen T. Infrequent dream recall associated with low performance but high overnight improvement on mirror-tracing. J Sleep Res. 2015;24:372–82.
51. Wamsley EJ, Tucker M, Payne JD, Benavides JA, Stickgold R. Dreaming of a learning task is associated with enhanced sleep-dependent memory consolidation. Curr Biol. 2010;20:850–5.
52. Llewellyn S. Such stuff as dreams are made on? Elaborative encoding, the ancient art of memory, and the hippocampus. Behav Brain Sci. 2013;36:589–607.
53. Wamsley EJ. Dreaming and offline memory consolidation. Curr Neurol Neurosci Rep. 2014;14:433.
54. Murri L, Arena R, Siciliano G, Mazzotta R, Muratorio A. Dream recall in patients with focal cerebral lesions. Arch Neurol. 1984;41:183–5.
55. Lavie P, Pratt H, Scharf B, Peled R, Brown J. Localized pontine lesion: nearly total absence of REM sleep. Neurology. 1984;34:118–20.
56. Solms M. Dreaming and REM sleep are controlled by different brain mechanisms. Behav Brain Sci. 2000;23:843–50.
57. Solms M. The neuropsychology of dreaming: a clinico-anatomical study. Mahwah: Lawrence Erlbaum Associates; 1997.
58. Carroll C, Landau ME. Effects of pontine lesions on REM sleep. Curr Neurol Neurosci Rep. 2014;14:460.
59. Leu-Semenescu S, Uguccioni G, Golmard J-L, Czernecki V, Yelnik J, Dubois B, et al. Can we still dream when the mind is blank? Sleep and dream mentations in auto-activation deficit. Brain. 2013;136:3076–84.
60. Landolt HP, Raimo EB, Schnierow BJ, Kelsoe JR, Rapaport MH, Gillin JC. Sleep and sleep electroencephalogram in depressed patients treated with phenelzine. Arch Gen Psychiat. 2001;58:268–76.
61. Greenberg MS, Farah MJ. The laterality of dreaming. Brain Cogn. 1986;5:307–21.
62. Doricchi F, Violani C. Dream recall in brain-damaged patients: a contribution to the neuropsychology of dreaming through a review of the literature. In: Antrobus JS, Bertini M, editors. The neuropsychology of sleep and dreaming. Hillsdale: Lawrence Erlbaum Associates; 1992. p. 99–129.
63. Kosslyn SM, Ganis G, Thompson WL. Neural foundations of imagery. Nat Rev Neurosci. 2001;2:635–42.
64. Kosslyn SM, Thompson WL, Alpert NM. Neural systems shared by visual imagery and visual perception: a positron emission tomography study. NeuroImage. 1997;6:320–34.

65. Bischof M, Bassetti CL. Total dream loss: a distinct neuropsychological dysfunction after bilateral PCA stroke. Ann Neurol. 2004;56:583–6.
66. Poryazova R, Huber R, Khatami R, Werth E, Brugger P, Barath K, et al. Topographic sleep EEG changes in the acute and chronic stage of hemispheric stroke. J Sleep Res. 2015;24:54–65.
67. Murri L, Mancino M, Massetani R, Canapicchi R, Puglioli M, Rossi G. Effect of acute and chronic brain damage on dreaming. Res Comm Psychol Psychiatry Behav. 1989;14:121–42.
68. Schwartz S, Maquet P. Sleep imaging and the neuro-psychological assessment of dreams. Trends Cogn Sci. 2002;6:23–30.
69. Dang-Vu TT, Schabus M, Desseilles M, Sterpenich V, Bonjean M, Maquet P. Functional neuroimaging insights into the physiology of human sleep. Sleep. 2010;33:1589–603.
70. Siclari F, Tononi G. Local aspects of sleep and wakefulness. Curr Opin Neuorbiol. 2017;44:222–7.
71. Fell J, Fernández G, Lutz MT, Kockelmann E, Burr W, Schaller C, et al. Rhinal-hippocampal connectivity determines memory formation during sleep. Brain. 2006;129:108–14.
72. Fell J, Klaver P, Lehnertz K, Grunwald T, Schaller C, Elger CE, et al. Human memory formation is accompanied by rhinal-hippocampal coupling and decoupling. Nat Neurosci. 2001;4:1259–64.
73. Fell J, Staedtgen M, Burr W, Kockelmann E, Helmstaedter C, Schaller C, et al. Rhinal-hippocampal EEG coherence is reduced during human sleep. Eur J Neurosci. 2003;18:1711–6.
74. Magnin M, Bastuji H, Garcia-Larrea L, Mauguière F. Human thalamic medial pulvinar nucleus is not activated during paradoxical sleep. Cereb Cortex. 2004;14:858–62.
75. Massimini M, Ferrarelli F, Murphy M, Huber R, Riedner B, Casarotto S, et al. Cortical reactivity and effective connectivity during REM sleep in humans. Cogn Neurosci. 2010;1:176–83.
76. Takeuchi T, Ogilvie RD, Murphy TI, Ferrelli AV. EEG activities during elicited sleep onset REM and NREM periods reflect different mechanisms of dream generation. Clin Neurophysiol. 2003;114:210–20.
77. Esposito MJ, Nielsen TA, Paquette T. Reduced alpha power associated with the recall of mentation from stage 2 and stage REM sleep. Psychophysiology. 2004;41:288–97.
78. Marzano C, Ferrara M, Mauro F, Moroni F, Gorgoni M, Tempesta D, et al. Recalling and forgetting dreams: theta and alpha oscillations during sleep predict subsequent dream recall. J Neurosci. 2011;31:6674–83.
79. Scarpelli S, Marzano C, D'Atri A, Gorgoni M, Ferrara M, De Gennaro L. State- or trait-like individual differences in dream recall: preliminary findings from a within-subjects study of multiple nap REM sleep awakenings. Front Psychol. 2015;6:928.
80. Mölle M, Marshall L, Fehm HL, Born J. EEG theta synchronization conjoined with alpha desynchronization indicate intentional encoding. Eur J Neurosci. 2002;15:923–8.
81. Klimesch W. EEG alpha and theta oscillations reflect cognitive and memory performance: a review and analysis. Brain Res Brain Res Rev. 1999;29:169–95.
82. Osipova D, Takashima A, Oostenveld R, Fernández G, Maris E, Jensen O. Theta and gamma oscillations predict encoding and retrieval of declarative memory. J Neurosci. 2006;26:7523–31.
83. Sederberg PB, Kahana MJ, Howard MW, Donner EJ, Madsen JR. Theta and gamma oscillations during encoding predict subsequent recall. J Neurosci. 2003;23:10809–14.
84. Anderson KL, Rajagovindan R, Ghacibeh GA, Meador KJ, Ding M. Theta oscillations mediate the interaction between prefrontal cortex and medial temporal lobe in human memory. Cereb Cortex. 2010;20:1604–12.
85. Chellappa SL, Frey S, Knoblauch V, Cajochen C. Cortical activation patterns herald successful dream recall after NREM and REM sleep. Biol Psychol. 2011;87:251–6.
86. Chellappa SL, Munch M, Knoblauch V, Cajochen C. Age effects on spectral electroencephalogram activity prior to dream recall. J Sleep Res. 2012;21:247–56.
87. Mangiaruga A, Scarpelli S, Bartolacci C, De Gennaro L. Spotlight on dream recall: the ages of dreams. Nat Sci Sleep. 2018;10:1–12.
88. Nielsen TA. Ultradian, circadian, and sleep-dependent features of dreaming. In: Kryger M, Roth T, Dement WC, editors. Principles and practice of sleep medicine. 5th ed. New York: Elsevier; 2010. p. 576–84.

89. Scarpelli S, D'Atri A, Mangiaruga A, Marzano C, Gorgoni M, Schiappa C, et al. Predicting dream recall: EEG activation during NREM sleep or shared mechanisms with wakefulness? Brain Topogr. 2017;30:629–38.
90. Siclari F, Baird B, Perogamvros L, Bernardi G, LaRocque JJ, Riedner B, Boly M, et al. The neural correlates of dreaming. Nat Neurosci. 2017;20:872–8.
91. Maquet P, Péters J, Aerts J, Delfiore G, Degueldre C, Luxen A, et al. Functional neuroanatomy of human rapid-eye-movement sleep and dreaming. Nature. 1996;383:163–6.
92. Braun AR, Balkin TJ, Wesenten NJ, Carson RE, Varga M, Baldwin P, et al. Regional cerebral blood flow throughout the sleep-wake cycle. An H2(15)O PET study. Brain. 1997;120:1173–97.
93. Braun AR, Balkin TJ, Wesensten NJ, Gwadry F, Carson RE, Varga M, et al. Dissociated pattern of activity in visual cortices and their projections during human rapid eye movement sleep. Science. 1998;279:91–5.
94. Nir Y, Tononi G. Dreaming and the brain: from phenomenology to neurophysiology. Trends Cogn Sci. 2010;14:88–100.
95. Pace-Schott EF, Picchioni D. The neurobiology of dreaming. In: Kryger MH, Roth T, Dement WC, editors. Principles and practice of sleep medicine. 6th ed. Philadelphia: Elsevier; 2017. p. 529–38.
96. Christoff K, Gordon AM, Smallwood J, Smith R, Schooler JW. Experience sampling during fMRI reveals default network and executive system contributions to mind wandering. Proc Natl Acad Sci U S A. 2009;106:8719–24.
97. Foulkes D, Fleisher S. Mental activity in relaxed wakefulness. J Abnorm Psychol. 1975;84:66–75.
98. Domhoff GW. The neural substrate for dreaming: is it a subsystem of the default network? Conscious Cogn. 2011;20:1163–74.
99. Walker MP, Liston C, Hobson JA, Stickgold R. Cognitive flexibility across the sleep-wake cycle: REM-sleep enhancement and anagram problem solving. Brain Res Cogn Brain Res. 2002;14:317–24.
100. Koike T, Kan S, Misaki M. Connectivity pattern changes in default-mode network with deep non-REM and REM sleep. Neurosci Res. 2011;69:322–30.
101. Eichenlaub J-B, Bertrand O, Morlet D, Ruby P. Brain reactivity differentiates subjects with high and low dream recall frequencies during both sleep and wakefulness. Cereb Cortex. 2014a;24:1206–15.
102. Eichenlaub J-B, Nicolas A, Daltrozzo J, Redouté J, Costes N, Ruby P. Resting brain activity varies with dream recall frequency between subjects. Neuropsychopharmacology. 2014b;39:1594–602.
103. De Gennaro L, Cipolli C, Cherubini A, Assogna F, Cacciari C, Marzano C, et al. Amygdala and hippocampus volumetry and diffusivity in relation to dreaming. Hum Brain Mapp. 2011;32:1458–70.
104. De Gennaro L, Lanteri O, Piras F, Scarpelli S, Assogna F, Ferrara M, et al. Dopaminergic system and dream recall: an MRI study in Parkinson's disease patients. Hum Brain Mapp. 2016;37:1136–47.
105. Fox KCR, Spreng RN, Ellamil M, Andrews-Hanna JR, Christoff K. The wandering brain: meta-analysis of functional neuroimaging studies of mind-wandering and related spontaneous thought processes. NeuroImage. 2015;111:611–21.
106. Carr M, Nielsen T. Daydreams and nap dreams: content comparisons. Conscious Cogn. 2015;36:196–205.
107. Daltrozzo J, Claude L, Tillmann B, Bastuji H, Perrin F. Working memory is partially preserved during sleep. PLoS One. 2012;7:e50997.
108. Fisher C, Byrne J, Edwards A, Kahn E. A psychophysiological study of nightmares. J Am Psychoanal Assoc. 1970;18:747–82.
109. Kramer M. The selective mood regulatory function of dreaming: an update an revision. In: Moffitt A, Kramer M, Hoffmann R, editors. The functions of dreaming. Albany: State University of New York; 1993. p. 139–96.

110. Nardo D, Högberg G, Jonsson C, Jacobsson H, Hällström T, Pagani M. Neurobiology of sleep disturbances in PTSD patients and traumatized controls: MRI and SPECT findings. Front Psych. 2015;6:134.
111. Zhu Y, Du R, Zhu Y, Shen Y, Zhang K, Chen Y, et al. PET mapping of neurofunctional changes in a posttraumatic stress disorder model. J Nucl Med. 2016;57:1474–7.
112. Maquet P, Degueldre C, Delfiore G, Aerts J, Péters JM, Luxen A, et al. Functional neuroanatomy of human slow wave sleep. J Neurosci. 1997;17:2807–12.
113. Wamsley EJ, Antrobus JS. The expression of traces conditioning during non-REM sleep and its relation to subjective experience. Neurobiol Learn Mem. 2009;92:283–91.
114. Steinvorth S, Wang C, Ulbert I, Schomer D, Halgren E. Human entorhinal gamma and theta oscillations selective for remote autobiographical memory. Hippocampus. 2010;20:166–73.
115. Perogamvros L, Schwartz S. The roles of the reward system in sleep and dreaming. Neurosci Biobehav Rev. 2012;36:1934–51.
116. Jakobson AJ, Fitzgerald PB, Conduit R. Induction of visual dream reports after transcranial direct current stimulation (tDCS) during stage 2 sleep. J Sleep Res. 2012;21:369–79.
117. Bertini M, De Gennaro L, Ferrara M, Curcio G, Romei V, Fratello F, et al. Reduction of transcallosal inhibition upon awakening from REM sleep in humans as assessed by transcranial magnetic stimulation. Sleep. 2004;27:875–82.
118. Stumbrys T, Erlacher D, Schredl M. Testing the involvement of the prefrontal cortex in lucid dreaming: a tDCS study. Conscious Cogn. 2013;22:1214–22.
119. La Berge S, Nagel LE, DEment WC, Zarcone VP. Lucid dreaming verified by volitional communication during REM sleep. Perc Mot Skills. 1981;52:727–32.
120. Voss U, Holzmann R, Tuin I, Hobson JA. Lucid dreaming: a state of consciousness with features of both waking and non-lucid dreaming. Sleep. 2009;32:1191–200.
121. Voss U, Holzmann R, Hobson A, Paulus W, Koppehele-Gossel J, Klimke A, et al. Induction of self awareness in dreams through frontal low current stimulation of gamma activity. Nat Neurosci. 2014;17:810–2.
122. Dodet P, Chavez M, Leu-Semenescu S, Golmard JL, Arnulf I. Lucid dreaming in narcolepsy. Sleep. 2015;38:487–97.
123. Dresler M, Koch SP, Wehrle R, Spoormaker VI, Holsboer F, Steiger A, et al. Dreamed movement elicits activation in the sensorimotor cortex. Curr Biol. 2011;21:1833–7.
124. Dresler M, Wehrle R, Spoormaker VI, Koch SP, Holsboer F, Steiger A, et al. Neural correlates of dream lucidity obtained from contrasting lucid versus non-lucid REM sleep: a combined EEG/fMRI case study. Sleep. 2012;35:1017–20.
125. Axelrod V, Rees G, Lavidor M, Bar M. Increasing propensity to mind-wander with transcranial direct current stimulation. Proc Natl Acad Sci U S A. 2015;112:3314–9.
126. Maquet P, Laureys S, Peigneux P, Fuchs S, Petiau C, Phillips C, et al. Experience-dependent changes in cerebral activity during human REM sleep. Nat Neurosci. 2000;3:831–6.
127. Peigneux P, Laureys S, Fuchs S, Collette F, Perrin F, Reggers J. Are spatial memories strengthened in the human hippocampus during slow wave sleep? Neuron. 2004;44:535–45.
128. Haynes JD, Rees G. Decoding mental states from brain activity in humans. Nat Rev Neurosci. 2006;7:523–34.
129. Horikawa T, Tamaki M, Miyawaki Y, Kamitani Y. Neural decoding of visual imagery during sleep. Science. 2013;340:639–42.
130. Andrillon T, Nir Y, Cirelli C, Tononi G, Fried I. Single-neuron activity and eye movements during human REM sleep and awake vision. Nat Commun. 2015;6:7884.
131. Benedetti F, Poletti S, Radaelli D, Ranieri R, Genduso V, Cavallotti S, et al. Right hemisphere neural activations in the recall of waking fantasies and of dreams. J Sleep Res. 2015;24:576–82.

Neurobiology of Memory and Sleep

6

Michael Blechner

Introduction

It is well known that sleep is comprised of two distinct phases: non-REM (NREM) vs REM. NREM is further divided into lighter stages (stage 1 and stage 2) and a deeper stage (stage 3). Stage 3, also known as slow wave sleep, is predominantly seen in the earlier portion of the night, while REM sleep is more predominant in the latter portion. The role of sleep has been reviewed extensively and has been postulated to allow for repair of cell tissue and thermoregulation, facilitate immune function, and preserve energy; however, other roles have been written about the function of sleep through the ages.

In the late sixteenth century and early seventeenth century, Dr. Christopher Wirtzung, writer of *A General Practise of Physicke*, described sleep as the "warming" of the body's spirits, the vehicle of all the processes of life emanating from the soul [1], while John Jones discussed the need for sleep to refresh ones' senses and improve digestion [2]. This was echoed by none other than Shakespeare in his works of Macbeth and Henry IV. Within the last few centuries, psychologists, philosophers, and playwrights alike spoke of new roles of sleep. British psychologist David Hartley proposed in 1801 the link between dreaming and memory [3]. However, it was not until 1924 when the first systematic studies of sleep and memory were performed by Jenkins and Dallenbach to test Ebbinghaus's theory of memory decay [4]. The knowledge elucidating sleep's impact in memory processing and brain plasticity has grown over the last half century and has become a topic of interest of many researchers. It is now assumed that sleep facilitates consolidation of memory by strengthening new memories and acclimating them into preexisting long-term memories. This chapter will outline what is known, while investigating perceived

M. Blechner (✉)
Department of Pediatric Pulmonary and Sleep Medicine, St. Joseph's Children's Hospital, Paterson, NJ, USA

L. M. DelRosso, R. Ferri (eds.), *Sleep Neurology*,
https://doi.org/10.1007/978-3-030-54359-4_6

relationships of various sleep stages and their impact on the production, restructuring, consolidation, storage, and strengthening of memories, which is imperative to performing our daily activities/functions, while reviewing the neurophysiologic basis for this. This information, which continues to grow with time, has been gathered from various physiological and behavioral studies performed over the last several decades. Initially, we will review underlying memory processes and the notion of memory consolidation.

Memory Processes

There are three major sub-processes in memory production – encoding, consolidation, and retrieval. Encoding is the process of forming a new memory trace in response to the perception of a stimulus. This process is highly susceptible to forgetting. Therefore, consolidation occurs to stabilize the memory trace by strengthening and integrating the memory into preexisting knowledge networks. As explained by Dudai [5], consolidation is the process by which newly acquired and generally labile memories encoded during wakefulness are believed to be reprocessed and converted into a more stable form and then integrated into preexisting memory networks during sleep. Finally, the stored memory may then be accessed and recalled during retrieval. While encoding and retrieval occur frequently during wakefulness, it is believed that sleep plays a major role in the consolidation of newly encoded memory into long-term storage. It is believed that this process depends on multiple variables, including different processes of neuronal plasticity. There are believed to be two types of consolidation processes: the first is coined "synaptic consolidation" which leads to remodeling and subsequent more effective synapses [5–8], and the second is referred to as "system consolidation," which redistribute newly encoded representations to other neuronal circuitries for long-term storage [5, 9].

Neurobiology of Memory

Long-term memory is generally divided into two main types: declarative (explicit) and procedural (implicit). Declarative memory ("knowing what") is memory that is consciously recalled, i.e., declared, and includes facts and events. It is further divided into episodic and semantic memory. Procedural or non-declarative memory ("knowing how") relates to unconscious memory such as skills and "how to do things." Examples of this include riding a bike or playing piano. These memories are generally acquired through repetition and are composed of automatic sensorimotor behaviors. Figure. 6.1 illustrates memory types and the stages of memory production (Fig. 6.1).

The locations in the brain where formation of long-term memory is processed were illustrated following a procedure in 1953 which included removal of a patient's

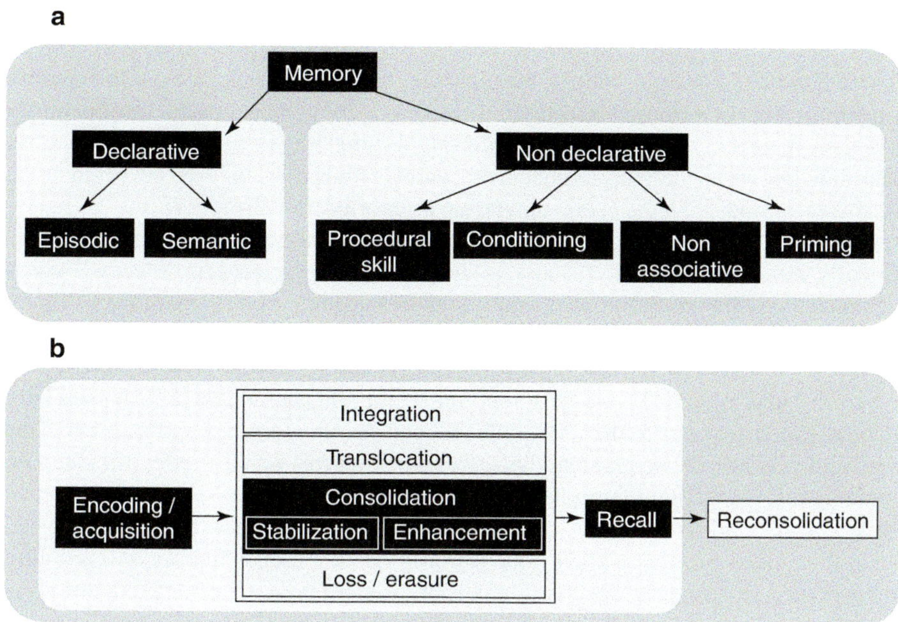

Fig. 6.1 Memory systems and memory stages. (**a**) Memory systems. Human memory is most commonly divided into declarative forms, with further subdivisions into episodic and semantic, and non-declarative forms, subdivided into an array of different types including procedural skill memory. (**b**) Developing stages of memory. Following initial encoding of a memory, several ensuing stages are proposed, beginning with consolidation, as well as integration of the memory representation, and translocation of the representation or erasure of the memory. Also, following later recall, the memory representation is believed to become unstable once again, requiring periods of reconsolidation. (Reprinted by permission from Springer: Spencer et al. [10])

medial temporal lobe, hippocampus, and amygdala to treat intractable seizures. After the surgery, the patient was still able to form procedural memories and short-term memories but demonstrated failure in converting immediate memory into stable long-term memory. As well, studies have shown that lesions of the hippocampus abolish the ability to acquire new declarative memory and produce retrograde amnesia where only older memories remain intact [9, 11].

Declarative memories require the involvement of medial temporal regions, specifically the hippocampus, with episodic memories rooted in temporal regions. Non-declarative memories such as perceptual skills originate from sensory cortices and procedural memories from the cerebellum, striatum, and motor areas. However, there is interaction between these systems, specifically seen in the early stages of new knowledge acquisition [12] and the transition from fast learning, seen in the hippocampus, to slow learning for long-term storage in the neocortex.

Role of Sleep in Memory

There have been several studies through the years reviewing barriers to memory retention and how sleep affects this process. Ebbinghaus performed memory research in 1885 and noticed that forgetting rapidly occurs in the first hours after learning, known as the "forgetting curve," and is reduced when sleep occurred in the retention period [13]. As well, others have reported impairment in the ability to remember following sleep deprivation [14]. There were two main beliefs that were initially considered when it came to forgetting, one related to "decay" which acknowledges the potential role time plays in the forgetting of memory [15] and another relating to "interference," an idea whereby new information interferes with the ability to retain older memories [16]. This is felt to have an even stronger effect when the new memories are quite similar to the old memories [17]. It was later demonstrated that sleep after learning reduced the amount of forgetting [4]. This was replicated on subsequent studies, strengthening this positive effect of sleep on memory and protecting memories from interference [18]. However, there appears to be a time dependency to how well memories may be protected from interference, such that, sleep occurring soon after learning has a stronger impact on inhibiting interference versus sleep occurring at a more remote time. One such example was seen in a study by Gais et al. who demonstrated that sleep within 3 hours of learning vocabulary had a more beneficial effect compared to sleep occurring 10 hours after [19]. Thus, not only does sleep have a positive effect on memory, but the timing when sleep occurs also plays an important role. However, it has additionally been illustrated that the stage of sleep also has an impact on how memory is protected from interference and how memory consolidation occurs.

Role of Specific Sleep Stages in Memory

Sleep stages are identified by characteristic brain activity rhythms: delta waves in slow wave sleep and spindle activity in stage 2 sleep.

The synaptic homeostasis theory describes the importance of not only upscaling synapses, which leads to their strengthening, but a system where downscaling may occur as well to simplify encoding of new information. This is believed to occur during slow wave sleep (SWS). One example is seen when the ability to encode pictures following a nap without slow wave activity was significantly diminished [20, 21]. With this impairment, there was a noted reduction in hippocampus activity during learning. However, neocortical regions were not impacted, and the ability to learn motor skill tasks was not hindered. Additionally, slow wave activity [22], slow wave amplitude, and the length of the up-state have been positively related to overnight memory consolidation [23]. SWS was also demonstrated to play a role in declarative memory consolidation by Rasch et al. [24]. Subjects were taught a visuospatial task of recognizing the location of images within a matrix. The task was performed while participants were exposed to an odor (the scent of roses), and then the odor was reintroduced during SWS for some participants (Fig. 6.2a). Those

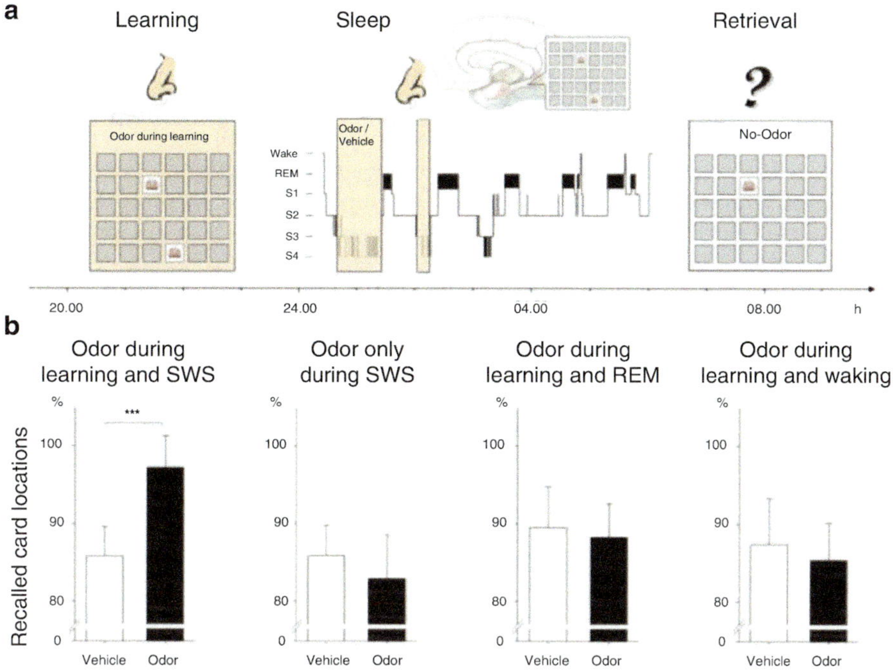

Fig. 6.2 Declarative memory reactivation during slow wave sleep. (**a**) Subjects performed a visuospatial learning task in the presence of an experimental odor (scent of roses). During subsequent slow wave sleep, this odor or a non-odor vehicle control was presented. Memory for the visuospatial task was tested following sleep. (**b**) Participants recalled more locations following sleep when the experimental odor was re-presented during slow wave sleep compared to vehicle (first panel). Such an effect was not observed if the odor present during slow wave sleep was not also present during learning (second panel), if the odor cue was re-presented during REM (third panel), or if the odor was re-presented during waking. (Reprinted by permission from Springer: Spencer et al. [10])

who were exposed to the odor during learning the task and subsequent SWS sleep had greater recall following sleep. This response was not seen with other sleep stages tested such as REM sleep. When the odor was presented during REM sleep, there was no benefit on subsequent recall compared to the no-odor (vehicle) control condition (Fig. 6.2b). This demonstrates the unique role of SWS on declarative memory consolidation.

Sleep spindles, generated in the thalamus and a component of stage 2 sleep, have been postulated to be associated with multiple forms of memory formation, consolidation, and storage. Spindle-associated discharges effectively triggered long-term potentiation (LTP) in neocortical synapses in in vitro models [25, 26]. It has been reported by several studies that there is a positive correlation between both time spent in stage 2 sleep and amount of spindle activity and motor skill improvement [27–31]. Thus, it is hypothesized that spindle activity induces synaptic potentiation

[25] and plays a role in procedural memory consolidation [27, 32–34]. As well, motor memory consolidation is thought to be impacted by spindle activity [28] with changes in striatal [35, 36] and hippocampal activity [37–39]. As an example, motor sequence learning tasks and the degree of overnight performance improvements positively correlate with increased spindle activity [29, 30, 33]. Sleep spindle promotion of skill consolidation is achieved by locally reactivating and functionally binding task-relevant cortical-cortical and subcortical regions, such as the hippocampus, putamen, thalamus, and motor-related cortical regions. Additionally, spindles are positively associated with declarative learning [40], declarative overnight performance changes [41], and intelligence [42].

Evidence shows that REM sleep also plays an important role in memory consolidation, specifically in regard to emotional memory [43–45]. Healthy students selectively deprived of REM sleep were found to exhibit impaired next-day recognition performance for negative, but not neutral, images encoded pre-sleep when compared to students deprived of NREM slow wave sleep [46]. Furthermore, 3 hours of late night REM-dominant sleep facilitate the consolidation of negative images [47]. This was reiterated by Payne, Chambers, and Kensinger in 2012 [48] who reported that greater negative memory consolidation is predicted by longer duration of REM sleep during retention intervals. As well, Gilson et al. [49] found in 2015 that greater REM density predicts greater selective consolidation of emotionally negative stories. With this in mind, both Walker and van der Helm in 2009 [43] and Harrington, Pennington, and Durant in 2017 [50] suggested that changes in REM sleep may increase vulnerability to the onset and recurrence of major depressive disorder by promoting the development of negative memory bias. Aside from emotional memory, there is also evidence that REM sleep plays a role in gross motor learning [51].

Role of Alternative Sleep Periods in Memory

Many studies have focused on full night sleep to investigate sleep and memory consolidation. However, it is also important to note that alternative sleep periods may play a role as well. This was achieved by van Schalkwijk et al. who, in 2017, studied 76 participants who were randomly assigned to a nap or wake group and performed a declarative word-pair association or procedural mirror tracing task. Performance changes from before compared to after a 90-minute retention interval filled with sleep or quiet wakefulness were evaluated between groups. Associations between performance changes, sleep architecture, spindles, and slow oscillations were investigated. They found a trend toward stronger forgetting in the declarative task across a wake period compared with a nap period while improved procedural task accuracy with a daytime nap compared to a period following daytime wakefulness [52]. This demonstrates that daytime napping prevented deterioration of performance for procedural mirror tracing, with a similar trend observed for declarative word-pair learning.

Conclusion

The knowledge surrounding the role of sleep and its impact on the body has grown over the centuries. However, only in recent history has the significance of sleep on memory been postulated and studied. Within just the last half century, we have learned how important sleep can be on multiple types of memory, including memory formation, consolidation, and storage. As well, it has been shown how various sleep stages and neural systems aid in this process. Yet, further work is needed to better understand this complex interaction and to determine how this information can be translated to patient care.

References

1. Wirtzung C. The general Practise of Physicke: Conteyning all inward and outward parts of the body, with all the accidents and infirmities that are incident unto them, Euen from the Crowne of the head to the sole of the Foote. London: Thomas Adams; 1617.
2. Jones J. The mysteries of opium Reveal'd: by Dr. John Jones. London: Richard Smith; 1701.
3. Hartley D. Observations on man, his frame, his duty, and his expectations. Warrington: J. Johnson; 1801.
4. Jenkins J, Dallenbach K. Obliviscence during sleep and waking. Am J Psychol. 1924;35:605–12.
5. Dudai Y. The neurobiology of consolidations, or how stable is the engram? Annu Rev Psychol. 2004;55:51–86.
6. Hofer SB. Structural traces of past experience in the cerebral cortex. J Mol Med. 2010;88:235–9.
7. Kandel ER. The molecular biology of memory storage: a dialogue between genes and synapses. Science. 2001;294:1030–8.
8. Redondo RL, Morris RGM. Making memories last: the synaptic tagging and capture hypothesis. Nat Rev Neurosci. 2011;12:17–30.
9. Frankland PW, Bontempi B. The organization of recent and remote memories. Nat Rev Neurosci. 2005;6:119–30.
10. Spencer R, Walker M, Stickgold R. Sleep and memory consolidation. In: Sleep disorders medicine: basic science, technical considerations and clinical aspects. 4th ed. New York: Springer; 2017. p. 205–23.
11. Corkin S. What's new with the amnesic patient HM? Nat Rev Neurosci. 2002;3:153–60.
12. Peigneux P, Laureys S, Delbeuck X, Maquet P. Sleeping brain, learning brain. The role of sleep for memory systems. Neruoreport. 2001;12:A111–24.
13. Van Ormer EB. Sleep and retention. Psychol Bull. 1933;30:415–39.
14. Patrick G. On the effects of loss of sleep. Psychol Rev. 1896;3:468–83.
15. Thorndike EL. Educational psychology, The psychology of learning, vol. 2. New York: Teachers College Press; 1913.
16. McGeoch JA. Forgetting and the law of disuse. Psychol Rev. 1932;39:352–70.
17. Keppel G. Consolidation and forgetting. In: Weingartner H, Parker ES, editors. Consolidation and forgetting. Hillsdale, NJ: Lawrence Erlbaum Associates; 1984.
18. Ellenbogen J, Payne J, Stickgold R. The role of sleep in declarative memory consolidation: passive, permissive, active or none? Curr Opin Neurobiol. 2006;16:716–22.
19. Gais S, Lucas B, Born J. Sleep after learning aids memory recall. Learn Mem. 2006;13:259–62.
20. Van der Werf Y, Altena E, Schoonheim M, Sanz-Arigita E, Vis J, de Rijke W, van Someren EJW. Sleep benefits subsequent hippocampal functioning. Nat Neurosci. 2009;12:122–3.
21. Van der Werf Y, Altena E, Vis J, Koene T, van Someren EJW. Reduction of nocturnal slow-wave activity affects daytime vigilance lapses and memory encoding but not reaction time or implicit learning. Prog Brain Res. 2011;193:245–55.

22. Holz J, Piosczyk H, Feige B, et al. EEG sigma and slow-wave activity during NREM sleep correlate with overnight declarative and procedural memory consolidation. J Sleep Res. 2012;21:612–9.
23. Heib DP, Hoedlmoser K, Anderer P, et al. Slow oscillation amplitudes and up-state lengths relate to memory improvement. PLoS One. 2013;8:e82049.
24. Rasch B, Buchel C, Gais S, Born J. Odor cues during slow-wave sleep prompt declarative memory consolidation. Science. 2007;315(5817):1426–9.
25. Rosanova M, Ulrich D. Pattern-specific associative long-term potentiation induced by a sleep spindle-related spike train. J Neurosci. 2005;25:9398–405.
26. Timofeev I, Grenier F, Bazhenov M, Houweling AR, Sejnowski TJ, Steriade M. Short-and medium-term plasticity associated with augmenting responses in cortical slabs and spindles in intact cortex of cats in vivo. J Physiol. 2002;542:583–98.
27. Fogel SM, Smith CT. Learning-dependent changes in sleep spindles and Stage 2 sleep. J Sleep Res. 2006;15(3):250–5.
28. Fogel SM, Smith CT. The function of the sleep spindle: a physiological index of intelligence and a mechanism for sleep-dependent memory consolidation. Neurosci Biobehav Rev. 2011;35(5):1154–65.
29. Laventure S, Fogel S, Lungu O, Albouy G, Sevigny-Dupont P, Vien C, et al. NREM2 and sleep spindles are instrumental to the consolidation of motor sequence memories. PLoS Biol. 2016;14(3):e1002429.
30. Nishida M, Walker MP. Daytime naps, motor memory consolidation and regionally specific sleep spindles. PLoS One. 2007;2(4):e341.
31. Walker MP, Brakefield T, Morgan A, Hobson JA, Stickgold R. Practice with sleep makes perfect: sleep-dependent motor skill learning. Neuron. 2002;35(1):205–11.
32. Ackermann S, Rasch B. Differential effects of non-REM and REM sleep on memory consolidation? Curr Neurol Neurosci Rep. 2014;14(2):430.
33. Barakat M, Doyon J, Debas K, Vandewalle G, Morin A, Poirier G, et al. Fast and slow spindle involvement in the consolidation of a new motor sequence. Behav Brain Res. 2011;217(1):117–21.
34. Barakat M, Carrier J, Debas K, Lungu O, Fogel S, Vandewalle G, et al. Sleep spindles predict neural and behavioral changes in motor sequence consolidation. Hum Brain Mapp. 2013;34(11):2918–28.
35. Debas K, Carrier J, Orban P, Barakat M, Lungu O, Vandewalle G, et al. Brain plasticity related to the consolidation of motor sequence learning and motor adaptation. Proc Natl Acad Sci U S A. 2010;107(41):17839–44.
36. Debas K, Carrier J, Barakat M, Marrelec G, Bellec P, Hadj Tahar A, et al. Off-line consolidation of motor sequence learning results in greater integration within a cortico-striatal functional network. NeuroImage. 2014;99:50–8.
37. Albouy G, Sterpenich V, Balteau E, Vandewalle G, Desseilles M, Dang-Vu T, et al. Both the hippocampus and striatum are involved in consolidation of motor sequence memory. Neuron. 2008;58(2):261–72.
38. Albouy G, Sterpenich V, Vandewalle G, Darsaud A, Gais S, Rauchs G, et al. Interaction between hippocampal and striatal systems predicts subsequent consolidation of motor sequence memory. PLoS One. 2013;8(3):e59490.
39. Albouy G, Fogel S, King BR, Laventure S, Benali H, Karni A, et al. Maintaining vs. enhancing motor sequence memories: respective roles of striatal and hippocampal systems. NeuroImage. 2015;108:423–34.
40. Gais S, Molle M, Helms K, Born J. Learning-dependent € increases in sleep spindle density. J Neurosci. 2002;22:6830–4.
41. Clemens Z, Fabo D, Halasz P. Overnight verbal memory retention correlates with the number of sleep spindles. Neuroscience. 2005;132:529–35.
42. Bodizs R, Gombos F, Ujma PP, Kovacs I. Sleep spindling and fluid intelligence across adolescent development: sex matters. Front Hum Neurosci. 2014;8:952.

43. Walker MP, van der Helm E. Overnight therapy? The role of sleep in emotional brain processing. Psychol Bull. 2009;135(5):731e748.
44. Bennion KA, Payne JD, Kensinger EA. Selective effects of sleep on emotional memory: what mechanisms are responsible? Trans Issu Psychol Sci. 2015;1(1):79e88.
45. Genzel L, Spoormaker V, Konrad BN, Dresler M. The role of rapid eye movement sleep for amygdala-related memory processing. Neurobiol Learn Mem. 2015;122:110e121.
46. Wiesner CD, Pulst J, Krause F, Elsner M, Baving L, Pedersen A, et al. The effect of selective REM-sleep deprivation on the consolidation and affective evaluation of emotional memories. Neurobiol Learn Mem. 2015;122:131e141.
47. Groch S, Zinke K, Wilhelm I, Born J. Dissociating the contributions of slow-wave sleep and rapid eye movement sleep to emotional item and source memory. Neurobiol Learn Mem. 2015;122:122e130.
48. Payne JD, Chambers AM, Kensinger EA. Sleep promotes lasting changes in selective memory for emotional scenes. Front Integr Neuros. 2012;6:108.
49. Gilson M, Deliens G, Leproult R, Bodart A, Nonclercq A, Ercek R, et al. REM-enriched naps are associated with memory consolidation for sad stories and enhance mood-related reactivity. Brain Sci. 2015;6(1):1e18.
50. Harrington MO, Pennington K, Durrant SJ. The "affect tagging and consolidation" (ATaC) model of depression vulnerability. Neurobiol Learn Mem. 2017;140:43e51.
51. Smith CT, Nixon MR, Nader RS. Posttraining increases in REM sleep intensity implicate REM sleep in memory processing and provide a biological marker of learning potential. Learn Mem. 2004;11(6):714e719.
52. Van Schalkwijk F, Sauter C, Hoedlmoser K, Heib D, et al. The effect of daytime napping and full-night sleep on the consolidation of declarative and procedural information. J Sleep Res. 2007;28(1).

Neurobiology of Insomnia

7

Célyne H. Bastien, Thierry Provencher, Jessica Lebel, and Roxanne Bolduc-Landry

Introduction

Insomnia disorder (ID) is defined by a difficulty of either initiating or maintaining sleep as well as waking up too early in the morning at least three times a week. These difficulties are also accompanied by daily consequences (e.g., reported attention and memory deficits). To be considered as insomnia, these difficulties must be present for at least 3 months [1]. Insomnia can be a symptom of another disorder (e.g., mental, physical) or a stand-alone disorder; the DSM-5 now uses the mention "comorbid" to diagnose insomnia co-occurring with another disorder. In addition, the ICSD-3 [2] does not use the subtype classification anymore but warrants study to explore neurophysiological markers between 11 common insomnia subtypes identified in the ISCD-2 [3], especially the two most important, psychophysiological insomnia (PSY-I) and sleep-state misperception or paradoxical insomnia (PARA-I).

While estimation of sleep in PSY-I tends to be quite accurate (subjective reports are similar to polysomnography (PSG)), PARA-I is generally associated with complaints of sleep difficulties while PSG recordings appear normal, where a constant and important gap occurs between objective and subjective measures of sleep [4]. Depending on the diagnostic criteria used, as much as 50% of individuals suffering from insomnia could be poor estimators and, consequently, classified as suffering from PARA-I [5, 6]. Subtypes of insomnia are not all mutually exclusive. An individual can show objective sleep-onset, sleep-maintenance, or mixed objective sleep difficulties but also present important discrepancies between reports and laboratory sleep observations. As such, an individual might be complaining of PSG-confirmed sleep difficulties over a long period of time, without showing any signs of

C. H. Bastien (✉) · T. Provencher · J. Lebel · R. Bolduc-Landry
School of Psychology, Laval University, Quebec, QC, Canada
e-mail: Celyne.bastien@psy.ulaval.ca; Thierry.Provencher.1@ulaval.ca; Jessica.lebel@psy.ulaval.ca; roxanne.bolduc-landry.1@ulaval.ca

L. M. DelRosso, R. Ferri (eds.), *Sleep Neurology*,
https://doi.org/10.1007/978-3-030-54359-4_7

depression or anxiety and, at the same time, greatly misestimating sleep or wake which may then correspond to PARA-I.

The prevalence of insomnia can reach up to 60% when present with another disorder, while as a stand-alone, it occurs chronically in about 12% of the population [7]. Many predisposing, precipitating, and maintaining factors have been linked to the disorder [8, 9]. From the neurocognitive models [10, 11] through the cognitive [12] and psychobiological inhibition ones [13], which have guided researchers and clinicians in their search to better understand insomnia, one approach gaining more and more attention to uncover specifics of the pathophysiology of this disorder is its neurobiology. In that regard, work on different neurotransmitters and other amino acids modulating wake and sleep has been the focus of many recent reviews. For example, orexin appears as an important candidate in insomnia neurobiology as well as the deficiency in benzodiazepine (BZD) receptors. As refined study techniques are rapidly developing, the neurobiology of insomnia is revealing itself.

This chapter will first present the sleep-wake regulation system to describe how insomnia might emerge. Second, we will introduce techniques used to study the neurobiology of insomnia and then will go on to discuss the different phenotypes of insomnia, including recent developments in our knowledge of short sleepers. Finally, the status of our understanding in daily functioning in insomnia and cortical areas involved will be briefly presented. Future directions will conclude this chapter. Note that this chapter deals only with insomnia not comorbid with any other disorders (psychiatric, mental, medical, etc.).

Sleep-Wake Regulation Systems

In the last decades, several review articles have brought together what we know concerning the sleep-wake regulatory systems. The 2005 review by Saper et al. [14] published in *Nature* was the first to compile such knowledge and remains the most cited review on the subject. We will briefly review cortical structures and neurotransmitters involved in sleep-wake regulation.

Wakefulness

The ascending reticular activating system or RAS has been shown to promote wakefulness. This system originates in the brain stem and releases a wide variety of neurotransmitters, mainly serotonin, dopamine, acetylcholine, norepinephrine, histamine, and glutamate. Its projections separate into two branches, which projects into the thalamus, hypothalamus, basal forebrain (BF), and cerebral cortex (see Fig. 7.1a).

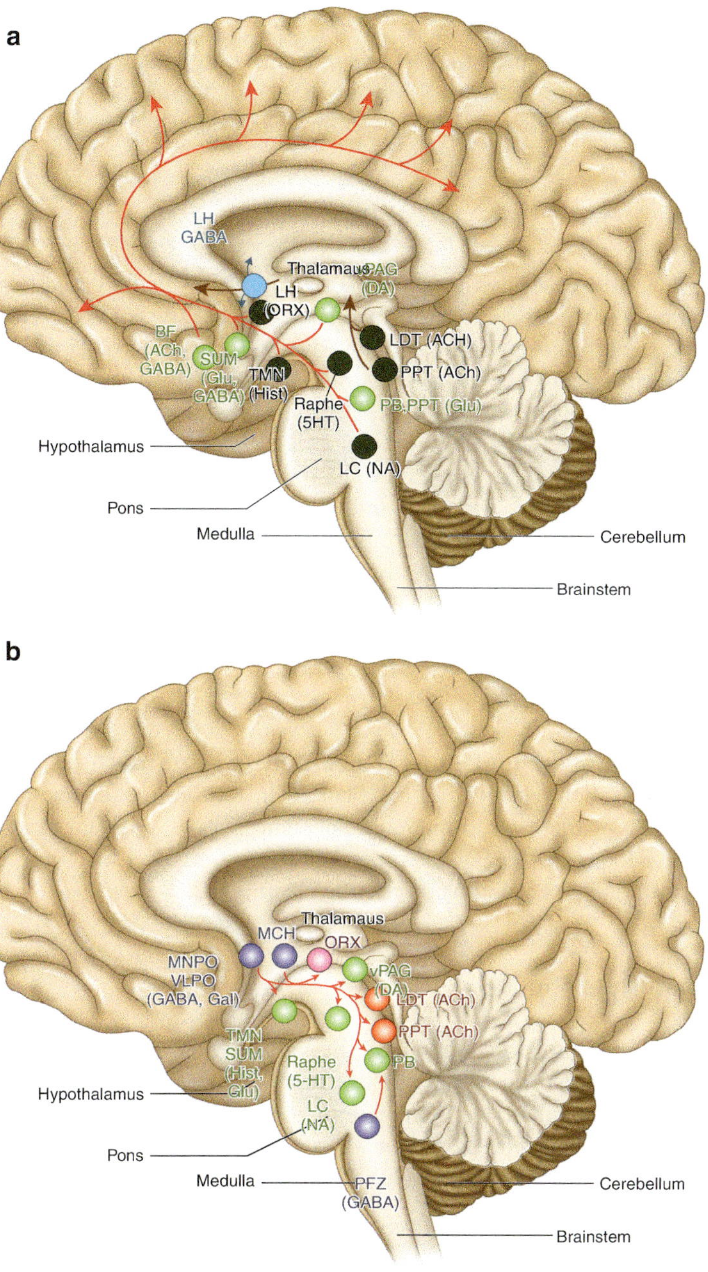

Fig. 7.1 (**a**) A schematic drawing showing the fast neurotransmitter systems that are at play in promoting wakefulness (i.e., the key components of the ascending arousal system). (**b**) A schematic drawing showing the fast neurotransmitter systems that contribute to sleep promotion (i.e., key projections of the VLPO to the main components of the ascending arousal system). (Reproduced with permission from Saper and Fuller [91]. With permission from Elsevier)

Sleep

The ventrolateral median preoptic (VLPO) nuclei, located in the hypothalamus, act as the sleep-promoting system. VLPO neurons release the inhibitory neurotransmitters GABA and galanin to inhibit wake-promoting systems (see Fig. 7.1b).

Flip-Flop Switch

The two previously described systems (i.e. wakefulness and sleep) are mutually inhibitory and create a feedback loop. When the arousal system is activated during wakefulness, it inhibits the sleep-promoting system; when neurons from the VLPO fire rapidly during sleep, they block the arousal system. This has been termed "flip-flop switch" due to the nature of the mutually inhibitory elements that produce sharp transitions between the two different states.

The neuropeptide orexin (hypocretin) is thought as a stabilizer of the switch. Orexin neurons project into the monoaminergic and cholinergic cell groups of the wake-promoting system, thus reinforcing wakefulness [15] (Fig. 7.2). A second neuropeptide, melanin-concentrating hormone (MCH), is involved in sleep-wake regulation, being active during sleep and becoming silent during wakefulness.

Homeostatic Regulation

In his two-process model, Borbély [16] described *process S* as the need for sleep or the accumulation of sleep-inducing substances in the brain. Adenosine has been proposed as one of these sleep-inducing substances. It was shown to accumulate mainly in the BF during the day, creating the homeostatic pressure for sleep, while also disinhibiting the VLPO, thus promoting sleep.

Circadian Regulation

Process C involves the 24-hour circadian rhythm known to act as a biological clock for sleep-wake cycles. It is located in the suprachiasmatic nucleus (SCN) of the hypothalamus and is modulated by melatonin, secreted by the pineal gland according to the light signal received by the photopigment melanopsin. The circadian clock is connected, via a multistage pathway, to the arousal system and the VLPO.

Alterations in Insomnia

A novel intermediate state has been suggested as being the result of a faulty flip-flop switch, blurring the boundaries between wake and sleep. Arousal systems and sleep-promoting systems were both shown to be hyperactive in INS, which could in part

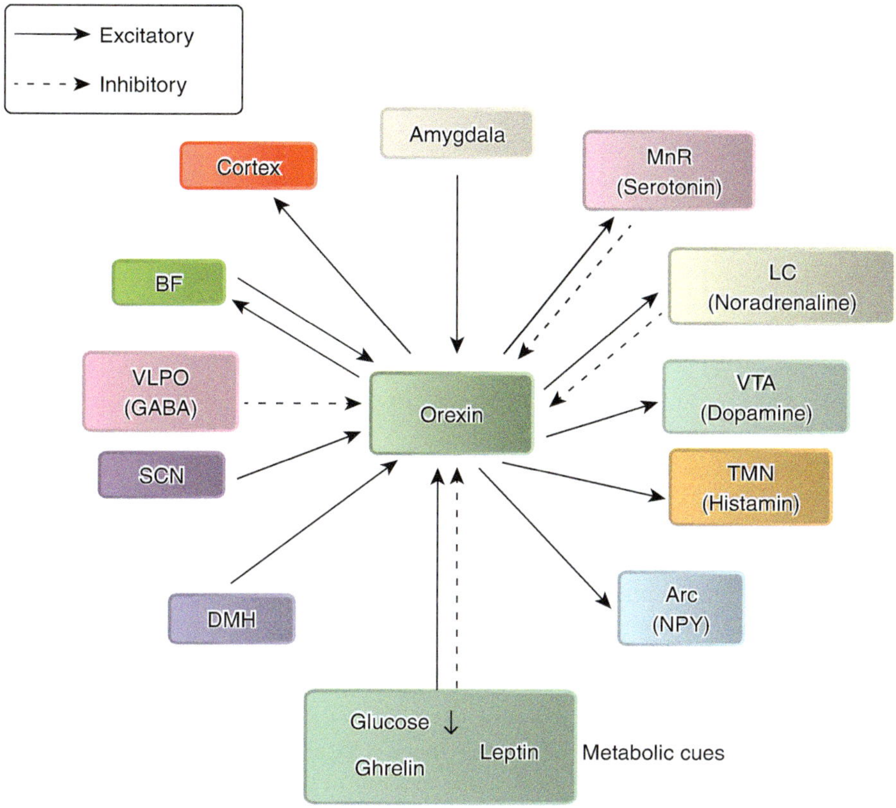

Fig. 7.2 Input and output pathways of orexin neurons in the lateral hypothalamus. Abbreviations: *BF* basal forebrain, *VLPO* ventrolateral preoptic nucleus, *SCN* suprachiasmatic nucleus, *DMH* dorsomedial hypothalamus, *Arc* arcuate nucleus, *NPY* neuropeptide Y, *TMN* tuberomammillary nucleus, *VTA* ventral tegmental area, *LC* locus coeruleus, *MnR* median raphe nucleus. (Reproduced with permission from Ono and Yamanaka [93]. With permission from Elsevier)

explain why some insomnia sufferers report worse sleep than they actually get (PARA-I). INS also suffer from fragmented sleep (frequent microarousals), which can be explained by reduced GABAergic activity or increased orexinergic activity. Furthermore, maladaptive behaviors often found in INS, like increased time in bed, might reduce the homeostatic processes needed to fall asleep. Indeed, on neurobiological and neurochemical levels, a decreased sleep drive in INS would lead to disinhibition of the arousal-promoting brain centers, thus promoting arousal over sleep.

As is discussed throughout this chapter, INS present a different activation pattern from good sleeper controls. Hyperactivation has been the main research focus for a long time, and in recent works, hypoactivation has been highlighted. It may be that sleep in insomnia is more of a localized phenomenon instead of a whole-brain phenomenon as it is in good sleepers; thus INS exhibit certain signs of hyper- and hypoactivation.

Methods and Techniques

Electroencephalography (EEG)

Electroencephalography (EEG) is a technique that allows the recording of low-intensity electrical potentials on specific regions of the scalp to measure the brain's neuronal activity. A power spectral analysis (PSA) can be performed on the resulting EEG recording to quantify the frequency and amplitude of brain waves[1] throughout the night [17, 18]. The power of those frequency bands can be expressed as absolute (i.e., raw power) or relative (i.e., percentage of power represented by each frequency band). Although the classification of frequency bands may differ, they can be defined as follows: slow waves (0–1 Hz), delta (1–4 Hz), theta (4–7 Hz), alpha (7–11 Hz), sigma (11–14 Hz), beta 1 (14–20 Hz), beta 2 (20–35 Hz), gamma (35–60 Hz), and omega (60–125 Hz) [17].

It is well recognized that, at sleep onset, insomnia sufferers (INS) have higher beta activity and reduced delta activity compared to good sleepers (GS) [19–21]. Higher beta wave activity is usually associated with alertness, whereas reduced delta wave activity may represent a decrease in drowsiness, both mechanisms thus decreasing the likelihood of falling asleep. This is not surprising as most INS report having difficulty falling asleep. These results support the main hypothesis of cerebral hyperactivation underlying several theoretical models developed over the past decades [10, 12, 13]. This hyperactivation documented by several studies is said to be even more present in non-paradoxical sleep (NREM) of INS [18, 22–26]. An increase in higher-frequency activity (i.e., beta) may also contribute to a higher rate of microarousals during REM sleep, now thought to be particularly fragmented in patients with insomnia [27–29].

Insomnia Subtypes

After six nights of EEG recording, Krystal and his colleagues [30] observed that the relative and absolute power of all frequency bands was similar for GS, PARA-I, and PSY-I during REM sleep. However, significant differences were found between groups during NREM sleep. PSY-I showed significantly higher relative power in the sigma frequency band compared to GS, whereas the NREM sleep of PARA-I was characterized by lower delta activity and higher alpha, beta, and sigma activity than GS. These results suggest that NREM sleep frequency is potentially related to sleep complaints and misperception. Although diagnosis of both these insomnia subtypes in research can be based on the work of Edinger and his colleagues [4], data indicates that we may theoretically be able to categorize insomnia sufferers in subtypes based solely on their brain activity; to this end, more research on the matter is still needed.

Our research group found significant differences between PARA-I and PSY-I during NREM sleep [25]. Results show higher absolute activity in slow waves, delta, theta, alpha, sigma, and gamma bands in PARA-I compared to PSY-I during

[1] Frequency and amplitude are measured, respectively, in hertz (Hz) and microvolts (μV).

NREM sleep. Surprisingly, there were no differences in beta activity between groups. Our results partly corroborated those obtained by Krystal et al. [30] by indicating that PARA-I exhibits a greater decrease in delta band activity in NREM sleep and an increase in alpha, sigma, and beta frequency band activity. In REM sleep, relative power in slow wave bands, theta, alpha, and sigma is higher in PSY-I than PARA-I [25]. Thus, in paradoxical insomnia, there is an increase in cortical activity (i.e., greater slow and fast EEG activity) during NREM sleep and decreased slow cortical activity in REM sleep, which corroborates the hypothesis that this mode of cortical activation, not necessarily seen in PSY-I, could be related to some extent to sleep misperception. Harvey and Tang [31] observed that in PARA-I, sleep tends to be misperceived as wake, along with greater displays of worry and selective attention toward sleep-related threats, and more brief awakenings, all those features being related to faster brain activity.

Structural Magnetic Resonance Imaging (MRI)

Structural magnetic resonance imaging (MRI) is a neuroimaging technique allowing the detection of abnormal volume in brain structures. A paper by Riemann et al. [11] gathered results from recent neuroimaging studies revealing that insomnia is accompanied by a reduction in gray matter volume in several brain regions such as the medial frontal lobes, the dorsolateral prefrontal cortex, the parietal cortex, and the anterior cingulate cortex. Research shows that the decrease in bilateral hippocampi gray matter volume seems to be predominant [32–34]. This is interesting since INS often report having memory issues and it is known that the hippocampus is critical for memory consolidation. Since this structure is affected by the amount and quality of sleep attained, a study by Noh et al. [35] aimed at clarifying the effect of chronic sleep deprivation on brain structure and cognition. Their results strongly suggested that chronic sleep deprivation experienced in insomnia tends to impair memory and frontal lobe function, which eventually may contribute to a reduction in hippocampal volume.

Functional Magnetic Resonance Imaging (fMRI)

In addition to structural alterations, functional MRI (fMRI) measures the hemodynamic response related to neuronal activity in the brain [17]. Increased blood flow in a brain structure is usually interpreted as increased activity in that same region. An fMRI study by Santarnecchi et al. [36] demonstrates that sleep deprivation or poor sleep quality induces a variety of changes in both good sleepers [37] and patients with sleep disorders [38–41]. Performance (or lack thereof) in neuropsychological tasks of attention and memory is often associated with hypoactivation (i.e., decreased blood flow) of task-related areas, particularly in frontal-subcortical networks [42]. An EEG study by Corsi-Cabrera et al. [26] corroborates these findings, their sample of INS patients presenting impaired frontal deactivation and disengagement of brain regions involved in executive control, attention, and self-awareness.

Regions of the *default mode network* (DMN), the network thought to be involved in awareness, mental representations of the self, processing of context-related information, and modulation of conscious processes (DMN; [43]), also present alterations in insomnia. One study revealed reduced deactivation in the DMN in INS during a working memory task, which correlated with a decreased performance in this task [40]. This deactivation in DMN may be indicative of increased worry and rumination among INS during the day [44]. In a study by Baglioni and colleagues [45], the activation of the amygdala was increased in INS compared to GS when exposed to images of individuals awake in bed at night. This suggests a strong link between the hyperactivation of the limbic system in insomnia and the high level of negative emotions experienced by INS [46, 47].

In 2018, Leerssen et al. [48] used both structural and resting-state fMRI to evaluate hippocampal volume and functional connectivity in insomnia. In their study, there were no differences in hippocampal volume between INS and GS. This contradicts results by Noh et al. [35], who showed that hippocampal volumes were different. However, Leerssen et al. [48] observed that, compared to GS, INS tend to have higher functional connectivity between bilateral hippocampi and left middle frontal gyrus (MFG) in insomnia. According to the authors, these structures are part of a circuit that is further activated by maladaptive rumination and deactivated with sleep. Thus, when the strength of this connectivity increases, insomnia severity tends to increase, while sleep efficiency decreases. The fMRI underpinnings related to the different subtypes of insomnia still remain to be clarified.

Positron Emission Tomography (PET)

Positron emission tomography (PET) is a functional nuclear medicine imaging technique that allows observing certain metabolic processes in the body. In the last decades, several studies using PET with insomnia patients have used fludeoxyglucose (FDG) (a glucose analogue) as an agent to investigate the metabolic activity of regions according to their glucose uptake [49].

Using FDG as a radiotracer, a PET study by Nofzinger et al. [50] revealed an increase in overall brain glucose metabolism in INS compared to GS during NREM sleep. Specifically, there is an inhibition of the usual decline in activity in patients with insomnia in several subsequent brain regions such as the ascending reticular activating system, hypothalamus, thalamus, amygdala, hippocampus, insula, and anterior and prefrontal cingulate cortex. According to these authors, these results suggest that insomnia is accompanied by increased activation of alertness, regulation of emotions, and several cognitive systems. For this reason, insomnia can be conceptualized as a corticolimbic hyperactivity disorder that can interfere with brain structures that promote sleep [51].

During wakefulness, INS show relative hypometabolism in the temporal, parietal, occipital, and prefrontal cortices, left hemisphere, thalamus, hypothalamus, and brain stem compared to GS. These results are in line with the study by Thomas et al. [52] who indicated that the hypometabolism of these similar brain structures follows sleep deprivation, even in GS. A paper by Buysse et al. [53] stipulates that INS present relatively greater metabolism in the frontoparietal cortex than GS during NREM sleep, most strikingly in the precuneus (i.e., a component of the *default mode network*). These results suggest that INS may be chronically deprived of sleep, hence the relative hypometabolism during the day, and that their sleep is associated with an increase in cerebral metabolism. The difficulty experienced by most INS to fall asleep therefore represents a central difficulty in inhibiting certain wakefulness-oriented mechanisms at the sleep onset.

Evoked Potentials (ERPs)

Evoked potentials or event-related potentials (ERPs)[2] are small voltage fluctuations generated in brain structures in response to specific events or stimuli [54]. A few decades ago, researchers were already beginning to take an interest in the relation between ERPs and insomnia. For example, one study in 1993 showed that N350 (i.e., sleep-specific ERP thought to reflect the inhibition of processing of potentially sleep-disrupting stimulus input; [55]) was of smaller amplitude for INS than GS at the beginning of sleep stage 2 [56]. In this same study, there was also a greater amplitude of P300 (i.e., ERP thought to reflect the updating of an active working memory; [57]) immediately before sleep onset and a shorter response latency in INS compared to GS during awakenings [56]. In other words, INS have trouble inhibiting wake processes, making it difficult for sleep mechanisms to be initiated, while they also present cognitive (i.e., memory) workload. These results suggest a state of hyperactivity in INS compared to GS, which is consistent with previous EEG research.

P1 (i.e., ERP thought to reflect the cognitive cost of attention; [58]) and N1 (i.e., ERP thought to reflect the outcome of a transient detector system; [59]) are significantly greater in amplitude in INS than in GS when measured during awakening [60, 61]. That said, during awakening, the amplitude of P2 (i.e., ERP thought to reflect an inhibitory process protecting sleep from irrelevant disturbance; [59]) tends to be lower in INS than in GS [60, 61]. This means that insomnia sufferers are less likely to fall or maintain sleep as they have trouble inhibiting certain diurnal processes somewhat incompatible with sleep.

In light of these results, our research group compared INS and GS in terms of ERPs [62]. The amplitude of N1 was greater in INS than GS when waking up in the morning which, again, indicates a certain state of cortical hyperarousal, whereas the

[2] Note that valence and peak latency are used to label the different components. For instance, "N" refers to a negative wave and "P" refers to a positive wave, whereas N100 and N350 (or N1 and N35) refer to negative waves peaking, respectively, at about 100 or 350 ms after stimulus onset, and P100, P200, and P300 (or P1, P2, and P3) refer to positive waves peaking, respectively, at about 100, 200, or 300 ms after stimulus onset.

amplitude of N350 was lower among INS at the beginning of the night. These results suggest that INS are more vigilant than GS during wakefulness, which also includes the period of time before sleep onset. This is consistent with the results of PET studies which observed INS cerebral hyperactivation at night, as well as their difficulty in inhibiting certain processes present during awakening, which decreases their likelihood of falling asleep [52].

Insomnia Subtypes

Currently few studies have examined the difference in ERPs between subtypes of insomnia. Preliminary results suggest that during the evening and early in the night, the amplitude of N1 and P2 is higher in PARA-I than PSY-I and GS and N350 amplitude is lower [63]. This may possibly indicate stronger cortical deficits of activation and inhibition in PARA-I. The preponderance of these neurophysiological mechanisms could have an impact on their sleep perception, perhaps leading them to falsely perceive that they sleep less than in reality, because their brain is hyperactive or more awake.

A study by Turcotte et al. [64] investigated N1 and P2 components between GS, PARA-I, and PSY-I in the evening, at sleep onset, and in early stage 2 sleep. They observed that during wakefulness and sleep onset, N1 was smaller for PSY-I and larger for PARA-I. At sleep onset, P2 was smaller for PSY-I than the other groups, but during wakefulness and stage 2 sleep, it was larger for PARA-I than GS. They concluded that these results may illustrate an inability for PSY-I to inhibit information processing during sleep onset and an overall enhanced attentional processing for PARA-I. Turcotte et al. [64] hypothesized that increased N1 may be a direct reflection of the obtrusiveness of stimulation. For PARA-I, this obtrusiveness may translate into a perception of "being more awake," especially during REM sleep.

Waking Aspects of Insomnia

Much evidence points toward a cortical and physiological arousal in insomnia as well as difficulties in the inhibitory processes in the evening and during sleep [11, 17, 62, 65, 66]. Although insomnia mainly concerns sleeping difficulties, manifestations of the disorder are also visible in the 24-hour sleep-wake period such as found in EEG, ERP, and MRI studies and in physiological measurement (e.g., HPA axis, cardiovascular, neurochemical). Sleep studies to date are not sufficient to fully understand the pathophysiology of this disorder. There is still limited literature about the physiological and cortical manifestation of insomnia during wake, and it mostly concerns wake at pre-sleep onset.

24-Hour Phenomenon

Over 30 years ago, a team of researchers suggested that there may be a central dysfunction in INS which would render this disorder a stable 24-hour phenomenon

[66]. In their study with 70 INS and 45 GS, they observed that individuals with difficulty sleeping at night showed signs of hyperarousal even before sleep onset. What is even more intriguing is that their sleep loss did not increase their daytime sleepiness, quite the contrary. INS showed significantly greater alertness during the day than asymptomatic sleepers, in line with the hyperarousal theory [65]. Those results suggest that insomnia and its cortical correlates may not be restricted to the presleep or sleep periods but may be present continuously, on a 24-hour basis [66].

In fact, in a recent study on daytime naps, it was observed that the brain activity of GS and INS not only differs during the night but also during the day [67]. For instance, GS had a longer total sleep time and better sleep efficiency than INS, while INS were awake significantly longer than GS. In this study, participants went through prolonged cognitive testing to induce sleepiness (i.e., mental fatigue). The fact that GS slept better than INS during naps suggests that hyperarousal predominates over mental fatigue resulting from cognitive testing.

Physiological Measurements and Cognitive Deficits

Thereby, insomnia is not only a night problem; diurnal manifestations are ubiquitous. In such, it is crucial to measure insomnia symptoms and consequences during wake. In their review, Vgontzas, Fernandez-Mendoza, Liao, and Bixler [68] discuss, among others, the HPA axis activity – which is associated with stress, arousal, and sleeplessness – in insomnia. Two studies [69, 70] have found that HPA activity was directly proportional to the severity of objective sleep disturbances. Short sleepers (sleepers getting less than 6 hours of sleep per night) suffering from insomnia exhibited higher amounts of cortisol, adrenocorticotropic hormone (ACTH), and norepinephrine (neurochemicals associated with high-level arousal) during the day, compared to normal/long sleepers. Vgontzas et al. [69] observed elevated level of ACTH and cortisol at any given time in INS when compared to GS. These results were not found in INS with subjective sleep complaints (group of individuals closely corresponding to PARA-I as defined by Bastien's group [17]) but only in those with high objective sleep difficulties (% sleep efficiency <70%; PSY-I as defined by Bastien's group [17]). The activity of the stress system could be hypothesized as being directly proportional to the degree of objective sleep disturbance in INS. Furthermore, short sleepers with insomnia were found to be significantly impaired on higher-order cognitive functions – poorer processing speed, lesser ability to flexibly switch attention, and poorer working visual memory as assessed by increased number of errors and omissions [71]. These impairments were associated with physiological arousal – hypercortisolemia and catecholaminergic and sympathetic activity – considering that it may lead to neurocognitive deficits [71]. Insomnia with objective short sleep duration is considered the most biologically severe phenotype of ID by some researchers [68]. Those findings on impaired high-order functions were replicated in INS with normal sleep duration [72]. Increased heart rate and blood pressure both during the day and night have also been reported [73]. Distinctions between subtypes or phenotypes of insomnia were overseen in the last studies.

Waking EEG

Results on wake EEG are somewhat conflicting. A majority of studies have shown higher-frequency power in the beta [74–77] and gamma range [74, 76] in INS, whereas another study did not observe any group differences [24]. Overall, higher beta and gamma frequency could be seen as indices of cortical hyperarousal.

Waking ERPs

As mentioned earlier, several studies have used ERPs in insomnia to depict cortical arousal in wake and sleep state. According to Colrain and Campbell [57], ERPs can be very useful in the assessment of daytime consequences following sleep disruption and, by extension, insomnia. Bastien, St-Jean, Morin, Turcotte, and Carrier [62] studied waking ERP amplitudes (evening and morning) in PSY-I and GS. During evening and the subsequent morning, PSY-I showed increased N1 and P2 amplitudes suggesting an increased sensibility to stimuli as well as an increased demand for inhibition. Thus, hyperarousal in PSY-I could be seen as pervasive not only during the night but also during daytime, further supporting the 24-hour conception of insomnia.

Cerebral Connectivity

In light of these results, INS seem to exhibit an overall difficulty to stop processing stimuli, even nonrelevant ones, as illustrated by a greater P2 amplitude during wake [63]. Similar results have been previously observed by Oades et al. [78] in attention-deficit hyperactivity disorder (ADHD) and Tourette disorder in children, suggesting that enhanced information processing of nonrelevant stimuli – which can be linked to attentional deficit – is reflected through an increase in P2 amplitude. According to this theory, increased P2 amplitude in this population could be a sign of increased effort to inhibit, even block, any stimulation. In addition, an MRI study by Killgore, Schwab, Kipman, DelDonno, and Weber [79] examined resting-state functional connectivity differences between individuals with or without insomnia-related sleep problems. They observed that difficulties in falling asleep were associated with increased functional connectivity between sensory regions and motor areas; and problems with sleep maintenance were associated with greater connectivity between the sensory regions themselves. According to these authors, their findings suggest that an increase in functional connectivity in those regions may contribute to the sustained sensory processing of environmental stimuli, delaying further sleep latency. Therefore, greater connectivity among the sensory and motor regions in insomnia may be conceivably seen as evidence of the hyperarousal theory of insomnia [9]. Greater beta EEG was also observed in those regions in INS [75] – and enhanced unwanted sensory awareness, leading to a difficulty in sleep initialization or maintenance. Even though these researchers did not relate their findings to ADHD, a review by Konrad and Eickhoff [80] reports the implication of similar brain connectivity arousal in sensory and motor regions in insomnia and ADHD.

Insomnia and ADHD

Common mechanisms seem to be at play in insomnia and ADHD, although the link between the two has yet to be fully understood. ADHD is commonly associated with sleep difficulties such as insomnia [81]. EEG studies in ADHD population typically report increased theta and delta activity and decreased alpha and beta activity (also called the theta/beta power ratio; [82, 83]) – which is quite the opposite of what is found in INS. At first, the theta/beta ratio was explained as cortical arousal, but lower skin conductance-level (SCL) results [82, 84] – skin conductance having a long history being a measure of CNS arousal – indicate that it is not arousal. Besides, when insomnia is concurrently examined in ADHD sufferers, hyperarousal does not seem to be at play, quite the contrary. Hypoarousal, or a lack of inhibition processes, rather, emerges. In a study with young ADHD children, Clarke et al. [85] highlight three clusters depicting different EEG profiles. The first two clusters are very similar and are characterized by increased theta activity and reduced beta activity – in accordance with the general ADHD EEG manifestations. The third cluster though shows an increase in relative beta, called spindling excessive beta (SEB), and decreased theta activity. Clarke et al. [85] first suggested that this third cluster was contradictorily associated with hyperarousal and might easily be associated with the hyperactive and inattentive problems found in ADHD. However, the same group [84] recently observed that SCL was reduced in ADHD children with excessive SEB and this cluster was then interpreted as showing hypoarousal. This hypoarousal might be connected to Espie's theorization [13] in which INS also lack inhibition processes. Thus we can appreciate why difficulties in concentration, sleep initiation, and maintenance may be seen in both ADHD and insomnia – and why these disorders are often comorbid.

Similarly, Arns, Swatzyn, Gunkelman, and Olbrich [86] also refer to hypoarousal as a diurnal manifestation present both in ADHD and insomnia. In 429 patients with heterogeneous psychiatric condition, a relative proportion of ADHD patient (12.1%) presented elevated SEB in the frontal region. Those participants also reported more insomnia complaints and hyperactivity/impulsivity symptoms. Arns et al. [86] thus postulated that SEB is associated to behavioral dimensions (insomnia and impulsivity/hyperactivity) and may reflect a problem with *de-arousal* processes. With those findings, SEB may be a sign of reduced vigilance associated with sleep problem which in turn is representative of hypoarousal. Hyperactivity/impulsivity symptoms and insomnia may thus both result from a lack of cortical inhibitory processes.

A New Way to See Insomnia?

In a recent study of high-density EEG comparing INS and GS in the evening, eyes opened or eyes closed, results pointed toward two principal evidences in the wake state for INS [75]. The first is a widespread enhanced broad beta power (16.3–40 Hz) which has been found in some earlier studies as well [74–77]. As presented earlier in this chapter, many studies have found elevated beta frequency power in sleep and have explained it as a manifestation of cortical hyperarousal [18, 22–26].

Accordingly, results from Colombo et al. [75] support the conception of an around the clock (24-hour) cortical hyperarousal in insomnia. There were also differences between INS and GS as to where the source of beta frequency power mostly originated. Those differences were maximal over sensorimotor cortices. Even though authors omitted to link hyperarousal in the sensorimotor cortices to clinical manifestations, we may perhaps stipulate that this arousal could be reflected in hyperactivity and impulsivity as seen in ADHD children and adults. Considering the interpretation of greater SEB in ADHD sufferers, it may lead to a new vision of this greater beta frequency power, a cortical attempt to inhibit arousal processes. This vision can be integrated in the second evidence elaborated by Colombo et al. [75] which is a widespread lower/upper alpha power (11–12.7 Hz). Authors explained the decreased alpha activity as a lack of inhibition in insomnia toward interfering cognitions. Normally, inhibition of non-pertinent cognition is shown with higher alpha frequency. This supports the hypothesis of Bastien et al. [62] about a lack of inhibition showed by greater P2 amplitude.

Riemann et al. [65] caution further researches to clearly understand the 24-hour insomnia theorization. Is it possible that the hyperarousal theory is misdeem and that there may exist two distinct phenomena as previously suggested by Espie's [13] biopsychosocial theory? Could there be hyperarousal and also hypoarousal of specific cortical processes at different times? Although there is only a paucity of results studying both the awake and sleep periods of INS, available data points toward this hypothesis.

Conclusion and Future Directions

This chapter has evidenced that INS do display signs of physiological hyperarousal albeit through the default mode network, with PET and other imagery techniques. Although a full portrait of the different subtypes of insomnia sufferers (PSY-I, PARA-I, and short-sleepers) remains to be documented in regard of nighttime cortical processes, the daytime consequences of these different subtypes are not well understood at all. It is also quite possible that misperception indexes be more indicated to study the different subtypes. As suggested in the last section, most likely both hyperarousal and hypoarousal mechanisms are at play at different times within the 24-hour cycle. Furthermore, while our INS subclassification is not optimal, little is known about the impact of each different symptoms of insomnia (sleep onset and sleep maintenance) on daytime functioning. In fact, studies are more and more using total sleep time as the "main" symptoms, while we still do not know if sleep-onset insomnia and sleep-maintenance insomnia are physiologically similar/different.

To further decipher the neuropathophysiology of insomnia, future studies must try to incorporate most recent techniques (PET, tDCS, etc.) when providing treatment. For example, more than a decade ago, both Cervena, Dauvilliers, Espa et al. [87] and Altena, Van Der Merf et al. [88] observed EEG (less beta and increased delta in NREM sleep) and fMRI changes (hypometabolism) at post-cognitive behavioral therapy for insomnia (CBT-I) compared to pretreatment. A very recent study has highlighted that INS are more subject to stage shifts than good sleepers, and this is in light of NREM sleep – S2 – which is often unstudied in any population [89], showing that the sleep of INS is more shallow and fragmented than the one of good sleepers in lighter sleep as it is in REM sleep [28]. The study of phasic events and other EEG measures during sleep can thus also offer insights as to what is happening cortically during the night in insomnia individuals. In addition, it would be really interesting to study homeostatic pressure in INS as this one might entertain a direct link with treatment (e.g., be involved in lack of response).

Further Readings

For reviews on the subject of sleep-wake regulation, see [14, 90–92]. On the same topic, specifically for insomnia, please see [11]. For a meta-analysis on the subject of theta/beta ratio in ADHD, see [83].

References

1. American Psychiatric Association. Diagnostic and statistical manual of mental disorders. 5th ed. Arlington: American Psychiatric Publishing; 2013.
2. American Academy of Sleep Medicine. International classification of sleep disorders: diagnostic and coding manual. 3rd ed. Westchester: Author; 2014.
3. American Academy of Sleep Medicine. International classification of sleep disorders: diagnostic and coding manual. 2nd ed. Westchester: Author; 2005.
4. Edinger JD, Bonnet MH, Bootzin RR, Doghramji K, Dorsey CM, Espie CA, et al. Derivation of research diagnostic criteria for insomnia: report of an American Academy of sleep medicine work group. Sleep. 2004;27(8):1567–96.
5. Edinger JD, Krystal AD. Subtyping primary insomnia: is sleep state misperception a distinct clinical entity? Sleep Med Rev. 2003;7(3):203–14.
6. St-Jean G, Bastien CH. Classification of insomnia sufferers based on laboratory PSG recordings and subjective sleep reports. Sleep. 2009;32:A284.
7. Morin CM, LeBlanc M, Daley M, Gregoire JP, Merette C. Epidemiology of insomnia: prevalence, self-help treatments, consultations, and determinants of help-seeking behaviors. Sleep Med. 2006;7(2):123–30.
8. Bastien CH, Vallieres A, Morin CM. Precipitating factors of insomnia. Behav Sleep Med. 2004;2(1):50–62.
9. Bastien CH, Vallieres A, Morin CM. Validation of the insomnia severity index as an outcome measure for insomnia research. Sleep Med. 2001;2(4):297–307.
10. Perlis ML, Giles DE, Mendelson WB, Bootzin RR, Wyatt JK. Psychophysiological insomnia: the behavioural model and a neurocognitive perspective. J Sleep Res. 1997;6(3):179–88.
11. Riemann D, Nissen C, Palagini L, Otte A, Perlis ML, Spiegelhalder K. The neurobiology, investigation, and treatment of chronic insomnia. Lancet Neurol. 2015;14(5):547–58.
12. Harvey AG. A cognitive model of insomnia. Behav Res Ther. 2002;40(8):869–93.
13. Espie CA. Insomnia: conceptual issues in the development, persistence, and treatment of sleep disorder in adults. Annu Rev Psychol. 2002;53:215–43.
14. Saper CB, Scammell TE, Lu J. Hypothalamic regulation of sleep and circadian rhythms. Nature. 2005;437(7063):1257–63.
15. Saper CB, Chou TC, Scammell TE. The sleep switch: hypothalamic control of sleep and wakefulness. Trends Neurosci. 2001;24(12):726–31.

16. Borbély AA. A two process model of sleep regulation. Hum Neurobiol. 1982;1(3):195–204.
17. Bastien CH. Insomnia: neurophysiological and neuropsychological approaches. Neuropsychol Rev. 2011;21(1):22–40.
18. Spiegelhalder K, Regen W, Feige B, Holz J, Piosczyk H, Baglioni C, et al. Increased EEG sigma and beta power during NREM sleep in primary insomnia. Biol Psychol. 2012;91(3):329–33.
19. Freedman RR. EEG power spectra in sleep-onset insomnia. Electroencephalogr Clin Neurophysiol. 1986;63(5):408–13.
20. Merica H, Blois R, Gaillard JM. Spectral characteristics of sleep EEG in chronic insomnia. Eur J Neurosci. 1998;10(5):1826–34.
21. Merica H, Gaillard JM. The EEG of the sleep onset period in insomnia: a discriminant analysis. Physiol Behav. 1992;52(2):199–204.
22. Maes J, Verbraecken J, Willemen M, De Volder I, van Gastel A, Michiels N, et al. Sleep misperception, EEG characteristics and autonomic nervous system activity in primary insomnia: a retrospective study on polysomnographic data. Int J Psychophysiol. 2014;91(3):163–71.
23. Israel B, Buysse DJ, Krafty RT, Begley A, Miewald J, Hall M. Short-term stability of sleep and heart rate variability in good sleepers and patients with insomnia: for some measures, one night is enough. Sleep. 2012;35(9):1285–91.
24. Wu YM, Pietrone R, Cashmere JD, Begley A, Miewald JM, Germain A, et al. EEG power during waking and NREM sleep in primary insomnia. J Clin Sleep Med. 2013;9(10):1031–7.
25. St-Jean G, Turcotte I, Perusse AD, Bastien CH. REM and NREM power spectral analysis on two consecutive nights in psychophysiological and paradoxical insomnia sufferers. Int J Psychophysiol. 2013;89(2):181–94.
26. Corsi-Cabrera M, Figueredo-Rodriguez P, del Rio-Portilla Y, Sanchez-Romero J, Galan L, Bosch-Bayard J. Enhanced frontoparietal synchronized activation during the wake-sleep transition in patients with primary insomnia. Sleep. 2012;35(4):501–11.
27. Feige B, Al-Shajlawi A, Nissen C, Voderholzer U, Hornyak M, Spiegelhalder K, et al. Does REM sleep contribute to subjective wake time in primary insomnia? A comparison of polysomnographic and subjective sleep in 100 patients. J Sleep Res. 2008;17(2):180–90.
28. Riemann D, Spiegelhalder K, Nissen C, Hirscher V, Baglioni C, Feige B. REM sleep instability--a new pathway for insomnia? Pharmacopsychiatry. 2012;45(5):167–76.
29. Kryger MH, Roth T, Dement WC. Principles and practice of sleep medicine. 4th ed. Philadelphia: Elsevier; 2005.
30. Krystal AD, Edinger JD, Wohlgemuth WK, Marsh GR. NREM sleep EEG frequency spectral correlates of sleep complaints in primary insomnia subtypes. Sleep. 2002;25(6):630–40.
31. Harvey AG, Tang NK. (Mis)perception of sleep in insomnia: a puzzle and a resolution. Psychol Bull. 2012;138(1):77–101.
32. Riemann D, Voderholzer U, Spiegelhalder K, Hornyak M, Buysse DJ, Nissen C, et al. Chronic insomnia and MRI-measured hippocampal volumes: a pilot study. Sleep. 2007;30(8):955–8.
33. Joo EY, Kim H, Suh S, Hong SB. Hippocampal substructural vulnerability to sleep disturbance and cognitive impairment in patients with chronic primary insomnia: magnetic resonance imaging morphometry. Sleep. 2014;37(7):1189–98.
34. Neylan TC, Mueller SG, Wang Z, Metzler TJ, Lenoci M, Truran D, et al. Insomnia severity is associated with a decreased volume of the CA3/dentate gyrus hippocampal subfield. Biol Psychiatry. 2010;68(5):494–6.
35. Noh HJ, Joo EY, Kim ST, Yoon SM, Koo DL, Kim D, et al. The relationship between hippocampal volume and cognition in patients with chronic primary insomnia. J Clin Neurol. 2012;8(2):130–8.
36. Santarnecchi E, Del Bianco C, Sicilia I, Momi D, Di Lorenzo G, Ferrone S, et al. Age of insomnia onset correlates with a reversal of default mode network and supplementary motor cortex connectivity. Neural Plast. 2018;2018:3678534.
37. De Havas JA, Parimal S, Soon CS, Chee MW. Sleep deprivation reduces default mode network connectivity and anti-correlation during rest and task performance. NeuroImage. 2012;59(2):1745–51.

38. Altena E, Vrenken H, Van Der Werf YD, van den Heuvel OA, Van Someren EJ. Reduced orbitofrontal and parietal gray matter in chronic insomnia: a voxel-based morphometric study. Biol Psychiatry. 2010;67(2):182–5.
39. Stoffers D, Altena E, van der Werf YD, Sanz-Arigita EJ, Voorn TA, Astill RG, et al. The caudate: a key node in the neuronal network imbalance of insomnia? Brain. 2014;137(Pt 2):610–20.
40. Drummond SP, Walker M, Almklov E, Campos M, Anderson DE, Straus LD. Neural correlates of working memory performance in primary insomnia. Sleep. 2013;36(9):1307–16.
41. Santarnecchi E, Sicilia I, Richiardi J, Vatti G, Polizzotto NR, Marino D, et al. Altered cortical and subcortical local coherence in obstructive sleep apnea: a functional magnetic resonance imaging study. J Sleep Res. 2013;22(3):337–47.
42. Spiegelhalder K, Regen W, Prem M, Baglioni C, Nissen C, Feige B, et al. Reduced anterior internal capsule white matter integrity in primary insomnia. Hum Brain Mapp. 2014;35(7):3431–8.
43. Cavanna AE, Trimble MR. The precuneus: a review of its functional anatomy and behavioural correlates. Brain. 2006;129(Pt 3):564–83.
44. Nofzinger EA. Advancing the neurobiology of insomnia, a commentary on: "functional imaging of the sleeping brain" by Drummond et al. Sleep Med Rev. 2004;8(3):243–7.
45. Baglioni C, Spiegelhalder K, Regen W, Feige B, Nissen C, Lombardo C, et al. Insomnia disorder is associated with increased amygdala reactivity to insomnia-related stimuli. Sleep. 2014;37(12):1907–17.
46. Baglioni C, Spiegelhalder K, Lombardo C, Riemann D. Sleep and emotions: a focus on insomnia. Sleep Med Rev. 2010;14(4):227–38.
47. Espie CA, Broomfield NM, MacMahon KM, Macphee LM, Taylor LM. The attention-intention-effort pathway in the development of psychophysiologic insomnia: a theoretical review. Sleep Med Rev. 2006;10(4):215–45.
48. Leerssen J, Wassing R, Ramautar JR, Stoffers D, Lakbila-Kamal O, Perrier J, et al. Increased hippocampal-prefrontal functional connectivity in insomnia. Neurobiol Learn Mem. 2019;160:144–50.
49. Bailey DL, Townsend DW, Valk PE, Maisey MN. Positron emission tomography: basic sciences. London: Springer; 2005.
50. Nofzinger EA, Buysse DJ, Germain A, Price JC, Miewald JM, Kupfer DJ. Functional neuroimaging evidence for hyperarousal in insomnia. Am J Psychiatry. 2004;161(11):2126–8.
51. Spiegelhalder K, Nissen C, Riemann D. Clinical sleep-wake disorders II: focus on insomnia and circadian rhythm sleep disorders. Handb Exp Pharmacol. 2019;253:261–76.
52. Thomas M, Sing H, Belenky G, Holcomb H, Mayberg H, Dannals R, et al. Neural basis of alertness and cognitive performance impairments during sleepiness. I. Effects of 24 h of sleep deprivation on waking human regional brain activity. J Sleep Res. 2000;9(4):335–52.
53. Buysse DJ, Germain A, Hall M, Monk TH, Nofzinger EA. A neurobiological model of insomnia. Drug Discov Today Dis Model. 2011;8(4):129–37.
54. Blackwood DH, Muir WJ. Cognitive brain potentials and their application. Br J Psychiatry Suppl. 1990;9:96–101.
55. Muller-Gass A, Campbell K. The processing of infrequently-presented low-intensity stimuli during natural sleep: an event-related potential study. Noise Health. 2010;12(47):120–8.
56. Hull JS. Event-related potentials during the wake/sleep transition in adults with and without primary insomnia [Dissertation]. University of Southern Mississippi; 1993.
57. Colrain IM, Campbell KB. The use of evoked potentials in sleep research. Sleep Med Rev. 2007;11(4):277–93.
58. Luck SJ, Hillyard SA, Mouloua M, Woldorff MG, Clark VP, Hawkins HL. Effect of spatial cueing on luminance detectability: psychophysical and electrophysiological evidence for early selection. J Exp Psychol Hum Percept Perform. 1994;20(4):887–904.
59. Campbell K. Event-related potentials as a measure of sleep disturbance: a tutorial review. Noise Health. 2010;12(47):137–53.
60. Loewy DH, Bootzin RR. Event-related potentials measures of information processing in insomniacs at bedtime and during sleep. Sleep. 1998;21(Suppl. 1):98.

61. Loewy DH, Burdik RS, Al-Shajlawi A, Franzen PL, Bootzin RR. Enhanced information processing at the peri-sleep onset period in insomniacs as measured by event-related potentials. Sleep. 1999;22(Suppl. 1):S152.
62. Bastien CH, St-Jean G, Morin CM, Turcotte I, Carrier J. Chronic psychophysiological insomnia: hyperarousal and/or inhibition deficits? An ERPs investigation. Sleep. 2008;31(6):887–98.
63. Bastien CH. ERP measures during wakefulness and sleep-onset in psychophysiological and paradoxical insomnia sufferers. J Sleep Res. 2008;17:65.
64. Turcotte I, St-Jean G, Bastien CH. Are individuals with paradoxical insomnia more hyperaroused than individuals with psychophysiological insomnia? Event-related potentials measures at the peri-onset of sleep. Int J Psychophysiol. 2011;81(3):177–90.
65. Riemann D, Spiegelhalder K, Feige B, Voderholzer U, Berger M, Perlis M, et al. The hyperarousal model of insomnia: a review of the concept and its evidence. Sleep Med Rev. 2010;14(1):19–31.
66. Stepanski E, Zorick F, Roehrs T, Young D, Roth T. Daytime alertness in patients with chronic insomnia compared with asymptomatic control subjects. Sleep. 1988;11(1):54–60.
67. Perusse AD, Turcotte I, St-Jean G, Ellis J, Hudon C, Bastien CH. Types of primary insomnia: is hyperarousal also present during napping? J Clin Sleep Med. 2013;9(12):1273–80.
68. Vgontzas AN, Fernandez-Mendoza J, Liao D, Bixler EO. Insomnia with objective short sleep duration: the most biologically severe phenotype of the disorder. Sleep Med Rev. 2013;17(4):241–54.
69. Vgontzas AN, Bixler EO, Lin HM, Prolo P, Mastorakos G, Vela-Bueno A, et al. Chronic insomnia is associated with nyctohemeral activation of the hypothalamic-pituitary-adrenal axis: clinical implications. J Clin Endocrinol Metab. 2001;86(8):3787–94.
70. Vgontzas AN, Tsigos C, Bixler EO, Stratakis CA, Zachman K, Kales A, et al. Chronic insomnia and activity of the stress system: a preliminary study. J Psychosom Res. 1998;45(1):21–31.
71. Fernandez-Mendoza J, Calhoun S, Bixler EO, Pejovic S, Karataraki M, Liao D, et al. Insomnia with objective short sleep duration is associated with deficits in neuropsychological performance: a general population study. Sleep. 2010;33(4):459–65.
72. Varkevisser M, Kerkhof GA. Chronic insomnia and performance in a 24-h constant routine study. J Sleep Res. 2005;14(1):49–59.
73. Covassin N, de Zambotti M, Sarlo M, De Min TG, Sarasso S, Stegagno L. Cognitive performance and cardiovascular markers of hyperarousal in primary insomnia. Int J Psychophysiol. 2011;80(1):79–86.
74. Kwan Y, Baek C, Chung S, Kim TH, Choi S. Resting-state quantitative EEG characteristics of insomniac patients with depression. Int J Psychophysiol. 2018;124:26–32.
75. Colombo MA, Ramautar JR, Wei Y, Gomez-Herrero G, Stoffers D, Wassing R, et al. Wake high-density electroencephalographic Spatiospectral signatures of insomnia. Sleep. 2016;39(5):1015–27.
76. Corsi-Cabrera M, Rojas-Ramos OA, del Rio-Portilla Y. Waking EEG signs of non-restoring sleep in primary insomnia patients. Clin Neurophysiol. 2016;127(3):1813–21.
77. Wolynczyk-Gmaj D, Szelenberger W. Waking EEG in primary insomnia. Acta Neurobiol Exp (Wars). 2011;71(3):387–92.
78. Oades RD, Dittmann-Balcar A, Schepker R, Eggers C, Zerbin D. Auditory event-related potentials (ERPs) and mismatch negativity (MMN) in healthy children and those with attention-deficit or Tourette/tic symptoms. Biol Psychol. 1996;43(2):163–85.
79. Killgore WD, Schwab ZJ, Kipman M, Deldonno SR, Weber M. Insomnia-related complaints correlate with functional connectivity between sensory-motor regions. Neuroreport. 2013;24(5):233–40.
80. Konrad K, Eickhoff SB. Is the ADHD brain wired differently? A review on structural and functional connectivity in attention deficit hyperactivity disorder. Hum Brain Mapp. 2010;31(6):904–16.
81. Wynchank D, Bijlenga D, Beekman AT, Kooij JJS, Penninx BW. Adult attention-deficit/hyperactivity disorder (ADHD) and insomnia: an update of the literature. Curr Psychiatry Rep. 2017;19(12):98.

82. Barry RJ, Clarke AR, Johnstone SJ, McCarthy R, Selikowitz M. Electroencephalogram theta/beta ratio and arousal in attention-deficit/hyperactivity disorder: evidence of independent processes. Biol Psychiatry. 2009;66(4):398–401.
83. Arns M, Conners CK, Kraemer HC. A decade of EEG theta/beta ratio research in ADHD: a meta-analysis. J Atten Disord. 2013;17(5):374–83.
84. Clarke AR, Barry RJ, Dupuy FE, McCarthy R, Selikowitz M, Johnstone SJ. Excess beta activity in the EEG of children with attention-deficit/hyperactivity disorder: a disorder of arousal? Int J Psychophysiol. 2013;89(3):314–9.
85. Clarke AR, Barry RJ, McCarthy R, Selikowitz M. EEG-defined subtypes of children with attention-deficit/hyperactivity disorder. Clin Neurophysiol. 2001;112(11):2098–105.
86. Arns M, Swatzyna RJ, Gunkelman J, Olbrich S. Sleep maintenance, spindling excessive beta and impulse control: an RDoC arousal and regulatory systems approach? Neuropsychiatric Electrophysiology. 2015;1(1):5.
87. Cervena K, Dauvilliers Y, Espa F, Touchon J, Matousek M, Billiard M, et al. Effect of cognitive behavioural therapy for insomnia on sleep architecture and sleep EEG power spectra in psychophysiological insomnia. J Sleep Res. 2004;13(4):385–93.
88. Altena E, Van Der Werf YD, Sanz-Arigita EJ, Voorn TA, Rombouts SA, Kuijer JP, et al. Prefrontal hypoactivation and recovery in insomnia. Sleep. 2008;31(9):1271–6.
89. Wei Y, Colombo MA, Ramautar JR, Blanken TF, van der Werf YD, Spiegelhalder K, et al. Sleep stage transition dynamics reveal specific stage 2 vulnerability in insomnia. Sleep. 2017;40(9) https://doi.org/10.1093/sleep/zsx117.
90. Schwartz JRL, Roth T. Neurophysiology of sleep and wakefulness: basic science and clinical implications. Curr Neuropharmacol. 2008;6(4):367–78.
91. Saper CB, Fuller PM. Wake-sleep circuitry: an overview. Curr Opin Neurobiol. 2017;44:186–92.
92. Scammell TE, Arrigoni E, Lipton JO. Neural circuitry of wakefulness and sleep. Neuron. 2017;93(4):747–65.
93. Ono D, Yamanaka A. Hypothalamic regulation of the sleep/wake cycle. Neurosci Res. 2017;118:74–81.

Narcolepsy and Central Nervous System Hypersomnias

8

Fabio Pizza and Giuseppe Plazzi

Introduction

Physiological sleepiness is the subjective sensation experienced by an individual in relation to the need to sleep. Sleepiness is modulated by the interaction between the homeostatic and circadian processes [1]. Accordingly, sleepiness physiologically fluctuates across the 24 hours reflecting highest sleep propensity during nighttime hours and, albeit of minor intensity, during early afternoons [2]. Research has demonstrated that subjective sleepiness perception in the general population shows gender-related differences with young women complaining a subjective feeling and older men a high self-reported sleep propensity in passive and active conditions [3].

In the clinical setting, physiological sleepiness should be distinguished from excessive daytime sleepiness/hypersomnolence (EDS), the latter being "defined as the inability to stay awake and alert during the major waking episodes of the day, resulting in periods of irrepressible need for sleep or unintended lapses into drowsiness or sleep" [4]. EDS severity has a variable range encompassing the extremes of falling asleep during monotonous conditions and up to having sleep attacks while performing an active task or being in an interactive situation. Up to one third of the general population (26%) reports sleepiness, but when considering associated symptoms (15.6%) and frequent sleepiness occurrence (4.7%), the prevalence decreases pointing to the need of a pathological definition of EDS [5].

In the clinical context, EDS can be evaluated using self-reported validated questionnaires such as the Epworth sleepiness scale [6], which minimizes interindividual differences, while polysomnographic in-laboratory tests are required to

F. Pizza · G. Plazzi (✉)
DIBINEM, Department of Biomedical and Neuromotor Sciences, University of Bologna, Bologna, Italy

IRCCS, Institute of Neurological Sciences of Bologna, UOC Neurological Clinic, Bellaria Hospital, Bologna, Italy
e-mail: giuseppe.plazzi@unibo.it

L. M. DelRosso, R. Ferri (eds.), *Sleep Neurology*,
https://doi.org/10.1007/978-3-030-54359-4_8

classify central disorders of hypersomnolence (CDH) (i.e., the multiple sleep latency test, MSLT) [7].

EDS is the core CDH complaint, a group of disorders characterized by prominent EDS despite normal nocturnal sleep and circadian rhythm features. Main CDH are narcolepsy (type 1, type 2), idiopathic hypersomnia, Kleine Levin syndrome, and insufficient sleep syndrome, as well as hypersomnias associated with medical or psychiatric disorders and with medication or substances. The diagnostic criteria defining CDH disorders have changed over time reflecting the increased pathophysiological knowledge on narcolepsy and the difficulties in defining the narcolepsy in specific patient populations such as children [4]. Narcolepsy is biologically linked to hypocretin neurotransmission, with narcolepsy type 1 being caused by hypocretin deficiency (i.e., a unique hypocretin deficiency syndrome) [8], whereas the other CDH disorders are poorly defined from the etiological standpoint. We will review the clinical/neurophysiological of CDH disorders (excluding the forms related to comorbid diseases or to medications/substances) in parallel to their current neurobiological knowledge.

Narcolepsy

Narcolepsy syndrome was firstly described by the clinicians Westphal and Gelineau [9, 10], and the psychoanalyst Vogel observed the rapid occurrence of REM sleep at sleep onset de facto identifying for the first time its neurophysiological disease marker [4, 11]. Narcolepsy is indeed characterized by a high sleep propensity during daytime coupled with the early occurrence of REM sleep at sleep onset [4], the so-defined sleep-onset REM periods (SOREMPs), which are the neurophysiological disease fingerprint on the MSLT to confirm narcolepsy diagnosis [7, 12].

Narcolepsy patients typically wake up refreshed in the morning, present with subsequent overwhelming daytime sleepiness and daytime sleep episodes with SOREMPs mirrored by self-reported dream occurrence during short naps. Patients with narcolepsy report feeling refreshed after awakening from daytime naps. Moreover, intense sleepiness in narcolepsy can be associated with occurrence of hypnagogic multisensory hallucinations, and sleep episodes can be linked to either hallucinatory content or sleep-related paralyses [13].

Narcolepsy most frequently arises in childhood or adolescence [14], and in young patients, EDS can present with daytime hyperactivity and irritability [15], features that contribute to wrong diagnoses (e.g., attention deficit hyperactivity disorder) and to diagnostic delay [16, 17].

Narcolepsy is further defined in narcolepsy type 1 (NT1) and narcolepsy type 2 (NT2), the former associated with cataplexy and hypocretin deficiency and the latter without cataplexy and, whenever available, with normal hypocretin levels [4]. NT1 is a defined clinical entity that has cataplexy (sudden loss of muscle strength triggered by emotions during wakefulness) as a pathognomonic symptom and a well-grounded neurobiological base in the hypocretin deficiency mirroring the loss of hypothalamic hypocretin-producing neurons [8]. NT1 is therefore the only CDH with a known neurobiological etiology. The hypocretin system damage underlying NT1 is most probably due to an autoimmune process, given the following biological evidences: (1) NT1 patients share a specific HLA predisposition carrying the HLA-DQB1*06:02 antigen [18] as well as other predisposing genetic features linked to the configuration of the immune system such as the T-cell receptor alpha [19]; (2) other polymorphisms of immune-related genes including P2RY11, CTSH, and TNFSF4 have been associated with NT1 [20, 21]; (3) environmental factors such as streptococcal infections have been supported by the finding of elevated antistreptococcal antibodies in NT1 patients close to disease onset [22] and by the association with physician-diagnosed strep throat [23]; and (4) a strong association linked NT1 to influenza A virus subtype H1N1 given the peaks of disease incidence after the influenza pandemia in China [24] as well as after the vaccination campaigns with Pandemrix in northern European countries [25]. The immunological disease mechanisms have not been fully elucidated, but a possible molecular mimicry between hypocretinergic neurons and a piece of the influenza hemagglutinin protein has been documented [26]. Moreover, recent data disclosed the presence of autoreactive T cells targeting antigens expressed by hypocretin neurons, the blood and cerebrospinal fluid of NT1 and NT2 patients [26–29]. Finally, indirect evidences pointing to an autoimmune etiology of narcolepsy have been provided also by studies on cytokines in the serum and cerebrospinal fluid that showed the signature of an activation of the immune system [30–33]. Further clinical evidences point to a possible link between narcolepsy phenotypes (NT1 and NT2). Indeed, several patients recall the onset of EDS and cataplexy years apart [34], thus suggesting that in a significant percentage of NT1 patients sleepiness appeared before cataplexy in a context of an NT2 phenotype. Cerebrospinal hypocretin deficiency can be a telltale for impending NT1 as proved by disease course in patients without cataplexy at the time of lumbar puncture [35], and an evolution of cerebrospinal hypocretin levels toward definite deficiency may also parallel the clinical disease course as proved in few cases evaluated longitudinally [36, 37]. Therefore, at least in a subset of patients, at early disease stage, NT1 can present with a narcolepsy without cataplexy (NT2) phenotype related to a partial hypocretinergic system defect [38]. Apart from the hypocretin system, other neurotransmitters have been involved in narcolepsy pathophysiology but bet with controversial results. Indeed cerebrospinal histamine has been found reduced in patients with narcolepsy and other CDH [39, 40], a finding not always replicated [41] and possibly related to adaptive mechanisms across disease course. Intriguingly however, EDS and also cataplexy respond to histaminergic treatment [42, 43].

Idiopathic Hypersomnia

Idiopathic hypersomnia (IH) was firstly described by Professor Roth in Prague [44, 45], as a condition of overwhelming EDS with prominent non-REM features, thus also called non-REM narcolepsy [4]. EDS in IH is characterized by "sleep drunkenness" that is an extreme awakening difficulty (i.e., sleep inertia) consisting of "confusion, disorientation, poor motor coordination and repeated return to sleep" and frequently by an abnormally prolonged sleep duration [44]. Several works highlighted a continuum in the CDH disease spectrum [46], and the IH phenotype was further differentiated considering the different presence of REM sleep-related symptoms (paralyses, hallucinations) and objective (SOREMP occurrence) evidences [47]. IH patients indeed suffer from prolonged nocturnal sleep duration, morning awakening difficulties, and EDS with automatic behaviors; however, a variable REM sleep propensity, sleep paralyses, and hallucinations can be overlapping symptoms rendering an IH disease spectrum encompassing pure non-REM and REM sleep-related complaints [47]. IH lacks a biological disease marker and is most often a diagnosis of exclusion of other sleep disorders causing EDS; therefore, IH diagnosis has been variably defined in subsequent international diagnostic manuals [48]. Indeed, prolonged nocturnal sleep duration casts difficulties in applying the classical nocturnal polysomnography plus MSLT approach for the need of a defined morning awakening to perform the MSLT contrasting with the utility of documenting nocturnal sleep duration in ecological conditions. IH was classified in two forms (with or without long sleep time defined as a documented sleep period >10 hours) both requiring high sleep propensity confirmation at the MSLT [49]. Recent works applying different continuous polysomnography protocols (free running, free running with invited napping, bed resting) evidenced the utility of documenting sleep duration across the 24 hours as a potential diagnostic marker [50–52], as well as of analyzing other features of the sleep-onset period [53]. To date IH diagnosis requires either the evidence of high non-REM sleep propensity at the MSLT or at least 11 hours of sleep across the 24 hours documented by polysomnography or wrist actigraphy [4]. The difficult morning awakening is mirrored by the evidence of high slow wave sleep representation across nighttime and, most notably, in early morning hours [50, 52]. EDS differs in IH and narcolepsy not only for the absence of dreaming but also for the unrefreshing potential of sleep, for the difficulties in performing short naps, and for the occurrence of slow wave sleep followed by sleep inertia at awakening.

The IH phenotype can forerun an NT1 as rarely documented [36] but can also have a benign disease course with a good response to stimulant treatment or a spontaneous improvement [52]. Given the intrinsic link between IH and non-REM sleep, different etiologies may however underlay IH and narcolepsy. An important observation was made by the group of Professor Rye who disclosed that the cerebrospinal fluid of IH patients was able to potentiate in vitro GABA neurotransmission [54], a finding reminding the previous description of the idiopathic recurring stupor as a disorder connected to the production of endogenous substances with properties similar to benzodiazepines [55]. Subsequent studies showed the possible therapeutic use of flumazenil [56] and of clarithromycin [57] further supporting a possible link between GABAergic transmission and IH. However, these findings were largely not replicated [58] making the relation between GABA receptor potentiation and IH elusive. Other studies analyzed the genetic background of IH patients and suggested a possible link between IH and the HLA system as known for the narcolepsy disease spectrum [59, 60]; however, despite a trend toward a higher representation of the HLA DQB1*0602, this association does not seem able to explain disease predisposition [61]. Clinical, neurophysiological, and pathophysiological research are therefore needed to define IH pathophysiology and its relations with the narcolepsy disease spectrum.

Kleine Levin Syndrome (KLS)

Kleine Levin syndrome is a peculiar CDH characterized by the sudden recurrence of attacks of EDS and prolonged sleep duration across the 24 hours persisting days and variably associated with altered cognition, perception, appetite (anorexia or hyperphagia), or disinhibited behavior (such as hypersexuality) with normal sleep patterns and daytime functioning in between the acute phases [4]. The syndrome was described at the beginning of the last century [62, 63], and the disorder is an extremely rare condition (prevalence below two cases per million) with onset most frequently during adolescence [64, 65]. Spontaneous disease course is benign with a progressive reduction in frequency and intensity of the acute attacks and a spontaneous resolution after a median period of 14 years most often before 30 years of age at least in patients with onset during adolescence [4, 64].

EDS in KLS has unique characteristics. Patients in the acute phase show an enormous, sudden, increase of 24-hour total sleep time and are difficult to awaken. Sleep drunkenness, dreaming, and hallucinations are also commonly reported [64]. Aside eating and sexual disorders, patients experience a wide range of cognitive dysfunctions encompassing amnesia, spatiotemporal disorientation, and altered perception of themselves and of the environment with dreamy state and derealization feelings. Even if by definition patients should have normal sleep patterns and daytime functioning outside of the attacks, recent data disclosed a common psychiatric comorbidity (mostly mood and anxiety disorders in up to 21% of cases) also during the asymptomatic periods [66] as well as subtle cognitive deficits [67].

KLS pathophysiology is to date unresolved. Few data are available on the potential role of the hypocretinergic system, and the patients who underwent lumbar puncture inside and outside the acute phases showed a significant reduction of hypocretin levels during the attacks of hypersomnolence, however, with levels that never reached the evidence of a definite hypocretin deficiency [68–70]. Similarly to narcolepsy, also in KLS, a potential role of triggering factors and most notably of infections (up to 42% of cases) has been reported [71]. Some studies focused on the potential presence of other clues pointing to immune-mediated etiology, and while some data suggested a potential role of the HLA, other data did not confirm the association [72, 73]. Other unconfirmed data suggest a possible role of the immune system in disease pathophysiology, such as the finding of increased antibodies against GAD65 [74]. Finally, a single study identified a genetic predisposition due to variants of the LMOD3 gene that was associated with KLS compared to controls and also linked to symptomatic cases in familial KLS [75].

Conclusions

EDS is the core complaint of CNS hypersomnias, and clinical assessment should at first distinguish physiological from pathological sleepiness during daytime. In this context, NT1 is the only disorder with a wide etiological knowledge that point to an immune attack to the hypocretinergic neurons of the hypothalamus in genetically predisposed individuals. The other CDH, namely, NT2, IH, and KLS, are poorly understood. There is an overall continuum in CDH features that is reflected by small pieces of evidence that may show an overlap with NT1 in terms of genetic predisposition and minor dysfunction of the hypocretinergic system. However, non-NT1 CDH suffer from a less definite clinical and neurophysiological definition that makes the investigation on disease etiology extremely difficult. Further clinical, neurophysiological, and biological studies are therefore warranted with the aim of coupling longitudinal disease evaluation with neurobiological assessments.

Financial Support The authors have no potential financial conflict of interest to disclose.

Conflict of Interest Giuseppe Plazzi participated in the advisory board for UCB Pharma, Jazz pharmaceuticals, Bioprojet, and Idorsia outside the submitted work. Fabio Pizza has no potential financial conflict of interest to disclose.

References

1. Borbély AA. A two process model of sleep regulation. Hum Neurobiol. 1982;1:195–204.
2. Monk TH, Buysse DJ, Reynolds CF, Kupfer DJ. Circadian determinants of the postlunch dip in performance. Chronobiol Int. 1996;13:123–33.
3. Kim H, Young T. Subjective daytime sleepiness: dimensions and correlates in the general population. Sleep. 2005;28:625–34.
4. American Academy of Sleep Medicine. International classification of sleep disorders. 3rd ed. Darien: American Academy of Sleep Medicine; 2014.
5. Ohayon MM, Dauvilliers Y, Reynolds CF. Operational definitions and algorithms for excessive sleepiness in the general population: implications for DSM-5 nosology. Arch Gen Psychiatry. 2012;69:71–9. https://doi.org/10.1001/archgenpsychiatry.2011.1240.
6. Johns MW. A new method for measuring daytime sleepiness: the Epworth sleepiness scale. Sleep. 1991;14:540–5.
7. Littner MR, Kushida C, Wise M, et al. Practice parameters for clinical use of the multiple sleep latency test and the maintenance of wakefulness test. Sleep. 2005;28:113–21.
8. Peyron C, Faraco J, Rogers W, et al. A mutation in a case of early onset narcolepsy and a generalized absence of hypocretin peptides in human narcoleptic brains. Nat Med. 2000;6:991–7.
9. Westphal C. Eigentümliche mit Einschlafen verbundene Anfälle. Arch Psych. 1877;7:631–5.
10. Gelineau JBE. De la Narcolepsy (I). Gaz des Hôp. 1880;54:626–8.
11. Vogel G. Studies in psychophysiology of dreams. III. The dream of narcolepsy. Arch Gen Psychiatry. 1960;3:421–8.
12. Carskadon MA, Dement WC, Mitler MM, et al. Guidelines for the multiple sleep latency test (MSLT): a standard measure of sleepiness. Sleep. 1986;9:519–24.
13. Plazzi G, Fabbri C, Pizza F, Serretti A. Schizophrenia-like symptoms in narcolepsy type 1: shared and distinctive clinical characteristics. Neuropsychobiology. 2015;71:218–24.

14. Postiglione E, Antelmi E, Pizza F, et al. The clinical spectrum of childhood narcolepsy. Sleep Med Rev. 2018;38:70–85.
15. Rocca FL, Pizza F, Ricci E, Plazzi G. Narcolepsy during childhood: an update. Neuropediatrics. 2015;46:181–98.
16. Thorpy MJ, Krieger AC. Delayed diagnosis of narcolepsy: characterization and impact. Sleep Med. 2014;15:502–7.
17. Plazzi G, Pizza F, Palaia V, et al. Complex movement disorders at disease onset in childhood narcolepsy with cataplexy. Brain. 2011;134:3477–89.
18. Honda Y, Asaka A, Tanaka Y, Juji T. Discrimination of narcolepsy by using genetic markers and HLA. Sleep Res. 1983;12:254.
19. Hallmayer J, Faraco J, Lin L, et al. Narcolepsy is strongly associated with the T-cell receptor alpha locus. Nat Genet. 2009;41:708–11.
20. Faraco J, Lin L, Kornum BR, et al. ImmunoChip study implicates antigen presentation to T cells in narcolepsy. PLoS Genet. 2013;9:e1003270.
21. Kornum BR, Kawashima M, Faraco J, et al. Common variants in P2RY11 are associated with narcolepsy. Nat Genet. 2011;43:66–71.
22. Aran A, Lin L, Nevsimalova S, et al. Elevated anti-streptococcal antibodies in patients with recent narcolepsy onset. Sleep. 2009;32:979–83.
23. Koepsell TD, Longstreth WT, Ton TG. Medical exposures in youth and the frequency of narcolepsy with cataplexy: a population-based case-control study in genetically predisposed people. J Sleep Res. 2010;19:80–6.
24. Han F, Lin L, Warby SC, Faraco J, et al. Narcolepsy onset is seasonal and increased following the 2009 H1N1 pandemic in China. Ann Neurol. 2011;70:410–7.
25. Partinen M, Kornum BR, Plazzi G, et al. Narcolepsy as an autoimmune disease: the role of H1N1 infection and vaccination. Lancet Neurol. 2014;13:600–13.
26. Luo G, Ambati A, Lin L, et al. Autoimmunity to hypocretin and molecular mimicry to flu in type 1 narcolepsy. Proc Natl Acad Sci U S A. 2018;115:E12323–32.
27. Latorre D, Kallweit U, Armentani E, et al. T cells in patients with narcolepsy target self-antigens of hypocretin neurons. Nature. 2018;562:63–8.
28. Lippert J, Young P, Gross C, et al. Specific T-cell activation in peripheral blood and cerebrospinal fluid in central disorders of hypersomnolence. Sleep. 2019;42. https://doi.org/10.1093/sleep/zsy223.
29. Pedersen NW, Holm A, Kristensen NP, et al. CD8(+) T cells from patients with narcolepsy and healthy controls recognize hypocretin neuron-specific antigens. Nat Commun. 2019;10(1):837.
30. Lecendreux M, Libri V, Jaussent I, et al. Impact of cytokine in type 1 narcolepsy: role of pandemic H1N1 vaccination? J Autoimmun. 2015;60:20–31.
31. Hartmann FJ, Bernard-Valnet R, Quériault C, et al. High-dimensional single-cell analysis reveals the immune signature of narcolepsy. J Exp Med. 2016;213:2621–33.
32. Lecendreux M, Churlaud G, Pitoiset F, et al. Narcolepsy type 1 is associated with a systemic increase and activation of regulatory T cells and with a systemic activation of global T cells. PLoS One. 2017;12:e0169836.
33. Ramberger M, Högl B, Stefani A, et al. CD4+ T-cell reactivity to orexin/hypocretin in patients with narcolepsy type 1. Sleep. 2017;40. https://doi.org/10.1093/sleep/zsw070
34. Luca G, Haba-Rubio J, Dauvilliers Y, et al. Clinical, polysomnographic and genome-wide association analyses of narcolepsy with cataplexy: a European Narcolepsy Network study. J Sleep Res. 2013;22:482–95.
35. Andlauer O, Moore H 4th, Hong SC, et al. Predictors of hypocretin (orexin) deficiency in narcolepsy without cataplexy. Sleep. 2012;35:1247–55F.
36. Pizza F, Vandi S, Liguori R, et al. Primary progressive narcolepsy type 1: the other side of the coin. Neurology. 2014;83:2189–90.
37. Lopez R, Barateau L, Evangelista E, et al. Temporal changes in the cerebrospinal fluid level of hypocretin-1 and histamine in narcolepsy. Sleep. 2017;40. https://doi.org/10.1093/sleep/zsw010.

38. Thannickal TC, Nienhuis R, Siegel JM. Localized loss of hypocretin (orexin) cells in narcolepsy without cataplexy. Sleep. 2009;32:993–8.
39. Kanbayashi T, Kodama T, Kondo H, et al. CSF histamine contents in narcolepsy, idiopathic hypersomnia and obstructive sleep apnea syndrome. Sleep. 2009;32:181–7.
40. Nishino S, Sakurai E, Nevsimalova S, et al. Decreased CSF histamine in narcolepsy with and without low CSF hypocretin-1 in comparison to healthy controls. Sleep. 2009;32:175–80.
41. Dauvilliers Y, Delallée N, Jaussent I, et al. Normal cerebrospinal fluid histamine and telemethylhistamine levels in hypersomnia conditions. Sleep. 2012;35:1359–66.
42. Dauvilliers Y, Bassetti C, Lammers GJ, et al. Pitolisant versus placebo or modafinil in patients with narcolepsy: a double-blind, randomised trial. Lancet Neurol. 2013;12:1068–75.
43. Szakacs Z, Dauvilliers Y, Mikhaylov V, et al. Safety and efficacy of pitolisant on cataplexy in patients with narcolepsy: a randomised, double-blind, placebo-controlled trial. Lancet Neurol. 2017;16:200–7.
44. Roth B, Nevsimalova S, Rechtschaffen A. Hypersomnia with "sleep drunkenness". Arch Gen Psychiatry. 1972;26:456–62.
45. Roth B. Idiopathic hypersomnia: a study of 187 personally observed cases. Int J Neurol. 1981;15:108–18.
46. Aldrich MS. The clinical spectrum of narcolepsy and idiopathic hypersomnia. Neurology. 1996;46:393–401.
47. Bassetti C, Aldrich MS. Idiopathic hypersomnia. A series of 42 patients. Brain. 1997;120(Pt 8):1423–35.
48. Billiard M. Diagnosis of narcolepsy and idiopathic hypersomnia. An update based on the international classification of sleep disorders, 2nd edition. Sleep Med Rev. 2007;11:377–88.
49. American Academy of Sleep Medicine. International classification of sleep disorders, Diagnostic and coding manual. 2nd ed. Westchester: American Academy of Sleep Medicine; 2005.
50. Vernet C, Arnulf I. Idiopathic hypersomnia with and without long sleep time: a controlled series of 75 patients. Sleep. 2009;32:753–9.
51. Pizza F, Moghadam KK, Vandi S, et al. Daytime continuous polysomnography predicts MSLT results in hypersomnias of central origin. J Sleep Res. 2013;22:32–40.
52. Anderson KN, Pilsworth S, Sharples LD, et al. Idiopathic hypersomnia: a study of 77 cases. Sleep. 2007;30:1274–81.
53. Pizza F, Vandi S, Detto S, et al. Different sleep onset criteria at the multiple sleep latency test (MSLT): an additional marker to differentiate central nervous system (CNS) hypersomnias. J Sleep Res. 2011;20:250–6.
54. Rye DB, Bliwise DL, Parker K, et al. Modulation of vigilance in the primary hypersomnias by endogenous enhancement of GABAA receptors. Sci Transl Med. 2012;4:161ra151.
55. Rothstein JD, Guidotti A, Tinuper P, et al. Endogenous benzodiazepine receptor ligands in idiopathic recurring stupor. Lancet. 1992;340(8826):1002–4.
56. Trotti LM, Saini P, Koola C, et al. Flumazenil for the treatment of refractory hypersomnolence: clinical experience with 153 patients. J Clin Sleep Med. 2016;12:1389–94.
57. Trotti LM, Saini P, Freeman AA, et al. Improvement in daytime sleepiness with clarithromycin in patients with GABA-related hypersomnia: clinical experience. J Psychopharmacol. 2014;28:697–702.
58. Dauvilliers Y, Evangelista E, Lopez R, et al. Absence of γ-aminobutyric acid-a receptor potentiation in central hypersomnolence disorders. Ann Neurol. 2016;80:259–68.
59. Coelho FM, Pradella-Hallinan M, Predazzoli Neto M, et al. Prevalence of the HLA-DQB1*0602 allele in narcolepsy and idiopathic hypersomnia patients seen at a sleep disorders outpatient unit in São Paulo. Braz J Psychiatry. 2009;31(1):10–4.
60. Pizza F, Vandi S, Iloti M, et al. Nocturnal sleep dynamics identify narcolepsy type 1. Sleep. 2015;38:1277–84.
61. Miyagawa T, Toyoda H, Kanbayashi T, et al. An association analysis of HLA-DQB1 with narcolepsy without cataplexy and idiopathic hypersomnia with/without long sleep time in a Japanese population. Hum Genome Var. 2015;2:15031.

62. Kleine W. Periodische Schlafsucht. Monatsschrift fur Psychiatrie und Neurologie. 1925;57:285–320.
63. Levin M. Periodic somnolence and morbid hunger: a new syndrome. Brain. 1936;59:494–504.
64. Arnulf I, Lin L, Gadoth N, et al. Kleine-Levin syndrome: a systematic study of 108 patients. Ann Neurol. 2008;63:482–93.
65. Lavault S, Golmard J-L, Groos E, et al. Kleine-Levin syndrome in 120 patients: differential diagnosis and long episodes. Ann Neurol. 2015;77:529–40.
66. Groos E, Chaumereuil C, Flamand M, et al. Emerging psychiatric disorders in Kleine-Levin syndrome. J Sleep Res. 2018;27:e12690.
67. Uguccioni G, Lavault S, Chaumereuil C, et al. Long-term cognitive impairment in Kleine-Levin syndrome. Sleep. 2016;39:429–38.
68. Podestá C, Ferreras M, Mozzi M, et al. Kleine-Levin syndrome in a 14-year-old girl: CSF hypocretin-1 measurements. Sleep. 2006;7:649–51.
69. Lopez R, Barateau L, Chenini S, et al. Preliminary results on CSF biomarkers for hypothalamic dysfunction in Kleine-Levin syndrome. Sleep Med. 2015;16:194–6.
70. Wang JY, Han F, Dong SX, et al. Cerebrospinal fluid orexin a levels and autonomic function in Kleine-Levin syndrome. Sleep. 2016;39:855–60.
71. Arnulf I, Zeitzer JM, File J, et al. Kleine-Levin syndrome: a systematic review of 186 cases in the literature. Brain. 2005;128:2763–76.
72. Huang CJ, Liao HT, Yeh GC, et al. Distribution of HLA-DQB1 alleles in patients with Kleine-Levin syndrome. J Clin Neurosci. 2012;19:628–30.
73. Al Shareef SM, Smith RM, BaHammam AS. Kleine-Levin syndrome: clues to aetiology. Sleep Breath. 2018a;22:613–23.
74. Rout UK, Michener MS, Dhossche DM. GAD65 autoantibodies in Kleine-Levin syndrome. J Neuropsychiatry Clin Neurosci. 2014;26:E49–51.
75. Al Shareef SM, Basit S, Li S, et al. Kleine-Levin syndrome is associated with LMOD3 variants. J Sleep Res. 2018b;28:e12718.

Neurobiology of Parasomnias

9

Tommaso Baldini, Giuseppe Loddo, Raffaele Ferri, and Federica Provini

Introduction

Parasomnias are defined as "undesirable physical events (movements, behaviors) or experiences (emotions, perceptions, dreams) that occur during entry into sleep, within sleep, or during arousals from sleep," over which there is no deliberate control [1]. The *International Classification of Sleep Disorders* (ICSD-3) groups parasomnias into three categories according to the stage to which they are usually associated: NREM-related parasomnias, REM-related parasomnias, and other parasomnias. Sleep talking is considered a normal variant and occurs at some time in most normal sleepers (see Table 9.1).

General Pathophysiology

Human consciousness consists of three essential states of being: wake, NREM sleep, and REM sleep, which cycle in a predictable fashion throughout a 24-hour period [1].Those states are defined by an identifiable and stable cluster of behavioral, neurophysiological, and autonomic descriptors observed for a period of time

T. Baldini · G. Loddo
Department of Biomedical and Neuromotor Sciences, University of Bologna, Bellaria Hospital, Bologna, Italy
e-mail: tommaso.baldini4@studio.unibo.it; giuseppe.loddo2@unibo.it

R. Ferri
Sleep Research Centre, Oasi Research Institute - IRCCS, Troina, Italy
e-mail: rferri@oasi.en.it

F. Provini (✉)
Department of Biomedical and Neuromotor Sciences, University of Bologna, Bologna, Italy

IRCCS Institute of Neurological Sciences, Bologna, Italy
e-mail: federica.provini@unibo.it

L. M. DelRosso, R. Ferri (eds.), *Sleep Neurology*,
https://doi.org/10.1007/978-3-030-54359-4_9

Table 9.1 Classification of parasomnias. (Adapted from the American Academy of Sleep Medicine [1])

NREM sleep-related parasomnias
Disorders of arousal (DOA)
Confusional arousals
Sleep terrors
Sleepwalking
Sleep-related eating disorder (SRED)
REM sleep-related parasomnias
REM sleep behavior disorder (RBD)
Recurrent isolated sleep paralysis
Nightmare disorder
Other parasomnias
Exploding head syndrome
Sleep-related hallucinations
Sleep enuresis
Parasomnia due to a medical disorder
Parasomnia due to a medication or substance
Parasomnia, unspecified
Isolated symptoms and normal variants
Sleep talking

[2]. Traditionally wake and sleep have been considered as global and discrete states of being: indeed, in physiological conditions and in most sleep disorders, these clusters are preserved, and possible changes involve their quantity rather than their basic features [2].

However, it has progressively emerged that wake and sleep are not mutually exclusive states and admixture of features of the different states can occur [3, 4]. State shifts do not always occur as an "on or off" switch phenomenon but involve transitions of several neuronal networks for an equivocal stage to declare itself [1]. During the transition periods from one state to another, a temporary pathological dissociation of states can occur across different brain structures, resulting in a state of altered consciousness manifesting as parasomnia [5]. Hence, most parasomnias are dissociated states that combine wake and sleep elements [1]. As an example, disorders of arousal (DOA) are an admixture of wakefulness and NREM sleep during which higher cognitive function is impaired, while motor behaviors are retained. REM sleep behavior disorder (RBD) is an admixture of REM sleep coupled with waking levels of EMG activity [1].

State dissociation is a universal, physiological property of the brain, as demonstrated by phylogeny and ontogeny: some aquatic mammals exhibit unihemispheric sleep, and humans in the neonatal period show phases of undetermined sleep with mixed REM and NREM features. From an evolutionary perspective, this probably serves a protective function, allowing sleep preservation but increasing the probability of survival by keeping a lower arousal threshold [6, 7]. In adulthood, some form of dissociation persists in normal human brains in the form of "local sleep," electrophysiologically characterized by coexistence of wake and sleep EEG patterns in different cortical areas [4, 8, 9]. During sleep, sleep spindles and slow wave activity can arise in specific brain areas and be locally modulated [8–11]; during wakefulness, subsets of cortical neurons produce local slow waves with an increasing incidence with wake duration, probably to selectively satisfy a local need for sleep [9, 12]. In physiologic conditions this exerts little or no effect on the prevalent behavioral state; however, an exaggeration of this mechanism might underlie the state dissociation of parasomnias [2].

Some of the behaviors observed in human parasomnias often resemble stereotyped, "primitive" behaviors noted in other animals, such as defensive postures, violent gestures, and feeding; those archaic behaviors result from the activation of shared, mostly subcortical brain structures, called central pattern generators (CPGs). CPGs in humans are largely under neocortical control; therefore, it is possible that some parasomnic behaviors may result from their inappropriate activation (i.e., their disinhibition) [13].

Parasomnias can result from very different etiologies. In most cases of DOA, no identifiable neuropathology is present, and the disorder results from functional changes in different brain areas during NREM sleep. In contrast, RBD can result from serious neuropathology and often is followed or accompanied by the development of neurodegenerative disorders, particularly synucleinopathies [1].

NREM Sleep-Related Parasomnias

NREM sleep-related parasomnias include disorders of arousal (DOA) and sleep-related eating disorder (SRED).

Disorders of Arousal (DOA)

DOA include confusional arousals, sleep terrors, and sleepwalking, which although classified as distinct entities actually represent a spectrum of manifestations of increasing complexity occurring as sudden partial awakenings from deep sleep (slow wave sleep, SWS, or stage 3 NREM sleep, N3) [14]. DOA typically have onset in childhood (probably related to the abundance of SWS at this age) and tend to cease with age. However, they may continue into adulthood or appear de novo in adults [15].

Clinical Features As N3 occurs mainly in the first part of the night, DOA usually occur within the first 3 hours of sleep or so after the individual falls asleep. DOA can vary in clinical features, but in all episodes, the patient seems to be simultaneously awake and asleep: the individual exhibits a behavior more reminiscent of the awake state, but he/she is usually unresponsive to the environment with typically complete or partial amnesia for the event.

Confusional arousals consist of confusion and disorientation during and following arousal from SWS, without accompanying major behavioral disorders or severe autonomic responses [1]. The episodes last a few minutes and are terminated by retrograde and anterograde amnesia. Confusional arousals affect from 20% to 50% of children and 4% of adults [16]. *Sleep-related abnormal sexual behaviors (sexsomnia/sleep sex)* are considered variants of confusional arousals. Sexsomnia consists of inappropriate sexual behavior (masturbation, sexual intercourse, or even sexual assault) occurring during sleep, followed by morning amnesia [1, 17].

Sleep terrors (night terrors) are the most extreme and dramatic forms of DOA. The episodes are characterized by a sudden arousal from deep sleep often accompanied by a piercing scream, intense fear, extreme panic, confusion, and heightened autonomic discharge (tachycardia, tachypnea, diaphoresis, mydriasis, and increased muscle tone) [18]. A universal feature is inconsolability. Although appearing alert, with eyes open, the child may not recognize his parents, and any attempt at consolation is fruitless and may serve only to prolong or even intensify the episode [15]. The episode usually lasts no more than a few minutes; afterward, the child usually relaxes and returns spontaneously to sleep. If the patient wakes up at the end of the episode (as may happen in older children), he may describe a feeling of primitive threat or danger but usually does not offer the extended sequence of mental images resembling a dream [15]. Sleep terrors occur in 3% of children and have a higher male predilection.

Sleepwalking (somnambulism) episodes are characterized by complex behaviors, ranging from playing with the sheets, to changes in body position, to sitting up in bed and walking [1]. Sometimes quite complex, coordinated, and semi-purposeful activities such as dressing, arranging articles on a desk, going to the bathroom, or opening drawers or doors may be seen. The episodes end with the patient awakening, appearing confused and amnestic for the events. If restrained, the patient may attempt to avoid the other person but does not usually put up aggressive resistance. Injuries during sleepwalking are uncommon but can occur as a result of falls and injuries following displacement, during attempts to "escape," or when walking downstairs or into windows. The prevalence of somnambulism (sleepwalking) in the general population is between 1% and 17% and approaches 4% in adults [16].

Diagnosis The initial approach to the complaint of unusual sleep-related behavior is to determine whether further evaluation is necessary. Careful history-taking is usually sufficient to establish a correct diagnosis of DOA, and most cases do not warrant further evaluation (see Table 9.2) [15, 18].

However, if the events present with atypical features (such as a late age of onset, brief duration, high frequency, progressive increase in frequency during life, presence of stereotypic movements or dystonic postures, or no response to treatment) or are associated with negative effects, such as violence or injuries, are extremely disruptive to other household members, or result in the complaint of excessive daytime sleepiness, further investigations are needed [15]. Remarkably, adult-onset sleepwalking presents with different characteristics from childhood-onset sleepwalking, such as a higher association with neurological comorbidities, EEG abnormalities, and violence [19]. In such cases, of particular importance is the distinction between DOA and sleep-related hypermotor epilepsy (SHE), a peculiar form of focal epilepsy in which motor seizures appear predominantly during sleep [20, 21].

Table 9.2 Diagnostic criteria for disorders of arousal. (Adapted from the American Academy of Sleep Medicine [1])

Diagnostic criteria (criteria A–E must be met)
A. Recurrent episodes of incomplete awakening from sleep
B. Inappropriate or absent responsiveness to efforts of others to intervene or redirect the person during the episode
C. Limited or no associated cognition or dream imagery
D. Partial or complete amnesia for the episode
E. The disturbance is not better explained by another sleep disorder, mental disorder, medication, or substance use

In these cases, video polysomnography (VPSG) is the exam of choice and allows a correct diagnosis in the majority of cases [22, 23]. DOA episodes typically begin with an arousal from SWS followed by a partial or complete persistence of sleep with diffuse rhythmic delta activity or diffuse delta and theta activity associated with tachycardia, tachypnea, and increased muscle tone (see Fig. 9.1) [22].

Prognosis and Management DOA are benign phenomena with a high prevalence in the general population, especially in children. They lack of psychological significance, and their natural history is to diminish over time; therefore, medications are rarely needed [5]. Management includes reassurance of the patient and relatives, a correct sleep hygiene, and measures to make the environment safe. Medications may be indicated when the episodes are extremely bothersome to the patient or other household members, result in symptoms of excessive daytime sleepiness, or are potentially injurious or violent. Options include imipramine or low-dose clonazepam (started at a dosage of 0.5 mg at bedtime) [15]. Nonpharmacologic treatment such as psychotherapy, progressive relaxation, or hypnosis is recommended for long-term management [5].

Pathophysiology of Disorders of Arousal

DOA are thought to arise from state dissociation during arousals from SWS, occurring due to abnormal neuronal excitability in different cortical areas [24, 25]. In addition, there are likely underlying physiologic phenomena that contribute to the appearance of complex motor actions during sleep, including disinhibition of central pattern generators (CPGs), dreams, and sleep-state instability [1].

State Dissociation

DOA consist of an NREM sleep-wake dissociated state, an admixture of wakefulness, and NREM sleep: the patient produces complex behaviors that are characteristic of wakefulness (such as eye opening, retained muscle tone, and ability to move and interact with the environment) but cognitive function fitting with sleep (inappropriate actions and lack of self-related awareness, insight, and recall).

Several data support this hypothesis. In around one third of cases, scalp EEG shows some evidence of state dissociation, with posterior dominant alpha rhythm (suggestive of resting wakefulness) but anterior theta activity and sometimes vertex waves or spindles (consistent with light NREM sleep) [22]. A SPECT study

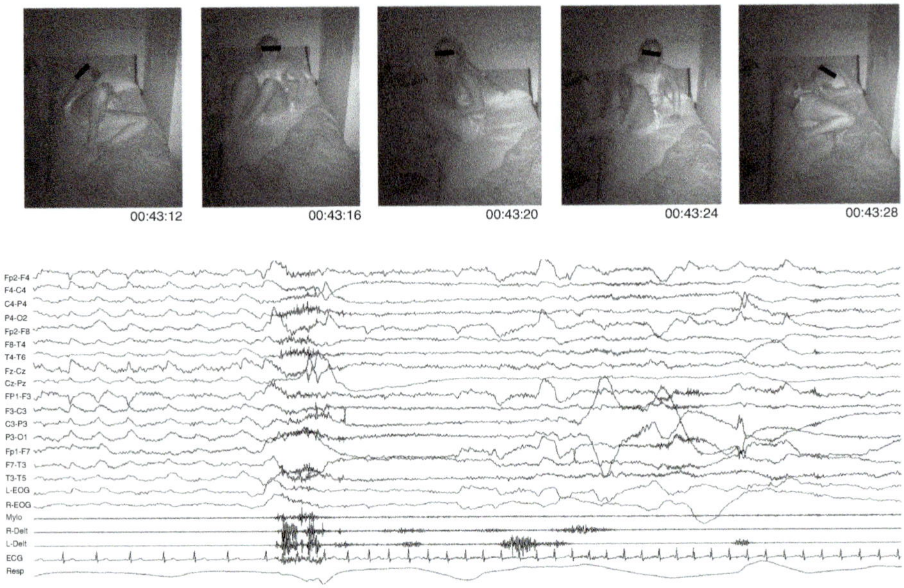

Fig. 9.1 Photo sequence and polysomnographic excerpt of a confusional arousal. The episode, arising from deep sleep in the first part of the night, began spontaneously with the patient abruptly setting upright on the bed with eyes open, turning to her right side with a frightened expression and moving her legs outside the bed with one foot on the floor. Subsequently she slowly lied down again. EEG at the episode onset was characterized by a diffuse delta activity; then theta activity appeared on anterior leads along with desynchronization on posterior leads. The episode was associated with tachycardia and modification in respiratory rate. *R, right; L, left; EOG, electrooculogram; Mylo, mylohyoid muscle; Delt, deltoid muscle; ECG, electrocardiogram; Resp, respirogram*

performed during sleepwalking found a decrease in regional blood flow in the frontoparietal associative cortices and an increase blood flow in the posterior cingulate cortex and in the anterior cerebellum [26]. As these structures are generally deactivated during normal slow wave sleep [27], a selective activation of thalamo-cingulate circuits with persisting inhibition of other thalamo-cortical arousal systems was suggested [26]. Stereo-EEG studies in epileptic patients with co-occurrence of disorders of arousal have recently confirmed these data, identifying the neurophysiological substrate of such blood perfusion changes. During DOA, different local activations occur, with local fast wake-like EEG activity on the motor, cingulate, insular, temporopolar, and amygdalar cortices and sleeplike EEG with increased delta activity on the frontoparietal associative cortices and in the hippocampus (see Fig. 9.2) [28, 29]. Additionally, a decrease in delta power and emergence of beta activity were observed in the ventral intermediate nucleus of the thalamus (VIM), which projects to the motor cortex, during a confusional arousal [6]. Taken together, these findings are consistent with the features of DOA during which motor and emotional activation are associated with a lack of insight and amnesia for the event (see Fig. 9.3) [6].

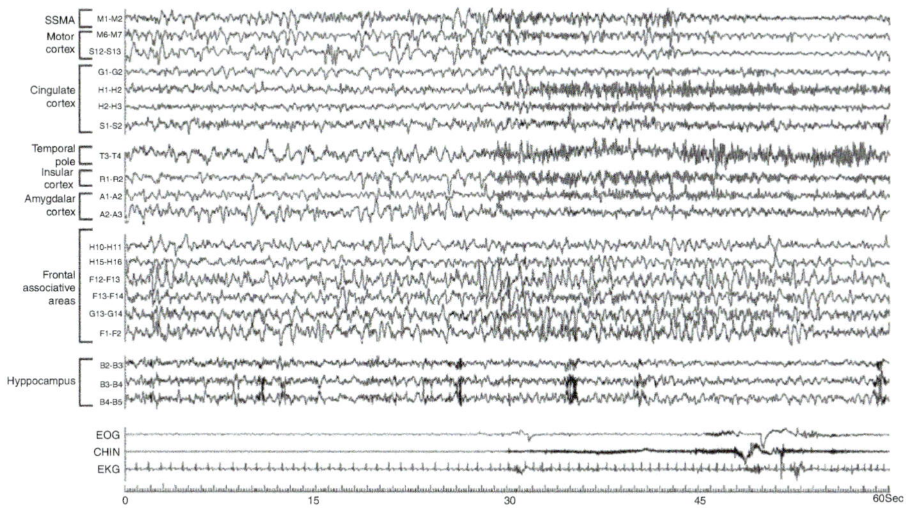

Fig. 9.2 Stereo-electroencephalogram (stereo-EEG) during a sleepwalking episode showing local dissociation. During the 30-s episode, the patient sat up in bed, uttering unintelligible words and failing to answer his mother's questions. EEG shows local activation in the motor, cingulate (frontal and central cingulate gyrus), insular, temporopolar, and amygdalar cortices in contrast to the presence of slow waves in the frontal and parietal dorsolateral cortices as well as persistent spindles in the hippocampal cortex. (Reproduced with courtesy from Terzaghi M et al., Dissociated local arousal states underlying essential clinical features of non-rapid eye movement arousal parasomnia: an intracerebral stereo-electroencephalographic study, J Sleep Res, 2012, John Wiley and Sons)

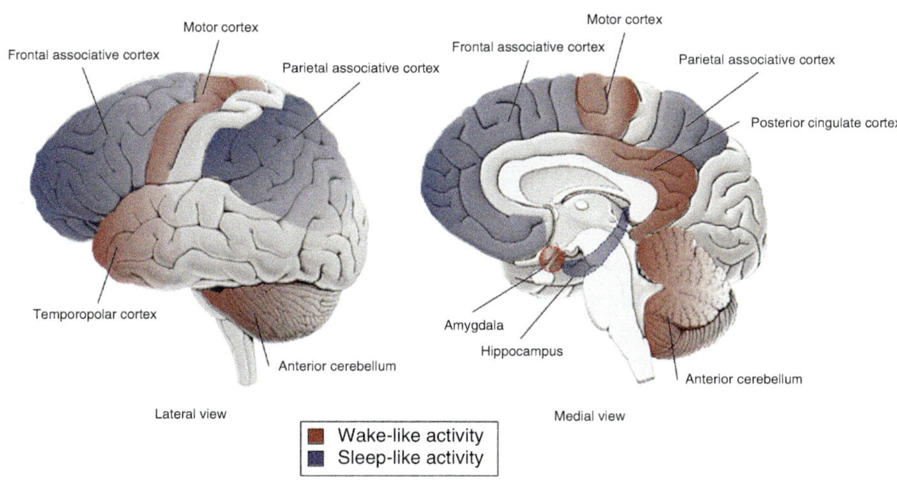

Fig. 9.3 State dissociation during disorders of arousal. Lateral and medial view of a human brain showing different local activity patterns (i.e., state dissociation) during disorders of arousal, as shown by SPECT and stereo-EEG studies. Specifically, wake-like activity of motor, temporopolar, anterior cerebellar, posterior cingulate cortices and amygdala (red) coexists with sleeplike activity of frontoparietal associative and hippocampal cortices (blue)

Abnormal Cortical Excitability and Precipitating Factors
The predisposition to state dissociation in affected subjects may originate from persistent, localized changes in neuronal excitability [25], specifically an increased arousability of motor and limbic areas in contrast with a reduced arousability of associative cortices [6]. A high-density EEG study found an all-night localized decrease in slow wave activity (SWA) in the cingulate, motor, and sensorimotor associative cortices, consistent with a lower arousal threshold of these areas [25]. This was also confirmed by an increased excitability of the motor cortex to transcranial magnetic stimulations (TMS) during wakefulness in sleepwalkers compared to healthy controls [30]. On the other hand, a SPECT study in ten sleepwalkers revealed a significant decrease of blood flow in frontal regions bilaterally during wakefulness, extending to parietal, temporal, and insular regions during SWS [31]. Hence, during arousals (spontaneous or evoked by stimuli), motor and limbic areas would easily be aroused, while associative areas would fail to fully transition into wakefulness, especially in conditions of increased SWS pressure. As a consequence, all the conditions increasing the amount of SWS, such as sleep deprivation, physical or emotional stress, fever, and medications affecting sleep (sedative-hypnotics – specifically z-drugs, neuroleptics, antidepressants, and beta-blockers), increase the likelihood of DOA episodes occurring [13, 15, 32–35]. Secondly, arousal by whatever mechanism (internal or external) can precipitate a DOA episode [1]. The list includes comorbid sleep disorders (particularly obstructive sleep apnea syndrome, OSAS, and periodic limb movement disorder, PLMD), bladder distention, physiological ending of a sleep cycle, mental activity, and noises [36]. Contrary to previous reports, alcohol does not seem to trigger DOA [1, 5].

Additionally, it has been suggested that downregulation (through synaptic pruning normally occurring during childhood) of cortical GABAergic inhibitory projections to brain stem or spinal cord CPGs could favor their disinhibition during the episodes [37]. This would explain the progressive decrease in episodes with age (due to maturation of these networks) and the resolution of NREM parasomnia upon treatment with GABA agonists such as clonazepam [7].

Central Pattern Generators (CPGs) and Dreams
CPGs are genetically determined or learned neural circuits, located in the spinal cord, the brain stem, or the cortex itself, that code for primitive, self-sustained patterns of behavior that are essential for survival. Their activation results in "basic drive states" such as locomotion, eating, aggression, and sex, similar to some behaviors occurring in DOA and more generally in NREM parasomnias [1]. In higher primates, CPGs during wakefulness are largely under neocortical control. Hence, it has been hypothesized that some behaviors of NREM parasomnias can result from disinhibition of these CPGs, favored by a transient dysfunction in the prefrontal cortex [38]. This could be true especially for the simpler manifestation of DOA and for "specialized" forms of NREM parasomnias, such as sexsomnia and SRED. On the other hand, there is increasing evidence that, in a number of cases of sleepwalking, complex and learned behaviors are performed, sometimes accompanied by congruent dreamlike mentation; this implies cortical engagement in DOA and

possibly a role of NREM sleep dreaming in the genesis of parasomnic behaviors that would represent "dream enactment behaviors" (DEB) in some cases [39–41].

Sleep Microstructure and Sleep-State Instability in DOA

Although the sleep macrostructure (i.e., the cycling NREM and REM sleep stage representation and distribution) is generally preserved in DOA, alterations in its "microstructure" have been reported [1]. Sleep microstructure includes EEG features and phasic phenomena occurring in the course of sleep, not belonging to the conventional 30-s scoring epoch and described as the cyclic alternating pattern (CAP) [42, 43]. The CAP is a periodic EEG pattern of NREM sleep, characterized by transient electrocortical events that are distinct from background EEG activity, recurring at up to 1-min intervals and lasting for 2–60 s [43]. A CAP cycle is composed of an A phase (represented by a recurring EEG feature, characterized by an abrupt frequency or amplitude shift from the background rhythm) and a B phase (represented by the intervening background that separates the repetitive elements). Compared to phase B, phase A can be composed of slower high-amplitude rhythms, faster lower-voltage rhythms, or by mixed patterns; examples include vertex sharp transients, K complex sequences with or without spindles, polyphasic bursts, and EEG arousals [43]. The rate at which CAP cycles recur is an index of sleep-state stability, where an increase in CAP rate associates with greater oscillations of state [44].

DOA patients have been found to have increases in CAP rate, in the number of CAP cycles, and in arousals with EEG synchronization [1, 5, 44]. Power spectral analysis of slow wave activity (SWA) in adult sleepwalkers has revealed several forms of SWS dysregulation, including high amount of SWS fragmentations, a significant increase in delta power just prior to an arousal, and increased SWA across all NREM sleep cycles [45]. These findings correlate with an increase in sleep instability and in arousal oscillations, potentially playing a role in triggering the parasomnic episodes in these patients [46].

Etiology of Disorders of Arousal

In conclusion, DOA occur when internal or external precipitating factors trigger the episodes in predisposed individuals. The precise cause of the "pathological" state dissociation and cortical abnormal excitability, however, is unknown but probably results from an interplay of genetic and environmental factors [7]. The overwhelming majority of individuals with DOA do not have underlying neurological or psychological pathology [1]. Recently, neuroanatomical alterations in the cingulate cortex of DOA patients have been reported, but it is unclear if they represent either a cause or an effect of chronic dysfunction [47]. A familial predisposition has been reported, suggesting a genetic base for DOA. First-degree relatives of sleepwalkers have a tenfold or higher increased likelihood of sleepwalking; and 57% of sleepwalkers have a positive familial history [48, 49]. A specific HLA gene (DQB1) appears to confer susceptibility to sleepwalking [50]; family studies have recognized several potential candidate genes, including the adenosine deaminase gene; however, no definite genes have been identified so far [51]. Interestingly, an increased frequency (nearly fivefold) of DOA was found in families of patients with sleep-related hypermotor epilepsy (SHE) [21]. In these patients, causative mutations have been identified in various subunits of the nicotine acetylcholine receptor gene and in the corticotrophin-releasing hormone gene [52]. These mutations confer a gain of function, with increased sensitivity to acetylcholine that may lead to changes in the excitability of cortical and subcortical networks and, at the same time, altering arousal mechanisms and destabilizing sleep [53, 54].

Sleep-Related Eating Disorder (SRED)
SRED is characterized by recurrent episodes of nonvoluntary nocturnal eating and drinking during partial arousals from sleep. Patients often display strange eating behavior (e.g., consumption of inedible or toxic substances such as frozen pizza, raw bacon, and cat food) with no or only partial recall at awakening. SRED generally occurs in women between the ages of 20 and 30; it is not usually associated with waking eating disorders, but a history of eating problems during childhood is common [1, 44]. SRED can be idiopathic but is usually associated with other sleep disorders (most commonly sleepwalking but also restless leg syndrome (RLS), OSAS, and irregular sleep-wake circadian rhythm disorders), use of psychotropic medications (such as triazolam, zolpidem, and quetiapine), or cessation of smoking or substance abuse [1, 5]. The disorder can cause adverse health consequences (weight gain and possibly metabolic syndrome), sleep disruption, injuries, and depression [1]. SRED must be distinguished from the night-eating syndrome (NES), characterized by excessive eating between dinner and bedtime and eating episodes during full awakenings from sleep [5]. SRED benefits most from treatment of the associated sleep disorders: in particular, RLS-related SRED is best treated with dopamine agonists such as pramipexole, while sleepwalking-related SRED benefits from low-dose benzodiazepines such as clonazepam [55–59].

Pathophysiology of SRED

The underlying pathophysiology is unclear, but it has been hypothesized that SRED is a sleepwalking variant with prominent features of an eating behavior; however, given its relatively homogeneous presentation, it has been classified as a separate disorder [1].

REM Sleep-Related Parasomnias

REM sleep-related parasomnias involve aberrations of REM sleep physiology (e.g., lack of atonia in REM sleep behavior disorder) or exaggeration of the features of REM sleep, as in nightmare disorder.

REM Sleep Behavior Disorder (RBD)

RBD is probably the best known example of state admixture and constitutes a critically important REM sleep parasomnia seen in older persons. The disorder is prevalent in males, with a typical age of onset between 40 and 70 years.

RBD is characterized by the appearance of abnormal behaviors during REM sleep that may cause injury or sleep disruption. This occurs due to the lack of REM sleep muscle atonia [2]. Because RBD emerges from REM sleep, which is maximally represented in the last half of the night, this is the time when events are most likely to occur. Episodes usually occur about once a week but may also occur as often as four times nightly over consecutive nights [56].

Clinical Features The central feature is the occurrence of diverse simple or complex motor behavior during REM sleep that mirrors the content of the patient's dreams (i.e., dream enactment behaviors, DEB). The behaviors include talking, laughing, swearing, shouting, reaching, grabbing, flailing, punching, hitting, or running; they are often violent in nature, causing self-injury or injury to the bed partner [18, 44]. In men the episodes appear to be more aggressive than in women, in whom vocalization and simple limb movements are more common [60]. Most patients view their dreams as nightmares, and the dream content often involves a chasing or attacking theme, where the patient is almost always the defender [61]. In contrast to NREM parasomnias, the sleeper awakens rapidly without confusion and often with good recall of a dream that corresponds with the enacted behaviors [56].

Etiology Two clinical forms of RBD are seen: acute/subacute and chronic. RBD is termed "idiopathic" (iRBD) when occurs as an isolated symptom and secondary when associated with an evident neurologic disorder (such as narcolepsy or a neurodegenerative disorder). In the majority of cases, RBD is a chronic, progressive disorder, associated with neurological disorders, such as narcolepsy or neurodegenerative diseases (predominantly synucleinopathies – Parkinson's disease (PD), multiple system atrophy (MSA), dementia with Lewy bodies (DLB), pure autonomic failure (PAF) – but also tauopathies, Alzheimer's disease and others) [61]. There is now evidence that most iRBDs represent the heralding sign of neurodegenerative disorders (see Prognosis) [61]. Chronic RBD may also be associated with other neurological disorders, such as narcolepsy. The acute/subacute form is associated with anticholinergics or REM suppressant medications (e.g., antidepressants such as selective serotonin reuptake inhibitors, SSRIs, and tricyclic antidepressants, TCAs) or in REM sleep rebound conditions (e.g., the first night on continuous positive airway pressure, CPAP, for OSAS, withdrawal from alcohol, and sedative-hypnotic agents) [44, 56]. Acute RBD has also been reported in the context of autoimmune diseases, such as encephalitis (e.g., anti-VGKC, Caspr2, and LGI1[1]) and Guillain-Barré syndrome (GBS) [2, 62, 63].

Diagnosis RBD can be suspected from the clinical history, which should include the bed partner's account; however, video polysomnography (VPSG) is required to confirm the diagnosis (see Table 9.3) [1]. The core PSG feature of RBD is the presence of REM sleep without atonia (RSWA or RWA), that is, increased EMG tone

[1] *VGKC, voltage-gated K+ channels; Caspr2, contactin-associated protein 2; LGI1, leucine-rich, glioma inactivated 1*

Table 9.3 Diagnostic criteria for REM sleep behavior disorder. (Adapted from the American Academy of Sleep Medicine [1])

Diagnostic criteria (criteria A–D must be met)
A. Repeated episodes of sleep-related vocalization and/or complex motor behaviors
B. These behaviors are documented by PSG to occur during REM sleep or are presumed to occur during REM sleep based on clinical history
C. PSG recording demonstrates RWA
D. The disturbance is not better explained by another sleep disorder, mental disorder, medication, or substance use

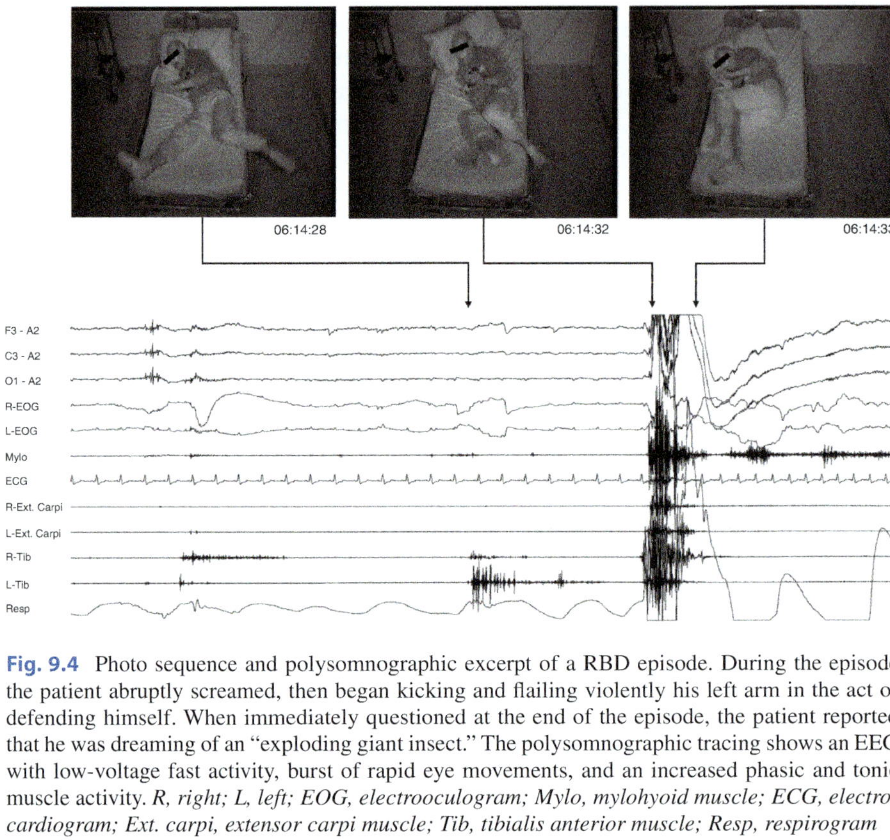

Fig. 9.4 Photo sequence and polysomnographic excerpt of a RBD episode. During the episode the patient abruptly screamed, then began kicking and flailing violently his left arm in the act of defending himself. When immediately questioned at the end of the episode, the patient reported that he was dreaming of an "exploding giant insect." The polysomnographic tracing shows an EEG with low-voltage fast activity, burst of rapid eye movements, and an increased phasic and tonic muscle activity. *R, right; L, left; EOG, electrooculogram; Mylo, mylohyoid muscle; ECG, electrocardiogram; Ext. carpi, extensor carpi muscle; Tib, tibialis anterior muscle; Resp, respirogram*

during REM sleep and/or twitches on EMG+/− movements that often appear to be dream enactment (DEB) during REM sleep (see Fig. 9.4) [44, 56, 61]. Of patients with RBD, more than 70% will also have periodic limb movements during sleep (PLMS) [64].

Prognosis and Management iRBD is a well-established risk factor for neurodegenerative diseases, especially synucleinopathies, and may precede the onset of the other symptoms by many decades [65, 66]. In a large iRBD cohort, the time of RBD diagnosis to neurodegenerative disorder phenoconversion was 33.1% at 5 years, 75.7% at 10 years, and 90.9% at 14 years, with a median conversion time of 7.5 years [67]. Therefore, iRBD may constitute a good therapeutic window to test for neuroprotective agents [44]. Spontaneous remissions of RBD are very rare, although DEBs may subside in the late stages of a neurodegenerative disorder [56].

Management includes measures to modify the sleep environment to reduce the risk of injuries (such as moving sharp and edged objects out of arms' way, placing a mattress on the floor adjacent to the bed, and placing a protective barrier on the side of the bed) and medications [61]. Clonazepam (0.25–0.5 mg/night) is the drug of choice in patients without dementia, gait disorders, or concomitant OSAS improving RBD frequency and severity [44, 61, 68]. The most troublesome adverse effects associated are morning confusion and memory impairment. Melatonin (3–12 mg), which has fewer side effects, can be used alone or in conjunction therapy. Most treated patients experience a marked improvement in the frequency and severity of RBD episodes [44, 61, 68]. The precise mechanisms of action of clonazepam and melatonin are unclear. Clonazepam, a GABA-A agonist, decreases phasic movements without restoring muscle atonia (i.e., RWA is still present), so it allegedly acts mainly by reducing the occurrence of unpleasant dreams. Conversely, melatonin reduces the occurrence of RWA without diminishing phasic movements, maybe through a direct inhibitory effect on motor neurons [61, 69].

Pathophysiology of RBD

As in DOA, state dissociation seems to have a central role in the pathogenesis of RBD. In this dissociated state, the mind is dreaming (REM dream mentation) while the body is awake (spinal motor neurons are still excitable) [2]. In RBD a failure of the mechanisms which normally maintain muscular atonia during REM sleep occurs. Additionally, other mechanisms are responsible for the appearance of dream enactment behaviors, specifically a modification in dream content and the activation of CPGs.

Normal Mechanisms of REM Sleep Atonia

The normal physiologic suppression of motor activity during REM sleep is the cumulative result of multiple neuronal circuits that predominantly originate in the dorsal pons and ultimately terminate on lower motor neurons (LMNs) in the spinal cord [61]. Normally the atonia of REM sleep is briefly interrupted by excitatory inputs that produce the rapid eye movements and the muscle jerks and twitches characteristic of REM sleep [70].

The REM atonia generator is located in the pontine reticular formation (PRF) and is composed mainly of glutamatergic neurons; also cholinergic and GABAergic neurons are involved [71–73]. This region has been called sublaterodorsal nucleus (SLD) in rats and peri-locus coeruleus α (peri-LC α) in cats; for the sake of simplicity, we will call it SLD for the rest of this chapter. This area is both necessary and sufficient to generate muscular atonia during REM sleep. Indeed, electrical stimulation of SLD produces bilateral loss of postural muscle tone, while SLD lesions produce REM without atonia [74, 75]. SLD neurons inhibit lower motor neurons (LMNs) by stimulation of the GABA/glycinergic bulbar gigantocellular ventral nucleus (GiV) and of glycinergic interneurons of the ventral horns, which in turn inhibit LMNs [72]. Additionally, PRF neurons also inhibit the noradrenergic locus coeruleus (LC) and serotoninergic dorsal raphe (DR) nuclei, which during wakefulness maintain a basal tone on LMNs [72, 76–79].

During wakefulness, contrariwise, SLD neurons are tonically inhibited by LC and DR neurons [72, 80–83]. These two REM-off nuclei are under the excitatory influence of the orexinergic neurons and inhibited by the melanin-concentrating hormone (MCH) neurons of the lateral hypothalamus [84–90]. Selective disruption of the orexinergic neurons is thought to be responsible for RBD occurring in narcolepsy. As orexin also promotes wakefulness and prevents inappropriate state shifts, a complex condition of wake-REM instability develops, with sleep attacks and two state dissociations that are opposite of RBD (i.e., muscle atonia appears during wakefulness): sleep paralysis and cataplexy [91].

RWA Results from Disruption of the Pontine REM Atonia Generator

The central pathophysiological feature of RBD is transient loss of the physiological REM sleep atonia (i.e., RWA), allowing the occurrence of motor acts. RWA seems to result from lesion or dysfunction of the brain stem areas which regulate atonia (particularly SLD) or their anatomic connections (see Fig. 9.5) [1, 61].

This can result from a variety of central nervous system pathologies. The clinical-anatomical correlation is quite evident in cases of acute/subacute RBD due to focal damage of the dorsal pons, including stroke, vasculitis, encephalitis, tumors, multiple sclerosis, and even spontaneous intracranial hypotension [92–97]. Neurodegenerative RBD may also be explained by this model. In synucleinopathies, the occurrence of RBD many years before the diagnosis is consistent with Braak staging system: according to this neuropathological model, in PD, α-synuclein pathology begins mainly in the medulla (and olfactory bulb) and gradually ascends to more rostral structures [98, 99]. This would explain why RBD, caused mainly by SLD degeneration, precedes by many years parkinsonism and cognitive decline at least in some patients [61]. Additionally, a combined loss of orexinergic and MCHergic neurons has been described in PD, possibly contributing to RBD [100]. Toxic RBD associated with antidepressants could be explained by either serotoninergic activation of REM-off nuclei or revealing of subclinical RBD in individuals with an underlying neurodegenerative disease [61, 101].

The Mechanism for Abnormal Behaviors and Dreams in RBD: Is It Really "Dream Enactment"?

In RBD speech and movements typically mirror the actual content of a dream; therefore, it has been proposed that dysfunction of the brain stem REM atonia generator would allow movements during REM sleep by unmasking cortically generated dreams (i.e., allowing dream enactment) [1, 69]. Moreover, during RBD episodes, patients can exhibit complex, learned behaviors (smoking, kissing, or speak with coherent syntax), so it has been argued that the motor cortex would be primarily responsible for

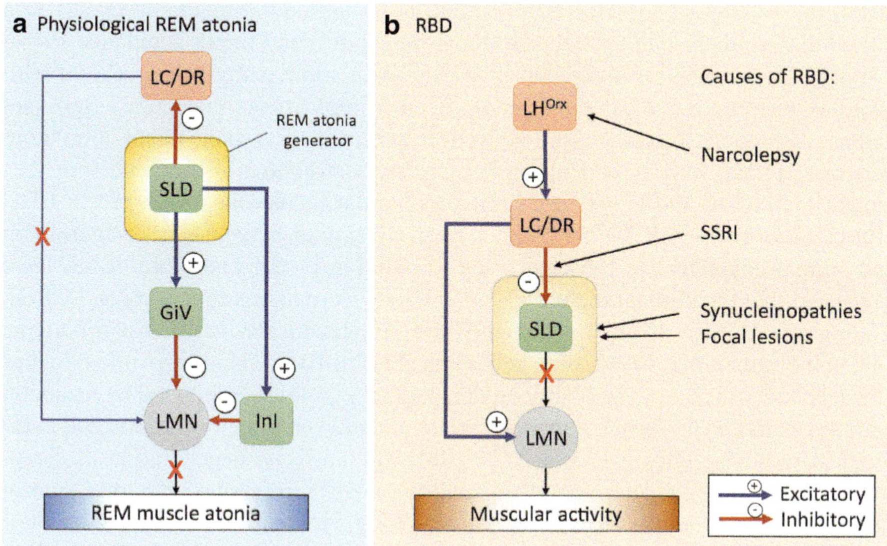

Fig. 9.5 Neuronal networks responsible for REM muscle atonia in physiologic conditions and in RBD. (**a**) Networks in physiological REM sleep. Green squares indicate nuclei active during REM sleep (REM-on), while red squares those inactive during REM sleep and active during wakefulness (REM-off). During REM sleep, the pontine sublaterodorsal nucleus (SLD) (i.e., the REM atonia generator) promotes muscle atonia by two mechanism: first, by stimulation of the bulbar gigantocellular ventral nucleus (GiV) and of the inhibitory interneurons of the spinal cord (InI), which in turn inhibit the lower motor neurons (LMNs), and second, by inhibition of locus coeruleus/dorsal raphe (LC/DR), which during wakefulness maintains a basal tone on LMNs, hence reducing facilitation of motor inputs. (**b**) Networks in RBD. The intermediate connections of the SLD are not shown for sake of simplicity. RBD results from disruption of the REM atonia generator or its connections, allowing an increased LMN tone by LC/DR. Synucleinopathies and focal lesions directly affect the SLD. SSRIs inhibit SLD by enhancing LC/DR activity. In narcolepsy, selective loss of orexinergic neurons is thought to be responsible for RBD to occur, but the precise mechanism is unknown

all of the pathological movements of RBD [1, 69, 102]. Of note, however, in patients with severe bradykinesia and rest tremor during wake, parkinsonism disappears during RBD behaviors, implying bypassing of the basal ganglia [13, 102].

This model, however, does not explain an intriguing feature of RBD, which is the elevated occurrence of nightmares with a chasing or attacking theme in front of a variety of individual psychological profiles [102, 103]. Indeed, RBD appears to be a disorder involving both behavior and dream synthesis, with a central role of the brain stem [103].

According to Hobson-McCarley's "activation-synthesis" model of dream generation, brain stem CPGs simultaneously produce ascending outputs to the cortex to activate cognitive, affective, and amnestic circuits that are synthesized in dreams and descending outputs to the spinal cord for motor behavioral expression, in a parallel manner [102, 104]. So both violent motor behaviors and abnormal dreams in RBD may result from hyperactivity of a common neuronal generator in the brain stem, occasionally with cortical engagement [70, 103, 105]. According to Revonsuo

[106], the biological function of dreaming is to simulate threatening events in a virtual environment and to rehearse threat perception and threat avoidance for an evolutionary purpose of higher survival. Although some patients can present with complex movements, most show jerky, repeated movements that mimic fight and defense behavior, suggesting the activation of CPGs of archaic threat-simulating behaviors [102]. Stimulation of some brain stem structures in rats and monkeys induces defensive and aggressive behaviors similar to the ones observed in RBD [105], and dreams in RBD are characterized by similar percentages of aggression and animal characters to those reported in children [103]. Therefore, it has been suggested that the clinical picture of RBD results from disinhibition of such CPGs, leading to a release of phylogenetically and ontogenetically early dream patterns and, at the same time, isomorphic behaviors [13, 70, 103]. The therapeutic effect of clonazepam, which acts especially on dream content, may be mediated by its action on brain stem CPGs, further supporting the link between dream synthesis and pathological motor activity [61, 104, 107, 108]. Alternatively, degeneration of sleep-related neural circuits may lead to disinhibition of brain stem CPGs that control motor behavior, resulting in exaggerated twitching, violent limb movements, and/or complex behavior; sensory feedback from moving limbs could then provide a source for abnormal dream mentation to the cortex [105]. Therefore, until future studies confirm one or the other model, it may be more precise to talk about "dream isomorphic behaviors" (DIB) instead of "dream enactment behaviors."

Parasomnia Overlap Disorder and Status Dissociatus

This group of complex conditions results from more extensive state dissociation, occurring not only between wake and sleep but also between NREM and REM sleep.

Parasomnia overlap disorder is a subtype of RBD in which patients have both RBD and either a NREM parasomnia (DOA, SRED, sexsomnia) or a rhythmic movement disorder [1]. Therefore, motor disinhibition occurs during both NREM and REM sleep [70]. Most cases begin during childhood or adolescence, idiopathic or symptomatic of a broad set of disorders (e.g., narcolepsy, multiple sclerosis, brain tumors, psychiatric disorders, and their pharmacotherapies) [1, 109].

Status dissociatus (SD) is an extreme form of state dissociation due to the complete breakdown of state-determining boundaries [3, 4]. Clinically, patients appear to be neither awake nor asleep; their behavioral sleep is very atypical, characterized by nearly continuous motor and verbal behaviors and reports of dreamlike mentation upon spontaneous or forced awakening; however, "sleep" can be perceived as normal and restorative by the patient. Polygraphically, there is simultaneous admixture of elements of wakefulness, REM sleep, and NREM sleep, and conventional stages are no longer identifiable. An underlying neurologic condition is virtually always present: protracted withdrawal from alcohol abuse, narcolepsy, MSA, and prior open-heart surgery.

A distinct condition is *agrypnia excitata (AE)*, characterized by the inability to generate sleep, autonomic and motor hyperactivity characterized by *oneiric stupor*, which consists of dream enactment gestures mimicking simple daily life activities, persisting throughout 24 h [1, 110, 111]. Polygraphically, conventional wakefulness, REM, and NREM sleep stages are no longer identifiable: a state of subwakefulness, characterized by a N1 stage with interspersed very short REM sleep periods, becomes the predominant diurnal and nocturnal pattern [1, 2, 70]. AE has been described in four different conditions: fatal familial insomnia (FFI), Morvan's syndrome, delirium tremens, and Mulvihill-Smith syndrome [1, 110, 111, 112].

Recurrent Isolated Sleep Paralysis

This disorder is characterized by recurrent episodes of inability to perform voluntary movements at sleep onset or on waking from sleep, in the absence of a diagnosis of narcolepsy. Consciousness is preserved, and full recall is present. An episode can last from seconds to a few minutes. Sleep deprivation and irregular sleep-wake schedules have been proposed as predisposing factors. The episodes appear to arise from REM sleep. The pathophysiology involves a state dissociation which is the opposite of RBD: elements of REM sleep (muscular atonia) persist into wakefulness just before sleep onset or on waking from sleep [1].

Nightmares

Nightmares are fearful, vivid, often frightening dreams, most often associated with REM sleep and therefore usually occurring during the second half of the night. Nightmares are mostly a normal phenomenon; they are present in up to 50% of children and progressively decrease with age. In a small percentage of individuals, however, recurrent and dysphoric nightmares continue to occur during adulthood, causing significant distress. Predisposing factors are anxious-depressive types of personality and significant psychopathology. The episodes can be precipitated by withdrawal of REM sleep-suppressant drugs (e.g., antidepressants such as SSRIs and TCAs, beta-blockers, or benzodiazepines) due to REM sleep rebound but also by administration of several medications (dopamine-receptor agonists, alcohol, REM sleep suppressants) [1, 113]. Of note, the effect of a medication on REM sleep does not always predict its propensity to cause nightmares. One possible explanation for this effect is that REM sleep, although suppressed, is delayed until later in the night and associated with cholinergic rebound and more intense dream activity [114]. Nightmares generally do not require any treatment except reassurance. In patients with recurring and fearful nightmares, however, combined behavioral or psychotherapy and REM sleep-suppressant medications may be helpful [44].

Other Parasomnias

This heterogeneous group includes parasomnias that bear no specific relationship to sleep stage.

Exploding head syndrome is characterized by a sudden loud noise or sense of explosion in the head, sometimes associated with a flash of light or a myoclonic jerk, occurring as the patient is falling asleep or upon waking during the night. The episode may vary from many on a single night to one every several weeks or months. This condition appears to be a sensory variant of hypnic jerks, but the neurophysiology is unknown [1].

Sleep-related hallucinations are hallucinatory experiences (mainly visual, but also auditory, tactile, or kinetic) that occur at sleep onset or on awakening from sleep. They usually take the form of complex, vivid, relatively immobile images of people or animals, sometimes distorted in shape or size, and can persist for many minutes but usually disappear if ambient illumination is increased. They are believed to result from a state dissociation in which dream ideation of REM sleep intrudes into wakefulness [1].

Isolated Symptoms and Normal Variants: Sleep Talking

Sleep talking (somniloquy) involves talking, with varying levels of comprehensibility, during REM or NREM sleep [115]. It has a high lifetime prevalence of 66% and a current prevalence of 17%, and it has a familial tendency. The episodes are generally brief, infrequent, and devoid of signs of emotional stress. The course is usually self-limited and benign [1].

Conclusions

Parasomnias are a heterogeneous group of disorders that in most cases result from a dissociation of the primary states of being. We are slowly beginning to understand the pathophysiological alterations that lie underneath, but a comprehensive understanding of these disorders remains elusive. Future research directions may include functional neuroimaging and genetic studies. An enhanced understanding of parasomnias could provide a better insight on the functioning of the brain and give us tools to improve the treatment of these conditions.

Acknowledgments The authors thank Francesco Mignani for providing excerpts of video polysomnographic recordings and Cecilia Baroncini for English revision.

References

1. American Academy of Sleep Medicine. International classification of sleep disorders. 3rd ed. Darien: American Academy of Sleep Medicine; 2014.
2. Antelmi E, Ferri R, Iranzo A, Arnulf I, Dauvilliers Y, Bhatia KP, et al. From state dissociation to status dissociatus [Internet]. Sleep Med Rev. W.B. Saunders; 2016 [cited 2017 Nov 21];1–13. Available from: http://www.sciencedirect.com/science/article/pii/S1087079215000945

3. Mahowald MW, Schenck CH. Status dissociatus – a perspective on states of being. Sleep. [Internet]. 1991 [cited 2018 Jan 30];14(1):69–79. Available from: http://www.ncbi.nlm.nih.gov/pubmed/1811323
4. Mahowald MW, Bornemann MAC, Schenck CH. State dissociation, human behavior, and consciousness. Curr Top Med Chem [Internet] 2011;11(19):2392–402. Available from: http://www.eurekaselect.com/openurl/content.php?genre=article&issn=1568-0266&volume=11&issue=19&spage=2392
5. Mahowald MW, Bornemann MAC. Non-REM arousal parasomnias. In: Principles and practice of sleep medicine [Internet]. Elsevier; 2011 [cited 2017 Nov 3];1075–82. Available from: http://linkinghub.elsevier.com/retrieve/pii/B9781416066453000943
6. Gibbs SA, Proserpio P, Terzaghi M, Pigorini A, Sarasso S, Lo Russo G, et al. Sleep-related epileptic behaviors and non-REM-related parasomnias: insights from stereo-EEG. Sleep Med Rev. 2016;25 [cited 2017 Nov 8]:4–20. Available from: http://www.sciencedirect.com/science/article/pii/S1087079215000726
7. Kothare SV, Ivanenko A. Parasomnias: clinical characteristics and treatment. New York: Springer; 2013.
8. Nir Y, Staba RJ, Andrillon T, Vyazovskiy VV, Cirelli C, Fried I, et al. Regional slow waves and spindles in human sleep. Neuron [Internet]. 2011 [cited 2016 Jan 26];70(1):153–69. Available from: http://www.pubmedcentral.nih.gov/articlerender.fcgi?artid=3108825&tool=pmcentrez&rendertype=abstract
9. Vyazovskiy VV, Olcese U, Hanlon EC, Nir Y, Cirelli C, Tononi G. Local sleep in awake rats. Nature [Internet]. 2011 [cited 2016 Feb 9];472(7344):443–7. Available from: http://www.pubmedcentral.nih.gov/articlerender.fcgi?artid=3085007&tool=pmcentrez&rendertype=abstract
10. Huber R, Ghilardi MF, Massimini M, Tononi G. Local sleep and learning. Nature. [Internet]. 2004. [cited 2015 Dec 9];430(6995):78–81. Available from: http://www.ncbi.nlm.nih.gov/pubmed/15184907
11. Krueger JM, Rector DM, Roy S, Van Dongen HPA, Belenky G, Panksepp J. Sleep as a fundamental property of neuronal assemblies. Nat Rev Neurosci [Internet]. 2008 [cited 2016 Feb 9];9(12):910–9. Available from: http://www.pubmedcentral.nih.gov/articlerender.fcgi?artid=2586424&tool=pmcentrez&rendertype=abstract
12. Bonnet MH. Acute sleep deprivation. In: Principles and practice of sleep medicine [Internet]. 5th ed. Elsevier Inc.; 2011. p. 54–66. Available from: https://doi.org/10.1016/B978-1-4160-6645-3.00005-0
13. Howell MJ. Parasomnias: an updated review. Neurotherapeutics [Internet]. Springer; 2012 [cited 2018 Jan 16];9(4):753–75. Available from: http://www.ncbi.nlm.nih.gov/pubmed/22965264
14. Broughton RJ. Sleep disorders: disorders of arousal? Enuresis, somnambulism, and nightmares occur in confusional states of arousal, not in "dreaming sleep". Science [Internet]. 1968 [cited 2018 Jan 15];159(3819):1070–8. Available from: http://www.ncbi.nlm.nih.gov/pubmed/4865791
15. Provini F, Tinuper P, Bisulli F, Lugaresi E. Arousal disorders. Sleep Med. [Internet]. Elsevier; 2011 [cited 2017 Nov 9];12(Suppl. 2):S22–6. Available from: http://www.sciencedirect.com/science/article/pii/S1389945711003182
16. Ohayon MM, Guilleminault C, Priest RG. Night terrors, sleepwalking, and confusional arousals in the general population: their frequency and relationship to other sleep and mental disorders. J Clin Psychiatry [Internet]. 1999 [cited 2018 Jan 11];60(4):268–76; quiz 277. Available from: http://www.ncbi.nlm.nih.gov/pubmed/10221293
17. Schenck CH, Arnulf I, Mahowald MW. Sleep and sex: what can go wrong? A review of the literature on sleep related disorders and abnormal sexual behaviors and experiences. Sleep. [Internet]. 2007. [cited 2018 Jan 11];30(6):683–702. Available from: http://www.ncbi.nlm.nih.gov/pubmed/17580590
18. Avidan A. Parasomnias and movement disorders of sleep. Semin Neurol. [Internet]. 2009. [cited 2018 Jan 11];29(04):372–92. Available from: http://www.ncbi.nlm.nih.gov/pubmed/19742413

19. Bargiotas P, Arnet I, Frei M, Baumann CR, Schindler K, Bassetti CL. Demographic, clinical and polysomnographic characteristics of childhood- and adult-onset sleepwalking in adults. Eur Neurol. [Internet]. 2017. [cited 2018 Jan 25];78(5–6):307–11. Available from: http://www.ncbi.nlm.nih.gov/pubmed/29073634
20. Tinuper P, Provini F, Bisulli F, Vignatelli L, Plazzi G, Vetrugno R, et al. Movement disorders in sleep: guidelines for differentiating epileptic from non-epileptic motor phenomena arising from sleep. Sleep Med Rev. [Internet]. 2007. [cited 2017 May 24];11(4):255–67. Available from: http://linkinghub.elsevier.com/retrieve/pii/S1087079207000196
21. Tinuper P, Bisulli F, Cross JH, Hesdorffer D, Kahane P, Nobili L, et al. Definition and diagnostic criteria of sleep-related hypermotor epilepsy. Neurology. [Internet]. 2016. [cited 2017 Nov 27];86(19):1834–42. Available from: http://www.ncbi.nlm.nih.gov/pubmed/27164717
22. Derry CP, Harvey AS, Walker MC, Duncan JS, Berkovic SF. NREM arousal parasomnias and their distinction from nocturnal frontal lobe epilepsy: a video EEG analysis. Sleep. [Internet]. 2009. [cited 2017 Nov 8];32(12):1637–44. Available from: http://www.ncbi.nlm.nih.gov/pubmed/20041600
23. Bisulli F, Vignatelli L, Provini F, Leta C, Lugaresi E, Tinuper P. Parasomnias and nocturnal frontal lobe epilepsy (NFLE): Lights and shadows – Controversial points in the differential diagnosis. Sleep Med [Internet]. 2011 [cited 2017 Nov 8];12:S27–32. Available from: http://www.ncbi.nlm.nih.gov/pubmed/22136895
24. Ferri R, Rundo F, Bruni O, Terzano MG, Stam CJ. Regional scalp EEG slow-wave synchronization during sleep cyclic alternating pattern A1 subtypes. Neurosci Lett. 2006;404(3):352–7.
25. Castelnovo A, Riedner BA, Smith RF, Tononi G, Boly M, Benca RM. Scalp and source power topography in sleepwalking and sleep terrors: a high-density EEG study. Sleep. [Internet]. 2016;39(10):1815–25. Available from: https://academic.oup.com/sleep/article-lookup/doi/10.5665/sleep.6162
26. Bassetti C, Vella S, Donati F, Wielepp P, Weder B. SPECT during sleepwalking. Lancet. [Internet]. 2000. [cited 2018 Jan 8];356(9228):484–5. Available from: http://www.ncbi.nlm.nih.gov/pubmed/10981896
27. Braun a R, Balkin TJ, Wesenten NJ, Carson RE, Varga M, Baldwin P, et al. Regional cerebral blood flow throughout the sleep-wake cycle. An H2(15)O PET study. Brain. 1997;120(Pt 7):1173–97.
28. Terzaghi M, Sartori I, Tassi L, Didato G, Rustioni V, LoRusso G, et al. Evidence of dissociated arousal states during NREM parasomnia from an intracerebral neurophysiological study. Sleep. [Internet]. 2009. [cited 2018 Jan 8];32(3):409–12. Available from: http://www.ncbi.nlm.nih.gov/pubmed/19294961
29. Terzaghi M, Sartori I, Tassi L, Rustioni V, Proserpio P, Lorusso G, et al. Dissociated local arousal states underlying essential clinical features of non-rapid eye movement arousal parasomnia: an intracerebral stereo-electroencephalographic study. J Sleep Res. [Internet]. 2012. [cited 2018 Jan 17];21(5):502–6. Available from: http://www.ncbi.nlm.nih.gov/pubmed/22329925
30. Oliviero A, Della Marca G, Tonali PA, Pilato F, Saturno E, Dileone M, et al. Functional involvement of cerebral cortex in adult sleepwalking. J Neurol. 2007;254(8):1066–72.
31. Desjardins M, Baril A, Desautels A, Marquis L, Soucy J, Montplaisir J, et al. Effects of sleep deprivation on brain perfusion patterns in sleepwalkers' wakefulness and slow wave sleep. Sleep [Internet]. Oxford University Press; 2017 [cited 2018 Jan 25];40(suppl_1):A273–4. Available from: https://academic.oup.com/sleep/article-lookup/doi/10.1093/sleepj/zsx050.737
32. Dang-Vu TT, Zadra A, Labelle M-A, Petit D, Soucy J-P, Montplaisir J. Sleep deprivation reveals altered brain perfusion patterns in somnambulism. Ferri R, editor. PLoS One [Internet]. 2015 [cited 2018 Jan 25];10(8):e0133474. Available from: http://www.ncbi.nlm.nih.gov/pubmed/26241047
33. Ferentinos P, Paparrigopoulos T. Zopiclone and sleepwalking. Int J Neuropsychopharmacol [Internet]. Oxford University Press; 2009 [cited 2018 Jan 25];12(01):141. Available from: https://academic.oup.com/ijnp/article-lookup/doi/10.1017/S1461145708009541

34. Joncas S, Zadra A, Paquet J, Montplaisir J. The value of sleep deprivation as a diagnostic tool in adult sleepwalkers. Neurology [Internet]. American Academy of Neurology; 2002 [cited 2018 Jan 25];58(6):936–40. Available from: http://www.ncbi.nlm.nih.gov/pubmed/11914411
35. Stallman HM, Kohler M, White J. Medication induced sleepwalking: a systematic review. Sleep Med Rev [Internet]. Elsevier; 2018 [cited 2018 Jan 25];37:105–13. Available from: http://www.ncbi.nlm.nih.gov/pubmed/28363449
36. Guilleminault C, Palombini L, Pelayo R, Chervin RD. Sleepwalking and sleep terrors in prepubertal children: what triggers them? Pediatrics. [Internet]. 2003. [cited 2018 Jan 11];111(1):e17–25. Available from: http://www.ncbi.nlm.nih.gov/pubmed/12509590
37. Huttenlocher P, de Courten C. The development of synapses in striate cortex of man. Hum Neurobiol. 1987;6(1):1–9.
38. Tassinari CA, Gardella E, Cantalupo G, Rubboli G. Relationship of central pattern generators with parasomnias and sleep-related epileptic seizures. Sleep Med Clin [Internet]. Elsevier; 2012 [cited 2018 Jan 11];7(1):125–34. Available from: http://linkinghub.elsevier.com/retrieve/pii/S1556407X12000045
39. Pillmann F. Complex dream-enacting behavior in sleepwalking. Psychosom Med. [Internet]. 2009. [cited 2018 Jan 25];71(2):231–4. Available from: http://www.ncbi.nlm.nih.gov/pubmed/19073753
40. Nielsen TA. A review of mentation in REM and NREM sleep: "Covert" REM sleep as a possible reconciliation of two opposing models. Behav Brain Sci [Internet]. Cambridge University Press; 2000 [cited 2018 Jan 25];23(6):S0140525X0000399X. Available from: http://www.journals.cambridge.org/abstract_S0140525X0000399X
41. Zadra A, Desautels A, Petit D, Montplaisir J. Somnambulism: clinical aspects and pathophysiological hypotheses. Lancet Neurol [Internet]. Elsevier; 2013 [cited 2018 Jan 25]:285–94. Available from: https://www.sciencedirect.com/science/article/pii/S1474442212703228
42. Muzet A. Alteration of sleep microstructure in psychiatric disorders. Dialogues Clin Neurosci. 2005;7:315–21.
43. Terzano MG, Parrino L, Sherieri A, Chervin R, Chokroverty S, Guilleminault C, et al. Atlas, rules, and recording techniques for the scoring of cyclic alternating pattern (CAP) in human sleep. Sleep Med [Internet]. 2001 [cited 2018 Jan 16];2(6):537–53. Available from: http://www.ncbi.nlm.nih.gov/pubmed/14592270
44. Daroff RB, Jankovic J, Mazziotta JC, Pomeroy SL. Chapter 102: Sleep and its disorders. In: Bradley's neurology in clinical practice. 7th ed. London: Elsevier; 2016. p. 1615–85.
45. Perrault R, Carrier J, Desautels A, Montplaisir J, Zadra A. Electroencephalographic slow waves prior to sleepwalking episodes. Sleep Med. 2014;15(12):1468–72.
46. Zucconi M, Oldani A, Ferini-Strambi L, Smirne S. Arousal fluctuations in non-rapid eye movement parasomnias: the role of cyclic alternating pattern as a measure of sleep instability. J Clin Neurophysiol. [Internet]. 1995. [cited 2018 Jan 16];12(2):147–54. Available from: http://www.ncbi.nlm.nih.gov/pubmed/7797629
47. Heidbreder A, Stefani A, Brandauer E, Steiger R, Kremser C, Gizewski ER, et al. Gray matter abnormalities of the dorsal posterior cingulate in sleep walking. Sleep Med [Internet]. Elsevier; 2017 [cited 2018 Jan 25];36:152–5. Available from: https://www.sciencedirect.com/science/article/pii/S1389945717302162
48. Lopez R, Jaussent I, Scholz S, Bayard S, Montplaisir J, Dauvilliers Y. Functional impairment in adult sleepwalkers: a case-control study. Sleep [Internet]. 2013 [cited 2018 Jan 8];36(3):345–51. Available from: https://academic.oup.com/sleep/article-lookup/doi/10.5665/sleep.2446
49. Hublin C, Kaprio J, Partinen M, Heikkilä K, Koskenvuo M. Prevalence and genetics of sleepwalking: a population-based twin study. Neurology. [Internet]. 1997. [cited 2018 Jan 8];48(1):177–81. Available from: http://www.ncbi.nlm.nih.gov/pubmed/9008515
50. Lecendreux M, Bassetti C, Dauvilliers Y, Mayer G, Neidhart E, Tafti M. HLA and genetic susceptibility to sleepwalking. Mol Psychiatry [Internet]. 2003 [cited 2018 Jan 10];8(1):114–7. Available from: http://www.ncbi.nlm.nih.gov/pubmed/12556916
51. Licis AK, Desruisseau DM, Yamada KA, Duntley SP, Gurnett CA. Novel genetic findings in an extended family pedigree with sleepwalking. Neurology. [Internet]. 2011. [cited

2018 Jan 8];76(1):49–52. Available from: http://www.neurology.org/cgi/doi/10.1212/WNL.0b013e318203e964
52. Bisulli F, Vignatelli L, Naldi I, Licchetta L, Provini F, Plazzi G, et al. Increased frequency of arousal parasomnias in families with nocturnal frontal lobe epilepsy: a common mechanism? Epilepsia. [Internet]. 2010. [cited 2017 May 23];51(9):1852–60. Available from: http://www.ncbi.nlm.nih.gov/pubmed/20477848
53. Hoda J, Gu W, Friedli M, Phillips HA, Bertrand S, Antonarakis SE, et al. Human nocturnal frontal lobe epilepsy: pharmacogenomic profiles of pathogenic nicotinic acetylcholine receptor β-subunit mutations outside the ion channel pore. Mol Pharmacol. 2008;74(2):379–91.
54. Combi R, Dalprà L, Ferini-Strambi L, Tenchini ML. Frontal lobe epilepsy and mutations of the corticotropin-releasing hormone gene. Ann Neurol. 2005;58(6):899–904.
55. Howell MJ, Schenck CH, Crow SJ. A review of nighttime eating disorders. Sleep Med Rev. [Internet]. 2009. [cited 2018 Jan 11];13(1):23–34. Available from: http://www.ncbi.nlm.nih.gov/pubmed/18819825
56. Fleetham JA, Fleming JAE. Parasomnias. CMAJ [Internet]. Canadian Medical Association; 2014 [cited 2017 Nov 3];186(8):E273–80. Available from: http://www.ncbi.nlm.nih.gov/pubmed/24799552
57. Chiaro G, Caletti MT, Provini F. Treatment of sleep-related eating disorder. Curr Treat Options Neurol. 2015;17(8):361.
58. Howell MJ, Schenck CH. Restless nocturnal eating: a common feature of Willis-Ekbom syndrome (RLS). J Clin Sleep Med [Internet]. 2012 [cited 2018 Jan 11];8(4):413–9. Available from: http://www.ncbi.nlm.nih.gov/pubmed/22893772
59. Provini F, Antelmi E, Vignatelli L, Zaniboni A, Naldi G, Calandra-Buonaura G, et al. Association of restless legs syndrome with nocturnal eating: a case-control study. Mov Disord. 2009;24(6):871–7.
60. Mahale RR, Yadav R, Pal PK. Rapid eye movement sleep behaviour disorder in women with Parkinson's disease is an underdiagnosed entity. J Clin Neurosci. 2016;28:43–6.
61. Boeve BF. REM sleep behavior disorder: updated review of the core features, the RBD-neurodegenerative disease association, evolving concepts, controversies, and future directions. Ann N Y Acad Sci. 2010;1184:15–54.
62. Iranzo A, Graus F, Clover L, Morera J, Bruna J, Vilar C, et al. Rapid eye movement sleep behavior disorder and potassium channel antibody-associated limbic encephalitis. Ann Neurol. 2006;59(1):178–82.
63. Cornelius JR, Pittock SJ, McKeon A, Lennon VA, Aston PA, Josephs KA, et al. Sleep manifestations of voltage-gated potassium channel complex autoimmunity. Arch Neurol. 2011;68(6):733–8.
64. Fantini ML, Michaud M, Gosselin N, Lavigne G, Montplaisir J. Periodic leg movements in REM sleep behavior disorder and related autonomic and EEG activation. Neurology. [Internet]. 2002;59(12):1889–94. Available from: http://graphics.tx.ovid.com/ovftpdfs/FPDDNCJCOEEKIG00/fs047/ovft/live/gv024/00006114/00006114-200212240-00015.pdf
65. Postuma RB, Gagnon J-F, Montplaisir J. Rapid eye movement sleep behavior disorder as a biomarker for neurodegeneration: the past 10 years. Sleep Med. [Internet]. 2013;14(8):763–7. Available from: http://www.ncbi.nlm.nih.gov/pubmed/23058689
66. Claassen DO, Josephs KA, Ahlskog JE, Silber MH, Tippmann-Peikert M, Boeve BF. REM sleep behavior disorder preceding other aspects of synucleinopathies by up to half a century. Neurology. 2010;75(6):494–9.
67. Iranzo A, Fernández-Arcos A, Tolosa E, Serradell M, Molinuevo JL, Valldeoriola F, et al. Neurodegenerative disorder risk in idiopathic REM sleep behavior disorder: study in 174 patients. PLoS One. 2014;9(2):e89741.
68. Aurora RN, Zak RS, Maganti RK, Auerbach SH, Casey KR, Chowdhuri S, et al. Best practice guide for the treatment of REM sleep behavior disorder (RBD). J Clin Sleep Med [Internet]. 2010;6(1):85–95.
69. Luppi PH, Clement O, Valencia Garcia S, Brischoux F, Fort P. New aspects in the pathophysiology of rapid eye movement sleep behavior disorder: the potential role of glutamate,

gamma-aminobutyric acid, and glycine. Sleep Med. [Internet]. 2013;14(8):714–718. Available from: http://www.ncbi.nlm.nih.gov/pubmed/23790501

70. Mahowald MW, Schenck CH. REM sleep parasomnias. In: Principles and practice of sleep medicine [Internet]. Elsevier; 2011 [cited 2017 Nov 3]. p. 1083–97. Available from: http://linkinghub.elsevier.com/retrieve/pii/B9781416066453000955
71. Jones BE. Paradoxical sleep and its chemical/structural substrates in the brain. Neuroscience. [Internet]. 1991. [cited 2016 Jan 31];40(3):637–56. Available from: http://www.sciencedirect.com/science/article/pii/030645229190002б
72. Brown RE, Basheer R, McKenna JT, Strecker RE, McCarley RW. Control of sleep and wakefulness. Physiol Rev. 2012;92(3):1087–187.
73. Luppi PH, Clément O, Fort P. Paradoxical (REM) sleep genesis by the brainstem is under hypothalamic control. Curr Opin Neurobiol. 2013;23(5):786–92.
74. Arrigoni E, Chen MC, Fuller PM. The anatomical, cellular and synaptic basis of motor atonia during rapid eye movement sleep. J Physiol. 2016;594:5391–414.
75. Lu J, Sherman D, Devor M, Saper CB. A putative flip-flop switch for control of REM sleep. Nature. [Internet]. 2006. [cited 2015 Sep 21];441(7093):589–94. Available from: http://www.ncbi.nlm.nih.gov/pubmed/16688184
76. Siegel JM, Moore R, Thannickal T, Nienhuis R. A brief history of hypocretin/orexin and narcolepsy. Neuropsychopharmacology [Internet] 2001;25(5 Suppl):S14–20. Available from: https://doi.org/10.1016/S0893-133X(01)00317-7
77. White SR, Fung SJ, Jackson DA, Imel KM. Serotonin, norepinephrine and associated neuropeptides: effects on somatic motoneuron excitability. Prog Brain Res [Internet]. 1996 [cited 2016 Feb 19];107:183–99. Available from: http://www.ncbi.nlm.nih.gov/pubmed/8782520
78. John J, Wu M-F, Boehmer LN, Siegel JM. Cataplexy-active neurons in the hypothalamus. Neuron. [Internet]. 2004. [cited 2016 Feb 19];42(4):619–34. Available from: http://www.sciencedirect.com/science/article/pii/S0896627304002478
79. Siegel JM. Clues to the functions of mammalian sleep. Nature. 2005;437(7063):1264–71.
80. McCarley RW. Mechanisms and models of REM sleep control. Arch Ital Biol [Internet]. 2004 [cited 2016 Jan 31];142(4):429–67. Available from: http://www.ncbi.nlm.nih.gov/pubmed/15493547
81. Mitani A, Ito K, Hallanger AE, Wainer BH, Kataoka K, McCarley RW. Cholinergic projections from the laterodorsal and pedunculopontine tegmental nuclei to the pontine gigantocellular tegmental field in the cat. Brain Res. [Internet]. 1988. [cited 2016 Jan 31];451(1–2):397–402. Available from: http://www.ncbi.nlm.nih.gov/pubmed/3251602
82. McCarley RW, Hobson JA. Neuronal excitability modulation over the sleep cycle: a structural and mathematical model. Science. [Internet]. 1975. [cited 2016 Jan 31];189(4196):58–60. Available from: http://www.ncbi.nlm.nih.gov/pubmed/1135627
83. Hobson JA, McCarley RW, Wyzinski PW. Sleep cycle oscillation: reciprocal discharge by two brainstem neuronal groups. Science. 1975;189(4196):55–8.
84. Horvath TL, Peyron C, Diano S, Ivanov A, Aston-Jones G, Kilduff TS, et al. Hypocretin (orexin) activation and synaptic innervation of the locus coeruleus noradrenergic system. J Comp Neurol. [Internet]. 1999. [cited 2016 Jan 13];415(2):145–59. Available from: http://www.ncbi.nlm.nih.gov/pubmed/10545156
85. Liu R-J, van den Pol AN, Aghajanian GK. Hypocretins (orexins) regulate serotonin neurons in the dorsal raphe nucleus by excitatory direct and inhibitory indirect actions. J Neurosci. [Internet]. 2002. [cited 2015 Dec 1];22(21):9453–64. Available from: http://www.ncbi.nlm.nih.gov/pubmed/12417670
86. Clément O, Sapin E, Libourel P-A, Arthaud S, Brischoux F, Fort P, et al. The lateral hypothalamic area controls paradoxical (REM) sleep by means of descending projections to brainstem GABAergic neurons. J Neurosci. [Internet]. 2012;32(47):16763–74. Available from: http://www.ncbi.nlm.nih.gov/pubmed/23175830
87. Verret L, Goutagny R, Fort P, Cagnon L, Salvert D, Léger L, et al. A role of melanin-concentrating hormone producing neurons in the central regulation of paradoxical sleep. BMC Neurosci [Internet]. 2003 [cited 2016 Jan 27];4:19. Available from: http://www.

pubmedcentral.nih.gov/articlerender.fcgi?artid=201018&tool=pmcentrez&rendertype=abstract
88. Konadhode RR, Pelluru D, Blanco-Centurion C, Zayachkivsky A, Liu M, Uhde T, et al. Optogenetic stimulation of MCH neurons increases sleep. J Neurosci [Internet]. 2013;33(25):10257–63. Available from: http://www.pubmedcentral.nih.gov/articlerender.fcgi?artid=3685832&tool=pmcentrez&rendertype=abstract
89. Saito Y, Cheng M, Leslie FM, Civelli O. Expression of the melanin-concentrating hormone (MCH) receptor mRNA in the rat brain. J Comp Neurol. [Internet]. 2001. [cited 2016 Feb 3];435(1):26–40. Available from: http://www.ncbi.nlm.nih.gov/pubmed/11370009
90. Bittencourt JC. Anatomical organization of the melanin-concentrating hormone peptide family in the mammalian brain. Gen Comp Endocrinol [Internet]. Elsevier Inc.; 2011;172(2):185–97. Available from: http://www.ncbi.nlm.nih.gov/pubmed/21463631
91. Wu MF, Gulyani SA, Yau E, Mignot E, Phan B, Siegel JM. Locus coeruleus neurons: cessation of activity during cataplexy. Neuroscience. 1999;91(4):1389–99.
92. Mathis J, Hess CW, Bassetti C. Isolated mediotegmental lesion causing narcolepsy and rapid eye movement sleep behaviour disorder: a case evidencing a common pathway in narcolepsy and rapid eye movement sleep behaviour disorder. J Neurol Neurosurg Psychiatry [Internet]. 2007;78(4):427–9. Available from: http://www.pubmedcentral.nih.gov/articlerender.fcgi?artid=2077786&tool=pmcentrez&rendertype=abstract
93. Xi Z, Luning W. REM sleep behavior disorder in a patient with pontine stroke. Sleep Med. 2009;10(1):143–6.
94. St. Louis EK, McCarter SJ, Boeve BF, Silber MH, Kantarci K, Benarroch EE, et al. Lesional REM sleep behavior disorder localizes to the dorsomedial pons. Neurology. 2014;83:1871–3.
95. Plazzi G, Montagna P. Remitting REM sleep behavior disorder as the initial sign of multiple sclerosis. Sleep Med. 2002;3(5):437–9.
96. Zambelis T, Paparrigopoulos T, Soldatos CR. REM sleep behaviour disorder associated with a neurinoma of the left pontocerebellar angle. J Neurol Neurosurg Psychiatry. 2002;72:821–2.
97. Salsone M, Salvino D, Mazza MR, Mumoli L, Bombardieri C, Quattrone A, et al. Reversible deep brain swelling causing REM behavior disorder. Neurology. 2016;86:1360.
98. Braak H, Del Tredici K, Rüb U, De Vos RAI, Jansen Steur ENH, Braak E. Staging of brain pathology related to sporadic Parkinson's disease. Neurobiol Aging. 2003;24(2):197–211.
99. Braak H, Ghebremedhin E, Rüb U, Bratzke H, Del Tredici K. Stages in the development of Parkinson's disease-related pathology. Cell Tissue Res. 2004:121–34.
100. Thannickal TC, Lai Y-Y, Siegel JM. Hypocretin (orexin) cell loss in Parkinson's disease. Brain [Internet]. 2007 [cited 2016 Jan 27];130(Pt 6):1586–95. Available from: http://www.ncbi.nlm.nih.gov/pubmed/17491094
101. Postuma RB, Gagnon J-F, Tuineaig M, Bertrand J-A, Latreille V, Desjardins C, et al. Antidepressants and REM sleep behavior disorder: isolated side effect or neurodegenerative signal? Sleep [Internet]. 2013;36(11):1579–85. Available from: http://www.pubmedcentral.nih.gov/articlerender.fcgi?artid=3792373&tool=pmcentrez&rendertype=abstract
102. Arnulf I. REM sleep behavior disorder: motor manifestations and pathophysiology. Mov Disord. 2012;27:677–89.
103. Fantini ML, Corona A, Clerici S, Ferini-Strambi L. Aggressive dream content without daytime aggressiveness in REM sleep behavior disorder. Neurology. 2005;65(7):1010–5.
104. Ferri R, Marelli S, Ferini-Strambi L, Oldani A, Colli F, Schenck CH, et al. An observational clinical and video-polysomnographic study of the effects of clonazepam in REM sleep behavior disorder. Sleep Med. 2013;14(1):24–9.
105. Blumberg MS, Plumeau AM. A new view of "dream enactment" in REM sleep behavior disorder. Sleep Med Rev. 2016;30:34–42.
106. Revonsuo A. The reinterpretation of dreams: an evolutionary hypothesis of the function of dreaming. Behav Brain Sci. 2000;23(6):877–901.
107. Li SX, Lam SP, Zhang J, Yu MWM, Chan JWY, Liu Y, et al. A prospective, naturalistic follow-up study of treatment outcomes with clonazepam in rapid eye movement sleep behavior disorder. Sleep Med. 2016;21:114–20.

108. D'Agostino A, Manni R, Limosani I, Terzaghi M, Cavallotti S, Scarone S. Challenging the myth of REM sleep behavior disorder: no evidence of heightened aggressiveness in dreams. Sleep Med. 2012;13(6):714–9.
109. Limousin N, Dehais C, Gout O, Héran F, Oudiette D, Arnulf I. A brainstem inflammatory lesion causing REM sleep behavior disorder and sleepwalking (parasomnia overlap disorder). Sleep Med. [Internet]. 2009. [cited 2018 Jan 25];10(9):1059–62. Available from: http://www.ncbi.nlm.nih.gov/pubmed/19345142
110. Provini F. Agrypnia excitata. Curr Neurol Neurosci Rep. 2013;13(4):341.
111. Lugaresi E, Provini F. Agrypnia excitata: clinical features and pathophysiological implications. Sleep Med Rev. [Internet]. 2001. [cited 2018 Jan 30];5(4):313–22. Available from: http://www.ncbi.nlm.nih.gov/pubmed/12530995
112. Ferri R, Lanuzza B, Cosentino FII, Iero I, Russo N, Tripodi M, Bosco P. Agrypnia excitata in a patient with progeroid short stature and pigmented Nevi (Mulvihill-Smith syndrome). J Sleep Res. 2005;14:463–70.
113. Pagel JF, Helfter P. Drug induced nightmares – an etiology based review. Hum Psychopharmacol. 2003;18(1):59–67.
114. Pace-Schott EF, Gersh T, Silvestri R, Stickgold R, Salzman C, Hobson JA. SSRI treatment suppresses dream recall frequency but increases subjective dream intensity in normal subjects. J Sleep Res. 2001;10(2):129–42.
115. Arnulf I, Uguccioni G, Gay F, Baldayrou E, Golmard J-L, Gayraud F, et al. What does the sleeping brain say? Syntax and semantics of sleep talking in healthy subjects and in parasomnia patients. Sleep. [Internet]. 2017. [cited 2018 Jan 25];40(11). Available from: http://www.ncbi.nlm.nih.gov/pubmed/29029239

Neurobiology of Sleep-Related Movements

10

Michela Figorilli, Monica Puligheddu, Patrizia Congiu, Stefan Clemens, and Raffaele Ferri

Periodic Limb Movements

Periodic limb movements (PLM) are defined as repetitive and somewhat stereotyped movements occurring in sleep (PLMS) or wakefulness (PLMW). PLM have been described polysomnographically in patients suffering from restless legs syndrome (RLS), during both wakefulness and sleep, firstly by Lugaresi and colleagues in 1965 [1]. Usually, PLM involve the lower limbs with an extension of the big toe, often combined with partial flexion of the ankle, the knee, and sometimes the hip, involving several muscles; in 75% of cases, the tibialis anterior (TA) is activated and more rarely the muscles of the upper limbs [2]. These movements may be unilateral or bilateral and may be associated with cortical and autonomic arousal or an awakening [2, 3]. PLMS are defined as movements lasting from 0.5 to 10 s, with an increase in electromyographic (EMG) signal to ≥8 μV above baseline and an intermovement interval between two consecutive PLMS (onset-to-onset) of 10–90s [4], and finally PLMS sequences are identified by at least four consecutive limb movements with intermovement intervals within this range.

PLMS are found in up to 95% of patients with RLS, but they may occur in other sleep disorders, such as narcolepsy, REM sleep behavior disorder (RBD), and

M. Figorilli · M. Puligheddu · P. Congiu
Sleep Disorder Center, Department of Medical Sciences and Public Health, University of Cagliari, Cagliari, Italy
e-mail: puligheddu@unica.it

S. Clemens
Department of Physiology, Brody School of Medicine, East Carolina University, Greenville, NC, USA
e-mail: clemenss@ecu.edu

R. Ferri (✉)
Sleep Research Centre, Oasi Research Institute - IRCCS, Troina, Italy
e-mail: rferri@oasi.en.it

L. M. DelRosso, R. Ferri (eds.), *Sleep Neurology*,
https://doi.org/10.1007/978-3-030-54359-4_10

obstructive sleep apnea syndrome (OSAS) [2, 5–10]. Moreover, PLMS can be observed also in several medical conditions, like cardiovascular disease, end-stage renal disease, and depression [11–17]. PLMS have also been observed in patients with neurodegenerative diseases, such as Parkinson's disease (PD) and multiple system atrophy (MSA) [18–20]. PLMS are considered to be an identifiable sleep disorder, named periodic limb movement disorder (PLMD), when they exceed 15/hour in adults or 5/hour in children and lead to sleep disruption and daytime consequences and cannot be explained by another sleep disorder or condition [2]. Finally, PLMS can be found also in the general healthy population, with a prevalence of 5–8%, which increases with age [21–23].

The pathophysiology of PLMS is not still completely understood. Given the close association of PLMS with RLS and the responsiveness of RLS patients to dopamine agonist and iron treatments, dopaminergic impairment has been proposed together with iron status depletion, as possible pathological mechanisms [2]. However, there are major debates regarding the localization of the neural circuits and structures implicated in the pathophysiology of PLMS [24]. Additionally, there is increasing evidence of a genetic diathesis implicated in pathophysiology of PLMS [2].

Dopaminergic Dysfunction and Supraspinal Networks

Dopaminergic dysfunction has been implicated in the pathophysiology of PLMS associated to RLS and PLMD. Supporting this theory, PET and SPECT studies have shown a depletion of striatal D2-subtype dopamine receptor in patients with RLS and PLMD [25–27]. Moreover, low doses of dopamine agonists that target the inhibitory D2-like receptor subtypes are very effective in suppressing PLMS since the first night of treatment [28, 29]. Therefore, the efficacy of these dopamine agonists together with neuroimaging studies indicates that the dopaminergic networks in the central nervous system, rather than in the peripheral nervous system, may play a role in PLMS pathology [30]. Converging evidence has highlighted a possible role of the dopaminergic diencephalon-spinal pathway originating from the posterior hypothalamic A11 cell group [31–33]. In contrast, while the basal ganglia have often been suspected to be involved in the pathogenesis of PLMS/RLS, there is not strong evidence for a direct connection of the dopaminergic basal ganglia to the spinal circuits that ultimately generate the movement [34–36].

The original dopaminergic hypothesis for RLS/PLMS, focused mainly around a possible hypodopaminergic state in nigrostriatal and spinal cord circuits, has been supported by the presence of (1) circadian secretion of dopamine, (2) association with iron depletion, and (3) dramatic and immediate response to dopaminergic replacement therapy [37, 38]. However, the dopamine metabolites in the CSF have not been found to be different between patients and controls [39], while 3-orthymethyl dopamine (3-OMD) and homovanillic acid (HVA) have been found to be significantly increased in the CSF of RLS patients, suggesting an increased dopamine production [40].

Brain imaging studies have reported controversial findings. The aforementioned SPECT and PET studies have shown decreased D2-receptor binding, probably associated with both decreased striatal D2 receptors, rather than an increased synaptic dopamine associated with a downregulation of D2-receptor [25, 26]. Likewise, Fluoro-L-Dopa (fDopa) analysis has shown declined fDopa uptake that indicates a fast turnover of dopamine, compatible with an increased dopamine production [41]. Overall, neuroimaging studies have shown an increased, rather than decreased, dopamine metabolism, suggesting a hyperdopaminergic state behind RLS/PLMS [42, 43]. This hyperdopaminergic stimulation might lead to a postsynaptic adjustment that creates relative evening and nighttime dopamine deficits, despite the overall increased dopamine [43].

However, beyond the dopamine dysfunction theory, there is some evidence of leg movement activity associated with other sleep disorders, such as sleep-related breathing disorders, which not always respond to dopamine agonists, suggesting a different mechanism that have to be yet fully understood [44, 45].

Iron Status

Several studies suggest a role for brain iron in RLS and PLMS pathophysiology [2, 46]. However, the precise mechanism is far from being clearly understood. Iron deficiency in the brain may be implicated in cellular loss by oxidative stress and modification of lipids, carbohydrates, protein, and DNA, through hydroxyl radical production [46, 47]. In fact, iron metabolism in the central nervous system is a crucial factor for several proteins involved in various biochemical processes, such as oxidative phosphorylation, myelin production, and the synthesis and metabolism of neurotransmitters [47]. Iron is an indispensable cofactor for dopamine production via tyrosine hydroxylase [48]. Thus, decreased extracellular dopamine is found in iron depletion, suggesting that iron affects the brain dopaminergic transmission in different ways and might play a role in RLS and PLMS genesis, leading to spinal hyperexcitability [30, 49]. The role of brain iron in the pathogenesis of RLS/PLMS has been supported by an in vivo MRI phase analysis, showing reduced iron in the substantia nigra, putamen, caudate, and thalamus [50]. However, it is not clear if brain iron metabolism impairment is implicated in all cases of RLS/PLMS.

Striatal and Cortical Hypoadenosinergic State

Recent experimental evidence suggests that brain iron deficiency might be associated with downregulation of adenosine A1 receptors (A1R) in the striatum and cortex which might provide a link between the alterations of dopaminergic and glutamatergic systems in RLS [51]. In fact, adenosine is a main endogenous sleep-promoting factor and a powerful modulator of both glutamate and dopamine neurotransmission. This hypoadenosinergic state might also support a mechanism explaining PLMS, secondary to hypersensitivity of cortico-striatal terminals [52].

The adenosine theory has recently received preliminary confirmation from a prospective open-label clinical trial in 13 patients with idiopathic RLS treated with dipyridamole, a nonselective inhibitor of adenosine transporters [53]. This study showed a significant decrease in PLMS index and PLMS/arousal index after treatment with a single evening dose of 100–400 mg dipyridamole for 8 weeks.

Spinal Cord

Both spinal and supraspinal mechanisms have been considered for the genesis and modulation of PLMS [54–60]. In fact, PLMS can be recorded also in some patients with complete spinal transection, even though in these patients only a minority of leg movements is actually periodic [57]. Moreover, PLMS resemble the spinal defense reflex mechanism that results from enhanced excitability in the flexor-reflex arc [31]. PLMS have been regarded as the expression of a so-called central pattern generator (CPG) for locomotion, characterized by stereotyped motor patterns produced by subcortical networks and modulated by phylogenetically recent neocortical structure [61]. This locomotor CPG is located in lamina VII of the thoracolumbar segments of the spinal cord (T12-L2), an area that, together with the motoneurons as the final common output of the CNS, receives strong innervation by descending dopamine fibers, presumably from the A11 cluster (Fig. 10.1) [62–64]. Thus, supraspinal pathways might act in a complex and coordinated synergy, producing an excitatory effect onto the spinal CPG that controls locomotion [65].

Furthermore, some authors have proposed an alteration in the hypothalamic/diencephalospinal (A11 area) dopaminergic tract that modulates the excitability of different neuronal populations, such as neurons of the dorsal horns and motoneural sites, suggesting a spinal dopamine dysfunction [32, 33]. Therefore, it has been postulated that PLMS originate from a sleep-related dysfunction of descending inhibitory dopamine pathways, resulting in a spinal hyperexcitability, due to a lowered inhibition [33, 56]. The responsiveness of PLMS to dopaminergic treatment, together with the presence of dopaminergic terminals in the spinal dorsal and ventral horns, supports the theory of a role of dopamine in the spinal motor control mechanism [33]. However, it seems reasonable that spinal generators are modulated and integrated in a more complex network, involving also supraspinal and supratentorial structures. In fact, PLMS have been reported in patients with lesions interrupting the descending supraspinal pathways, like subcortical stroke or subcortical and demyelinating lesions [66–69]. PLMS are frequently synchronized with the sleep EEG oscillation known as cyclic alternating pattern (CAP), which is not modified after dopaminergic therapy [70, 71]. CAP is a physiological NREM sleep periodic EEG activity, characterized by repeated spontaneous sequences of transient events (phase A), separated by return to background activity intervals (phase B) with duration up to 60 s (Fig. 10.2) [72]. This synchronization can be lost in patients with abnormalities of the connection between supraspinal and spinal structures, such as in spinal cord traumatic injuries and neurodegenerative diseases [57, 73].

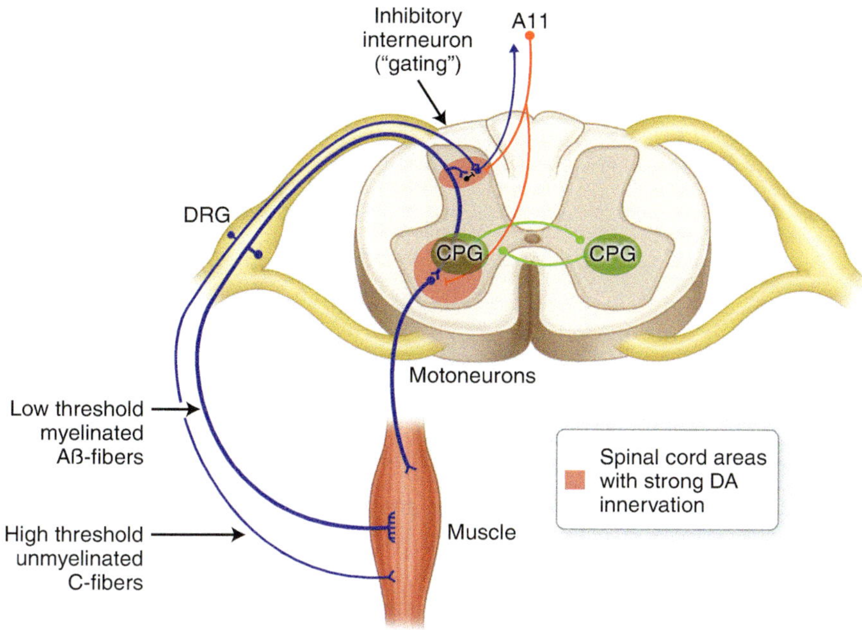

Fig. 10.1 Model of potential interactions of local neural circuits in a cross section of the thoraco-lumbar spinal cord that control locomotion and their descending modulation by dopaminergic fibers. The central pattern generator (CPG) for locomotion is located bilaterally in lamina VII of spinal segments T12–L2 (green ovals). Mutual inhibition of the underlying neurons across the midline (indicated by line-ball connections) organizes the activity patterns, and the CPG drives the motoneurons that control the leg movements. Both dorsal and ventral aspects of the spinal gray matter receive dopaminergic input descending from the A11 region of the posterior hypothalamus (projection areas indicated by red ovals). Within the dorsal horn, dopamine modulation controls the excitability of the high-threshold unmyelinated fibers that originate deep in the leg muscles and that likely carry the initial information that underlies the "urge to move" in RLS. These fibers are also negatively gated by activation of proprioceptive low-threshold Aß-fibers in the muscles and their associated inhibitory interneurons in the dorsal laminae, which is suspected to terminate the urge to move. Within the ventral horn, dopamine modulation can control both the locomotor CPG circuits and motoneuron excitability. *DA* dopamine; *DRG* dorsal root ganglion; *CPG* central pattern generator

Genetic Diathesis

The fact that a family history of RLS confers increased risk for PLMS and PLMD suggests a strong genetic predisposition. Genome-wide association studies (GWAS) have identified common risk variants in six genomic regions and single-nucleotide polymorphism (SNPs), namely, the disease-associated loci contained MEIS1, BTBD9, MAP2K5/SKOR1, PTPRD, and TOX3 [74–77]. Among these loci, MEIS1 has been found to be associated with the strongest risk, odds ratios (ORs) around 2.7

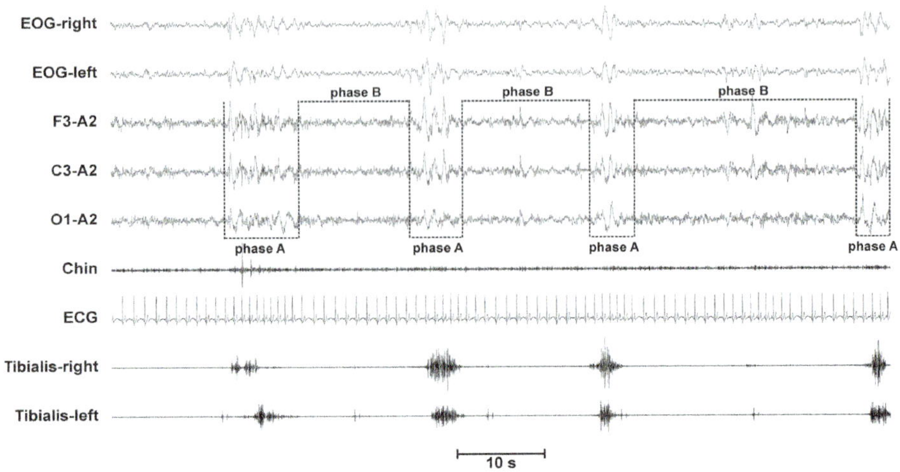

Fig. 10.2 Periodic limb movement occurring in synchronization with CAP phase A during sleep stage N2

[74–77]. Different studies have found a relationship between the common RLS-associated SNP rs3923809 to PLMS in the absence of RLS sensory symptoms, and also the authors found an association between PLMS and variants in BTBD9 and MEIS1 [78–80]. Conversely, PLMS alone were not related with rs3784709 (OR = 1.19) or rs3104767 (OR = 1.07), implying that MAP2K5/SKOR1 and TOX3 may be more profoundly associated with the sensory symptoms of RLS. Further studies are needed to asses SNPs for association to PLMS in other disorders or in a large cohort of subjects with PLMD.

Other Theories

Some authors have found that leg movement and leg discomfort are associated with functional magnetic resonance imaging (fMRI)-related activation in the thalamus and cerebellum [59, 81]. Moreover, PET studies have found enhanced dopaminergic and opioid binding in the thalamus [82, 83]. Another study found a positive relationship between GABA levels and both PLMS and RLS severity in the thalamus and a negative correlation of these measures in the cerebellum [84].

In conclusion, the pathophysiology of PLMS is far from being clearly understood. Converging evidence highlights the hypothesis that PLMS might be regulated in an oscillatory fashion by a complex mechanism involving interactions between the central nervous system, the respiratory system, and the cardiovascular system. The mutual relationships between these systems are also complex and cannot be explained by a simple cause/effect relationship. Thus a large variety of events, including PLMS, seem to be integrated in a multifaceted oscillatory mechanism that modulates their time structure in both physiology and pathology.

Other Leg Motor Activities During Sleep

There are three sleep-related motor events that are pretty similar to each other and may represent a different expression of the same phenomenon along a spectrum. In fact, they are characterized by small motor activation of the lower limbs, short-lasting bilateral alternating or unilateral activations (agonist/antagonist), happening during sleep or wakefulness, frequently associated with arousal and often in trains.

Alternating Leg Muscle Activation

Alternating leg muscle activation (ALMA) comprises brief alternating activation of the anterior tibialis muscles during sleep or, more frequently, in concomitance of an arousal from sleep (Fig. 10.3) [2]. ALMA seems to be a benign motor phenomenon without clinical impact, although it is not evident if it represents one of the nocturnal motor activities associated with RLS. The neurobiology of ALMA is not completely elucidated. Some authors speculated that ALMA might be an expression of a facilitated spinal central pattern generator for locomotion by descending fibers from brainstem serotoninergic nuclei [85]. Another study has shown a reduction in the rate of ALMA after treatment with pramipexole, suggesting that the spinal networks involved might be modulated by the inhibitory control of dopaminergic networks, like RLS and PLMS [86]. Moreover, the existence of ALMA, regardless of other sleep disorders, such as sleep apnea, PLMS, or use of antidepressant drugs, suggests that ALMA might be modulated by sleep instability, namely, by CAP, in an oscillatory fashion [86].

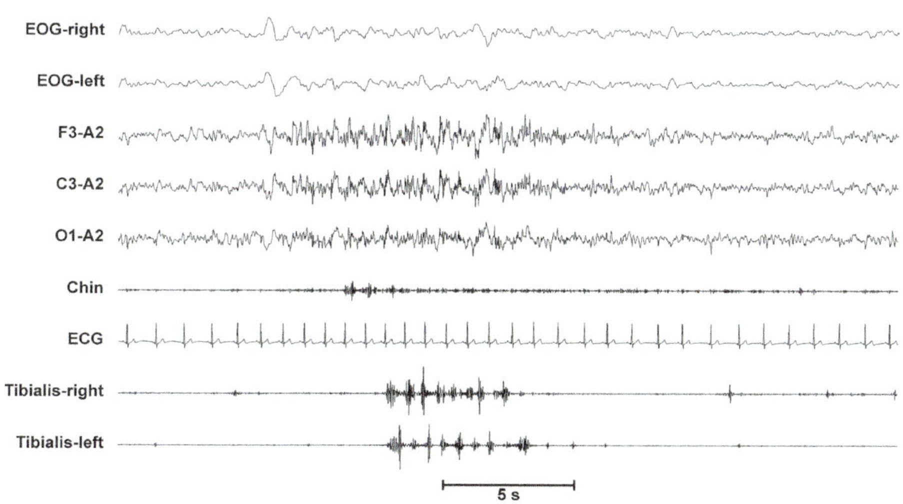

Fig. 10.3 Example of an ALMA event occurring during an arousal from sleep stage N2

Hypnagogic Foot Tremor

The hypnagogic foot tremor (HFT) is characterized by a rhythmic movement of feet or toes arising during the transition between wakefulness and sleep or during NREM sleep stage N1 and/or N2, with a frequency of 1–4 Hz [87, 88]. The pathophysiology of HFT is still unknown. Moreover, HFT may be considered to be a benign phenomenon without clinical significance.

High-Frequency Leg Movements

High-frequency leg movements (HFLM) are defined by at least four leg movements with 0.3–0.4 Hz frequency, typically unilateral, occurring in all sleep stages and also, frequently, during wakefulness [2, 89]. HFLM pathophysiology is yet to be explored, and further studies are needed to ascertain its neurobiology and clinical implications, such as sleep fragmentation, reduction of daytime performance, and cardiovascular and cerebrovascular consequences.

Excessive Fragmentary Myoclonus

Excessive fragmentary myoclonus (EFM) is characterized by muscle twitches that may occur asynchronously, symmetrically, and bilaterally, during wake-sleep transition or during all stages of sleep [90, 91]. EFM involves mostly distal muscles, such as fingers, toes, and corners of the mouth, and twitches are barely visible. EFM may be observed incidentally during PSG and resembles the REM twitches [88]. EFM pathophysiology and clinical significance is still debated; it is more common in the elderly and may occur in up 5–10% of patients with excessive daytime sleepiness and other sleep disorders, like sleep apnea [88]. Recently, Vetrugno et al. [92] have shown no neuromuscular abnormalities nor EEG arousal associated to EFM. One seminal work has postulated that EFM is originated and modulated by the reticulospinal tract [93]. More recently, Raccagni et al. [94] have found EMG abnormalities in patients with EFM, such as polyneuropathy, root lesions, and benign fasciculation, suggesting a probable peripheral nerve pathology behind EFM.

Finally, there are two main contradictory hypotheses for EFM pathophysiology that need to be clarified, namely, brainstem or peripheral nerve dysfunction. There is no therapy option available and, indeed, the need for treatment is unclear.

Propriospinal Myoclonus at Sleep Onset

Propriospinal myoclonus (PMS) at sleep onset is characterized by generalized and symmetric jerks, which arise first from spinal axial muscles of the trunk, neck, or abdomen and then propagate to more rostral and caudal muscles, through propriospinal polysynaptic pathways [95, 96]. PMS arises from relaxed wakefulness or drowsiness, associated with diffuse alpha EEG activity, and is usually inhibited by mental activation and sleep onset [2]. Some studies have postulated the existence of a spinal myoclonus generator at the thoracic level whose hyperexcitability might be associated with functional impairment of the spinal cord [97–99]. Moreover, the persistence of this phenomenon during the wake-sleep transition might be explained by the loss of supraspinal control over the spinal generator, namely, the sleep-related spinal inhibition from brainstem structures, possibly the reticular formation [100]. Finally, further studies are needed to determine if there are subtle spinal cord alterations behind PMS.

Sleep Bruxism

Sleep bruxism (SB) is a phenomenon defined by repetitive jaw muscle activity, such as clenching/grinding the teeth or bracing or thrusting the mandible, occurring during light sleep (NREM stages N1 and N2), and is generally associated with arousal from sleep [2]. The neurobiological background of SB remains controversial. Originally, SB was supposed to be generated by tooth occlusion factors, like disharmony or premature tooth contacts which trigger excessive jaw muscle activity [101–103]. However, further studies have shown no correlation between dental morphology and SB events and actually have shown decreased jaw muscle activity [104, 105]. Nowadays, SB is considered to be mainly modulated by the central autonomic nervous system, with psychological hyperarousal factors and marginal contribution from anatomical factors [106–108]. Another hypothesis of SB pathophysiology is based on the influence of stress and psychological factors [109, 110]. This theory is far from being confirmed; however, stress and psychological factors seem to represent a risk factor that may predispose, maintain, or exacerbate SB [101, 111]. Genetics predisposition for SB is not completely explored, and twin studies have found 20–50% of variance [112, 113]. Probably, SB is a polygenic pathology with several genes and their haplotypes interacting with environmental, psychological, and endogenous factors [24].

Recent converging evidence has pointed out the role of the central and autonomic nervous system in the genesis of SB which seems to be centrally mediated and modulated by the autonomic nervous system and brain arousal [24, 111]. Moreover, other studies have found a relationship between SB and CAP; in fact, SB events occur in clusters, are preceded by alpha EEG activity, and are associated with transient increases in heart rate and blood pressure [114–117]. The autonomic balance during sleep may influence SB events, and cardiovascular and respiratory factors may modulate SB [118–120]. Finally, SB has been found in some patients suffering from other sleep disorders, such as RLS, RBD, and sleep apnea.

Sleep-Related Rhythmic Movement Disorder

The sleep-related rhythmic movement disorder (RMD) is characterized by repetitive, stereotyped, and rhythmic motor behaviors occurring mainly during drowsiness or sleep [2]. RMD may involve large muscle groups like the head, neck, trunk, or limbs, and, depending on which muscles are involved, RMD may be differentiated in subtypes, namely, body rocking, head banging, and head rolling [2]. Different types of RMD may happen in the same subject during the same night [121, 122]. RMD is primarily observed in children, and it is considered to be a benign phenomenon that usually resolves spontaneously. However, patients with RMD may experience insomnia, parasomnia, and daytime sleepiness, but whether the movements cause sleep disturbances is not clear [24]. RMD may be present also during adulthood, generally associated with other primary sleep disorders, such as OSAS, RBD, narcolepsy, and RLS. The pathophysiology behind RMD is still uncertain and seems to be correlated to fluctuations in arousal modulated by central motor pattern generators of the brainstem [123].

Sleep-Related Leg Cramps

Sleep-related leg cramps are painful sensations usually in the calf or small muscles of the foot, caused by sudden and intense involuntary muscle contraction that leads to muscle spasm and hardness for several seconds [2]. Generally, sleep-related leg cramps are considered to be an idiopathic condition, but they may be associated with other disorders, such as vascular disease, lumbar canal stenosis, cirrhosis, hemodialysis, pregnancy, neuromuscular disease, and metabolic conditions. The neurobiological mechanism of sleep-related leg cramps is far from being understood. There are some theories that encompass abnormal spinal cord excitability, spinal disinhibition, abnormal terminal motor nerve excitability, and increased muscle contraction propagation through cross-activation of adjacent neurons [124, 125].

Sleep Starts

Sleep starts, also known as hypnic jerks, are a sleep-wake transition disorder and are characterized by sudden, brief, simultaneous contractions of the body or one or more body segments occurring at sleep onset [2]. This phenomenon may be spontaneous or induced by stimuli and is often accompanied by sensory, auditory, or visual sensations. Generally, it is considered to be a normal variant and a benign phenomenon without clinical consequences. However, when sleep starts are repetitive, intense, or frequent, it might represent a cause of distress and insomnia. The pathophysiology of sleep starts is uncertain, and it seems associated to the modulatory system in the brainstem reticular formation and forebrain and is triggered by instability of the system in the transition from wakefulness to sleep [88, 123].

Neck Myoclonus During Sleep

Neck myoclonus during sleep indicates the movement associated with a short stripe-shaped movement-induced artifact over the EEG leads during REM sleep. This phenomenon has been reported to occur frequently in routine polysomnography, being present in >50% of patients, but with a low frequency during the night (1 ± 2.7 per hour of REM sleep) and an age-related decline, from younger to older patients [126]. However, it has been found also in 35% of healthy sleepers with a rate up to 8.8/hour in REM sleep; this indicates the uncertain clinical significance of this phenomenon and confirms the probable physiological nature of this phenomenon that can be considered to be similar to other phasic motor events of REM sleep, such as muscle twitches [127].

However, in particular patient neck myoclonus or similar movements might reflect other conditions with a clinical significance. An elderly patient has been reported having a "facio-mandibular myoclonus during REM sleep" which resembles closely neck myoclonus, but this patient had nocturnal awakenings due to biting of the tongue and forceful jaw closings [128]. Another 3-year-old boy presenting with alternate brisk head turns to the right and left was found to have focal cortical myoclonus localized to the unilateral neck [129]. Finally, neck myoclonus has also been reported to occur in periodic sequences mainly during NREM sleep, in a fashion similar to that of PLMS; this neck myoclonus has probably a different meaning and is not correlated with neck myoclonus during REM sleep [130].

Conclusion

In this chapter, we have described a wide variety of sleep-related movements that should be clearly distinguished from sleep-related behaviors, typically seen in parasomnia or other disorders, which were not the focus here. Although many categories of sleep-related movements involve the lower limbs, it is not clear if this is simply due to the fact that leg movements are routinely recorded during night polysomnography while other muscle districts are not or if this reflects a real preponderance of leg movements over other movements. With respect to PLMS in RLS, however, it was reported that leg muscles were the muscles most frequently involved; in particular, the tibialis anterior muscle was activated in approx. 75% of movements [131]. Table 10.1 summarizes the features of the different sleep-related movement disorders listed in this chapter and their proposed neurobiological mechanisms; however, in most instances, no definite pathophysiological mechanism has been determined, and their clinical significance is still unclear. This points at a need for additional research in the future.

Table 10.1 Summary of sleep-related movement disorders and their neurobiological mechanisms

Disorder	Movement features	Proposed mechanism(s)
Periodic limb movements	Repetitive movements usually involving the lower limbs with an extension of the big toe, often combined with partial flexion of the ankle, the knee, and sometimes the hip, unilateral or bilateral, lasting from 0.5 to 10 s, at least four consecutive movements with interval (onset-to-onset) of 10–90 s	Role for dopamine dysfunction, iron, genetic diathesis, hypoadenosinergic state
Alternating leg muscle activation	Brief alternating activation (0.5–2 Hz) of the anterior tibialis muscles during sleep or, more frequently, in concomitance of an arousal from sleep	Spinal central pattern generator for locomotion facilitated by descending fibers from brainstem serotoninergic and/or dopaminergic nuclei
Hypnagogic foot tremor	Rhythmic movement of the feet or toes arising during the transition between wakefulness and sleep or during NREM sleep stage N1 and/or N2, at 1–4 Hz	Benign phenomenon without clinical significance
High-frequency leg movements	At least four leg movements with 0.3–0.4 Hz frequency, typically unilateral, occurring in all sleep stages and also, frequently, during wakefulness	Unknown
Excessive fragmentary myoclonus	Muscle twitches that may occur asynchronously, symmetrically, and bilaterally, during wake-sleep transition or during all stages of sleep	Brainstem or peripheral nerve dysfunction
Propriospinal myoclonus at sleep onset	Generalized and symmetric jerks, which arise first from spinal axial muscles of the trunk, neck, or abdomen and then propagate to more rostral and caudal muscles, through propriospinal polysynaptic pathways	Loss of sleep-related supraspinal control from brainstem structures over the spinal myoclonus generator
Sleep bruxism	Repetitive jaw muscle activity, such as clenching/grinding the teeth or bracing or thrusting the mandible, occurring during light sleep and generally associated with arousal from sleep	Combined role of central (arousal) and autonomic nervous systems
Sleep-related rhythmic movement disorder	Repetitive, stereotyped, and rhythmic motor behaviors occurring mainly during drowsiness or sleep	Fluctuations in arousal modulated by central motor pattern generators of the brainstem
Sleep-related leg cramps	Painful sensations usually in the calf or small muscles of the foot, caused by sudden and intense involuntary muscle contraction that leads to muscle spasm and hardness for several seconds	Abnormal spinal cord excitability, spinal disinhibition, abnormal terminal motor nerve excitability, and increased muscle contraction propagation through cross-activation of adjacent neurons
Sleep starts	Sudden, brief, simultaneous contractions of the body or one or more body segments occurring at sleep onset	Instability of brainstem mechanisms
Neck myoclonus during sleep	Movement associated with a short stripe-shaped movement-induced artifact over the EEG leads during REM sleep	Physiological REM sleep phasic motor phenomenon

References

1. Lugaresi E, Coccagna G, Tassinari CA, Ambrosetto C. Polygraphic data on motor phenomena in the restless legs syndrome. Riv Neurol. 1965;35(6):550–61.
2. ICSD-3 Online Version – American Academy of Sleep Medicine (AASM) [Internet]. AASM. 2015 [cited 2015 Jan 10]. Available from: http://www.aasmnet.org/store/product.aspx?pid=849
3. Sforza E, Nicolas A, Lavigne G, Gosselin A, Petit D, Montplaisir J. EEG and cardiac activation during periodic leg movements in sleep: support for a hierarchy of arousal responses. Neurology. 1999;52(4):786–91.
4. Ferri R, Fulda S, Allen RP, Zucconi M, Bruni O, Chokroverty S, et al. World Association of Sleep Medicine (WASM) 2016 standards for recording and scoring leg movements in polysomnograms developed by a joint task force from the International and the European Restless Legs Syndrome Study Groups (IRLSSG and EURLSSG). Sleep Med. 2016;26:86–95.
5. Allen RP, Picchietti DL, Garcia-Borreguero D, Ondo WG, Walters AS, Winkelman JW, et al. Restless legs syndrome/Willis–Ekbom disease diagnostic criteria: updated International Restless Legs Syndrome Study Group (IRLSSG) consensus criteria – history, rationale, description, and significance. Sleep Med. 2014;15(8):860–73.
6. Dauvilliers Y, Pennestri M-H, Petit D, Dang-Vu T, Lavigne G, Montplaisir J. Periodic leg movements during sleep and wakefulness in narcolepsy. J Sleep Res. 2007;16(3):333–9.
7. Fantini ML, Michaud M, Gosselin N, Lavigne G, Montplaisir J. Periodic leg movements in REM sleep behavior disorder and related autonomic and EEG activation. Neurology. 2002;59(12):1889–94.
8. Baran AS, Richert AC, Douglass AB, May W, Ansarin K. Change in periodic limb movement index during treatment of obstructive sleep apnea with continuous positive airway pressure. Sleep. 2003;26(6):717–20.
9. Chervin RD. Periodic leg movements and sleepiness in patients evaluated for sleep-disordered breathing. Am J Respir Crit Care Med. 2001;164(8 Pt 1):1454–8.
10. Al-Alawi A, Mulgrew A, Tench E, Ryan CF. Prevalence, risk factors and impact on daytime sleepiness and hypertension of periodic leg movements with arousals in patients with obstructive sleep apnea. J Clin Sleep Med. 2006;2(3):281–7.
11. Benz RL, Pressman MR, Hovick ET, Peterson DD. Potential novel predictors of mortality in end-stage renal disease patients with sleep disorders. Am J Kidney Dis. 2000;35(6):1052–60.
12. Espinar-Sierra J, Vela-Bueno A, Luque-Otero M. Periodic leg movements in sleep in essential hypertension. Psychiatry Clin Neurosci. 1997;51(3):103–7.
13. Yang C, White DP, Winkelman JW. Antidepressants and periodic leg movements of sleep. Biol Psychiatry. 2005;58(6):510–4.
14. Koo BB, Mehra R, Blackwell T, Ancoli-Israel S, Stone KL, Redline S, et al. Periodic limb movements during sleep and cardiac arrhythmia in older men (MrOS sleep). J Clin Sleep Med. 2014;10(1):7–11.
15. Koo BB, Sillau S, Dean DA, Lutsey PL, Redline S. Periodic limb movements during sleep and prevalent hypertension in the multi-ethnic study of atherosclerosis. Hypertens Dallas Tex 1979. 2015;65(1):70–7.
16. Coelho FMS, Georgsson H, Narayansingh M, Swartz RH, Murray BJ. Higher prevalence of periodic limb movements of sleep in patients with history of stroke. J Clin Sleep Med. 2010;6(5):428–30.
17. Yumino D, Wang H, Floras JS, Newton GE, Mak S, Ruttanaumpawan P, et al. Relation of periodic leg movements during sleep and mortality in patients with systolic heart failure. Am J Cardiol. 2011;107(3):447–51.
18. Puligheddu M, Figorilli M, Aricò D, Raggi A, Marrosu F, Ferri R. Time structure of leg movement activity during sleep in untreated Parkinson disease and effects of dopaminergic treatment. Sleep Med [Internet]. 2014 [cited 2015 Jan 24]. Available from: http://www.sciencedirect.com/science/article/pii/S1389945714001464

19. Wetter TC, Collado-Seidel V, Pollmächer T, Yassouridis A, Trenkwalder C. Sleep and periodic leg movement patterns in drug-free patients with Parkinson's disease and multiple system atrophy. Sleep. 2000;23(3):361–7.
20. Arnulf I, Konofal E, Merino-Andreu M, Houeto JL, Mesnage V, Welter ML, et al. Parkinson's disease and sleepiness: an integral part of PD. Neurology. 2002;58(7):1019–24.
21. Carrier J, Frenette S, Montplaisir J, Paquet J, Drapeau C, Morettini J. Effects of periodic leg movements during sleep in middle-aged subjects without sleep complaints. Mov Disord. 2005;20(9):1127–32.
22. Scofield H, Roth T, Drake C. Periodic limb movements during sleep: population prevalence, clinical correlates, and racial differences. Sleep. 2008;31(9):1221.
23. Ancoli-Israel S, Kripke DF, Klauber MR, Mason WJ, Fell R, Kaplan O. Periodic limb movements in sleep in community-dwelling elderly. Sleep. 1991;14(6):496–500.
24. Kryger MH, Roth T, Dement WC, editors. Principles and practice of sleep medicine. 6th ed. Philadephia: Elsevier; 2017. 1730 p. (ExpertConsult).
25. Turjanski N, Lees AJ, Brooks DJ. Striatal dopaminergic function in restless legs syndrome: 18F-dopa and 11C-raclopride PET studies. Neurology. 1999;52(5):932–7.
26. Michaud M, Soucy J-P, Chabli A, Lavigne G, Montplaisir J. SPECT imaging of striatal pre- and postsynaptic dopaminergic status in restless legs syndrome with periodic leg movements in sleep. J Neurol. 2002;249(2):164–70.
27. Staedt J, Stoppe G, Kögler A, Riemann H, Hajak G, Munz DL, et al. Nocturnal myoclonus syndrome (periodic movements in sleep) related to central dopamine D2-receptor alteration. Eur Arch Psychiatry Clin Neurosci. 1995;245(1):8–10.
28. Manconi M, Ferri R, Zucconi M, Oldani A, Fantini ML, Castronovo V, et al. First night efficacy of pramipexole in restless legs syndrome and periodic leg movements. Sleep Med. 2007;8(5):491–7.
29. Saletu M, Anderer P, Saletu B, Hauer C, Mandl M, Oberndorfer S, et al. Sleep laboratory studies in restless legs syndrome patients as compared with normals and acute effects of ropinirole. 2. Findings on periodic leg movements, arousals and respiratory variables. Neuropsychobiology. 2000;41(4):190–9.
30. Garcia-Borreguero D, Williams A-M. An update on restless legs syndrome (Willis-Ekbom disease): clinical features, pathogenesis and treatment. Curr Opin Neurol. 2014;27(4):493–501.
31. Rye DB, Trotti LM. Restless legs syndrome and periodic leg movements of sleep. Neurol Clin. 2012;30(4):1137–66.
32. Clemens S, Rye D, Hochman S. Restless legs syndrome: revisiting the dopamine hypothesis from the spinal cord perspective. Neurology. 2006;67(1):125–30.
33. Rijsman RM, Stam CJ, de Weerd AW. Abnormal H-reflexes in periodic limb movement disorder; impact on understanding the pathophysiology of the disorder. Clin Neurophysiol. 2005;116(1):204–10.
34. Kedia S, Moro E, Tagliati M, Lang AE, Kumar R. Emergence of restless legs syndrome during subthalamic stimulation for Parkinson disease. Neurology. 2004;63(12):2410–2.
35. Driver-Dunckley E, Evidente VGH, Adler CH, Hillman R, Hernandez J, Fletcher G, et al. Restless legs syndrome in Parkinson's disease patients may improve with subthalamic stimulation. Mov Disord. 2006;21(8):1287–9.
36. Ondo W. VIM deep brain stimulation does not improve pre-existing restless legs syndrome in patients with essential tremor. Parkinsonism Relat Disord. 2006;12(2):113–4.
37. Carlsson A, Svennerholm L, Winblad B. Seasonal and circadian monoamine variations in human brains examined post mortem. Acta Psychiatr Scand. 1980;61(S280):75–85.
38. Sowers JR, Vlachakis N. Circadian variation in plasma dopamine levels in man. J Endocrinol Investig. 1984;7(4):341–5.
39. Stiasny-Kolster K, Möller JC, Zschocke J, Bandmann O, Cassel W, Oertel WH, et al. Normal dopaminergic and serotonergic metabolites in cerebrospinal fluid and blood of restless legs syndrome patients. Mov Disord. 2004;19(2):192–6.

40. Allen RP, Connor JR, Hyland K, Earley CJ. Abnormally increased CSF 3-Ortho-methyldopa (3-OMD) in untreated restless legs syndrome (RLS) patients indicates more severe disease and possibly abnormally increased dopamine synthesis. Sleep Med. 2009;10(1):123–8.
41. Ruottinen HM, Partinen M, Hublin C, Bergman J, Haaparanta M, Solin O, et al. An FDOPA PET study in patients with periodic limb movement disorder and restless legs syndrome. Neurology. 2000;54(2):502–4.
42. Earley CJ, Kuwabara H, Wong DF, Gamaldo C, Salas RE, Brašić JR, et al. Increased synaptic dopamine in the putamen in restless legs syndrome. Sleep. 2013;36(1):51–7.
43. Allen RP. Restless leg syndrome/Willis-Ekbom disease pathophysiology. Sleep Med Clin. 2015;10(3):207–14, xi.
44. Hanly P, Zuberi N. Periodic leg movements during sleep before and after heart transplantation. Sleep. 1992;15(6):489–92.
45. Manconi M, Vitale G, Ferri R, Zucconi M, Ferini-Strambi L. Periodic leg movements in Cheyne-Stokes respiration. Eur Respir J. 2008 Dec;32(6):1656–62.
46. Guo S, Huang J, Jiang H, Han C, Li J, Xu X, et al. Restless legs syndrome: from pathophysiology to clinical diagnosis and management. Front Aging Neurosci. 2017;9:171.
47. Ward RJ, Zucca FA, Duyn JH, Crichton RR, Zecca L. The role of iron in brain ageing and neurodegenerative disorders. Lancet Neurol. 2014;13(10):1045–60.
48. Earley CJ, Connor J, Garcia-Borreguero D, Jenner P, Winkelman J, Zee PC, et al. Altered brain iron homeostasis and dopaminergic function in restless legs syndrome (Willis-Ekbom disease). Sleep Med. 2014;15(11):1288–301.
49. Dauvilliers Y, Winkelmann J. Restless legs syndrome: update on pathogenesis. Curr Opin Pulm Med. 2013;19(6):594–600.
50. Rizzo G, Manners D, Testa C, Tonon C, Vetrugno R, Marconi S, et al. Low brain iron content in idiopathic restless legs syndrome patients detected by phase imaging. Mov Disord. 2013;28(13):1886–90.
51. Ferré S, Quiroz C, Guitart X, Rea W, Seyedian A, Moreno E, et al. Pivotal role of adenosine neurotransmission in restless legs syndrome. Front Neurosci. 2017;11:722.
52. Yepes G, Guitart X, Rea W, Newman AH, Allen RP, Earley CJ, et al. Targeting hypersensitive corticostriatal terminals in restless legs syndrome. Ann Neurol. 2017;82(6):951–60.
53. Garcia-Borreguero D, Guitart X, Garcia Malo C, Cano-Pumarega I, José Granizo J, Ferré S. Treatment of restless legs syndrome/Willis-Ekbom disease with the non-selective ENT1/ENT2 inhibitor dipyridamole: testing the adenosine hypothesis. Sleep Med [Internet]. 2018 [cited 2018 Feb 26]. Available from: http://linkinghub.elsevier.com/retrieve/pii/S1389945718300492
54. Yokota T, Hirose K, Tanabe H, Tsukagoshi H. Sleep-related periodic leg movements (nocturnal myoclonus) due to spinal cord lesion. J Neurol Sci. 1991;104(1):13–8.
55. Dickel MJ, Renfrow SD, Moore PT, Berry RB. Rapid eye movement sleep periodic leg movements in patients with spinal cord injury. Sleep. 1994;17(8):733–8.
56. Bara-Jimenez W, Aksu M, Graham B, Sato S, Hallett M. Periodic limb movements in sleep: state-dependent excitability of the spinal flexor reflex. Neurology. 2000;54(8):1609–16.
57. Ferri R, Proserpio P, Rundo F, Lanza A, Sambusida K, Redaelli T, et al. Neurophysiological correlates of sleep leg movements in acute spinal cord injury. Clin Neurophysiol. 2015;126(2):333–8.
58. Tergau F, Wischer S, Paulus W. Motor system excitability in patients with restless legs syndrome. Neurology. 1999;52(5):1060–3.
59. Bucher SF, Seelos KC, Oertel WH, Reiser M, Trenkwalder C. Cerebral generators involved in the pathogenesis of the restless legs syndrome. Ann Neurol. 1997;41(5):639–45.
60. Briellmann RS, Rösler KM, Hess CW. Blink reflex excitability is abnormal in patients with periodic leg movements in sleep. Mov Disord. 1996;11(6):710–4.
61. Tassinari CA, Cantalupo G, Högl B, Cortelli P, Tassi L, Francione S, et al. Neuroethological approach to frontolimbic epileptic seizures and parasomnias: the same central pattern generators for the same behaviours. Rev Neurol (Paris). 2009;165(10):762–8.

62. van Dijken H, Dijk J, Voom P, Holstege JC. Localization of dopamine D2 receptor in rat spinal cord identified with immunocytochemistry and in situ hybridization. Eur J Neurosci. 1996;8(3):621–8.
63. Holstege JC. The ventro-medial medullary projections to spinal motoneurons: ultrastructure, transmitters and functional aspects. Prog Brain Res. 1996;107:159–81.
64. Holstege JC, Van Dijken H, Buijs RM, Goedknegt H, Gosens T, Bongers CM. Distribution of dopamine immunoreactivity in the rat, cat and monkey spinal cord. J Comp Neurol. 1996;376(4):631–52.
65. Ferri R. Two legs, one heart, one sleeping brain. Sleep Med. 2006;7(3):299–300.
66. Woo HG, Lee D, Hwang KJ, Ahn TB. Post-stroke restless leg syndrome and periodic limb movements in sleep. Acta Neurol Scand. 2017;135(2):204–210. https://doi.org/10.1111/ane.12582.
67. Manconi M, Rocca MA, Ferini-Strambi L, Tortorella P, Agosta F, Comi G, et al. Restless legs syndrome is a common finding in multiple sclerosis and correlates with cervical cord damage. Mult Scler. 2008;14(1):86–93.
68. Ferini-Strambi L, Filippi M, Martinelli V, Oldani A, Rovaris M, Zucconi M, et al. Nocturnal sleep study in multiple sclerosis: correlations with clinical and brain magnetic resonance imaging findings. J Neurol Sci. 1994;125(2):194–7.
69. Lee S-J, Kim J-S, Song I-U, An J-Y, Kim Y-I, Lee K-S. Poststroke restless legs syndrome and lesion location: anatomical considerations. Mov Disord. 2009;24(1):77–84.
70. Zucconi M, Ferri R, Allen R, Baier PC, Bruni O, Chokroverty S, et al. The official World Association of Sleep Medicine (WASM) standards for recording and scoring periodic leg movements in sleep (PLMS) and wakefulness (PLMW) developed in collaboration with a task force from the International Restless Legs Syndrome Study Group (IRLSSG). Sleep Med. 2006;7(2):175–83.
71. Parrino L, Boselli M, Buccino GP, Spaggiari MC, Di Giovanni G, Terzano MG. The cyclic alternating pattern plays a gate-control on periodic limb movements during non-rapid eye movement sleep. J Clin Neurophysiol. 1996;13(4):314–23.
72. Parrino L, Ferri R, Bruni O, Terzano MG. Cyclic alternating pattern (CAP): the marker of sleep instability. Sleep Med Rev. 2012;16(1):27–45.
73. Raggi A, Ferri R. Sleep disorders in neurodegenerative diseases. Eur J Neurol. 2010;17(11):1326–38.
74. Winkelmann J, Schormair B, Xiong L, Dion PA, Rye DB, Rouleau GA. Genetics of restless legs syndrome. Sleep Med. 2017;31:18–22.
75. Schormair B, Kemlink D, Roeske D, Eckstein G, Xiong L, Lichtner P, et al. PTPRD (protein tyrosine phosphatase receptor type delta) is associated with restless legs syndrome. Nat Genet. 2008;40(8):946–8.
76. Winkelmann J, Schormair B, Lichtner P, Ripke S, Xiong L, Jalilzadeh S, et al. Genome-wide association study of restless legs syndrome identifies common variants in three genomic regions. Nat Genet. 2007;39(8):1000–6.
77. Winkelmann J, Czamara D, Schormair B, Knauf F, Schulte EC, Trenkwalder C, et al. Genome-wide association study identifies novel restless legs syndrome susceptibility loci on 2p14 and 16q12.1. PLoS Genet. 2011;7(7):e1002171.
78. Stefansson H, Rye DB, Hicks A, Petursson H, Ingason A, Thorgeirsson TE, et al. A genetic risk factor for periodic limb movements in sleep. N Engl J Med. 2007;357(7):639–47.
79. Moore H, Winkelmann J, Lin L, Finn L, Peppard P, Mignot E. Periodic leg movements during sleep are associated with polymorphisms in BTBD9, TOX3/BC034767, MEIS1, MAP2K5/SKOR1, and PTPRD. Sleep. 2014;37(9):1535–42.
80. Winkelman JW. Periodic limb movements in sleep – endophenotype for restless legs syndrome? N Engl J Med. 2007;357(7):703–5.
81. Margariti PN, Astrakas LG, Tsouli SG, Hadjigeorgiou GM, Konitsiotis S, Argyropoulou MI. Investigation of unmedicated early onset restless legs syndrome by voxel-based morphometry, T2 relaxometry, and functional MR imaging during the night-time hours. AJNR Am J Neuroradiol. 2012;33(4):667–72.

82. Cervenka S, Pålhagen SE, Comley RA, Panagiotidis G, Cselényi Z, Matthews JC, et al. Support for dopaminergic hypoactivity in restless legs syndrome: a PET study on D2-receptor binding. Brain J Neurol. 2006;129(Pt 8):2017–28.
83. von Spiczak S, Whone AL, Hammers A, Asselin M-C, Turkheimer F, Tings T, et al. The role of opioids in restless legs syndrome: an [11C]diprenorphine PET study. Brain J Neurol. 2005;128(Pt 4):906–17.
84. Winkelman JW, Schoerning L, Platt S, Jensen JE. Restless legs syndrome and central nervous system gamma-aminobutyric acid: preliminary associations with periodic limb movements in sleep and restless leg syndrome symptom severity. Sleep Med. 2014;15(10):1225–30.
85. Chervin RD, Consens FB, Kutluay E. Alternating leg muscle activation during sleep and arousals: a new sleep-related motor phenomenon? Mov Disord. 2003;18(5):551–9.
86. Cosentino FII, Iero I, Lanuzza B, Tripodi M, Ferri R. The neurophysiology of the alternating leg muscle activation (ALMA) during sleep: study of one patient before and after treatment with pramipexole. Sleep Med. 2006;7(1):63–71.
87. Wichniak A, Tracik F, Geisler P, Ebersbach G, Morrissey SP, Zulley J. Rhythmic feet movements while falling asleep. Mov Disord. 2001;16(6):1164–70.
88. Walters AS. Clinical identification of the simple sleep-related movement disorders. Chest. 2007;131(4):1260–6.
89. Yang C, Winkelman JW. Clinical and polysomnographic characteristics of high frequency leg movements. J Clin Sleep Med. 2010;6(5):431–8.
90. Broughton R, Tolentino MA, Krelina M. Excessive fragmentary myoclonus in NREM sleep: a report of 38 cases. Electroencephalogr Clin Neurophysiol. 1985;61(2):123–33.
91. Broughton R, Tolentino MA. Fragmentary pathological myoclonus in NREM sleep. Electroencephalogr Clin Neurophysiol. 1984;57(4):303–9.
92. Vetrugno R, Plazzi G, Provini F, Liguori R, Lugaresi E, Montagna P. Excessive fragmentary hypnic myoclonus: clinical and neurophysiological findings. Sleep Med. 2002;3(1):73–6.
93. Gassel MM, Marchiafava PL, Pompeiano O. Phasic changes in muscular activity during desynchronized sleep in unrestrained cats. An analysis of the pattern and organization of myoclonic twitches. Arch Ital Biol. 1964;102:449–70.
94. Raccagni C, Löscher WN, Stefani A, Wanschitz J, Kraemer L, Heidbreder A, et al. Peripheral nerve function in patients with excessive fragmentary myoclonus during sleep. Sleep Med. 2016;22:61–4.
95. Brown P, Thompson PD, Rothwell JC, Day BL, Marsden CD. Axial myoclonus of propriospinal origin. Brain J Neurol. 1991;114(Pt 1A):197–214.
96. Montagna P, Provini F, Vetrugno R. Propriospinal myoclonus at sleep onset. Neurophysiol Clin. 2006;36(5–6):351–5.
97. Meinck HM, Piesiur-Strehlow B. Reflexes evoked in leg muscles from arm afferents: a propriospinal pathway in man? Exp Brain Res. 1981;43(1):78–86.
98. Pierrot-Deseilligny E. Propriospinal transmission of part of the corticospinal excitation in humans. Muscle Nerve. 2002;26(2):155–72.
99. Roze E, Bounolleau P, Ducreux D, Cochen V, Leu-Semenescu S, Beaugendre Y, et al. Propriospinal myoclonus revisited: clinical, neurophysiologic, and neuroradiologic findings. Neurology. 2009;72(15):1301–9.
100. Antelmi E, Provini F. Propriospinal myoclonus: the spectrum of clinical and neurophysiological phenotypes. Sleep Med Rev. 2015;22:54–63.
101. Rugh JD, Barghi N, Drago CJ. Experimental occlusal discrepancies and nocturnal bruxism. J Prosthet Dent. 1984;51(4):548–53.
102. Kardachi BJ, Bailey JO, Ash MM. A comparison of biofeedback and occlusal adjustment on bruxism. J Periodontol. 1978;49(7):367–72.
103. Tsukiyama Y, Baba K, Clark GT. An evidence-based assessment of occlusal adjustment as a treatment for temporomandibular disorders. J Prosthet Dent. 2001;86(1):57–66.

104. Lobbezoo F, Rompré PH, Soucy JP, Iafrancesco C, Turkewicz J, Montplaisir JY, et al. Lack of associations between occlusal and cephalometric measures, side imbalance in striatal D2 receptor binding, and sleep-related oromotor activities. J Orofac Pain. 2001;15(1):64–71.
105. Ommerborn MA, Giraki M, Schneider C, Fuck LM, Handschel J, Franz M, et al. Effects of sleep bruxism on functional and occlusal parameters: a prospective controlled investigation. Int J Oral Sci. 2012;4(3):141–5.
106. Manfredini D, Bucci MB, Sabattini VB, Lobbezoo F. Bruxism: overview of current knowledge and suggestions for dental implants planning. Cranio. 2011;29(4):304–12.
107. Lobbezoo F, Ahlberg J, Manfredini D, Winocur E. Are bruxism and the bite causally related? J Oral Rehabil. 2012;39(7):489–501.
108. Lobbezoo F, Naeije M. Bruxism is mainly regulated centrally, not peripherally. J Oral Rehabil. 2001;28(12):1085–91.
109. Bayar GR, Tutuncu R, Acikel C. Psychopathological profile of patients with different forms of bruxism. Clin Oral Investig. 2012;16(1):305–11.
110. Manfredini D, Lobbezoo F. Role of psychosocial factors in the etiology of bruxism. J Orofac Pain. 2009;23(2):153–66.
111. Klasser GD, Rei N, Lavigne GJ. Sleep bruxism etiology: the evolution of a changing paradigm. J Can Dent Assoc. 2015;81:f2.
112. Munechika T, Fujishiro H, Okuda M, Iwamoto K, Torii Y, Iritani S, et al. Rapid eye movement sleep without atonia may help diagnose Lewy body disease in middle-aged and older patients with somatic symptom disorder: RWA and somatic symptom disorder. Psychogeriatrics. 2016.
113. Lobbezoo F, Visscher CM, Ahlberg J, Manfredini D. Bruxism and genetics: a review of the literature. J Oral Rehabil. 2014;41(9):709–14.
114. Carra MC, Macaluso GM, Rompré PH, Huynh N, Parrino L, Terzano MG, et al. Clonidine has a paradoxical effect on cyclic arousal and sleep bruxism during NREM sleep. Sleep. 2010;33(12):1711–6.
115. Terzano MG, Mancia D, Salati MR, Costani G, Decembrino A, Parrino L. The cyclic alternating pattern as a physiologic component of normal NREM sleep. Sleep. 1985;8(2):137–45.
116. Mizumori T, Sumiya M, Kobayashi Y, Inano S, Yatani H. Prediction of sleep bruxism events by increased heart rate. Int J Prosthodont. 2013;26(3):239–43.
117. Schames SE, Schames J, Schames M, Chagall-Gungur SS. Sleep bruxism, an autonomic self-regulating response by triggering the trigeminal cardiac reflex. J Calif Dent Assoc. 2012;40(8):670–1, 674–6.
118. Huynh N, Kato T, Rompré PH, Okura K, Saber M, Lanfranchi PA, et al. Sleep bruxism is associated to micro-arousals and an increase in cardiac sympathetic activity. J Sleep Res. 2006;15(3):339–46.
119. Nashed A, Lanfranchi P, Rompré P, Carra MC, Mayer P, Colombo R, et al. Sleep bruxism is associated with a rise in arterial blood pressure. Sleep. 2012;35(4):529–36.
120. Khoury S, Rouleau GA, Rompré PH, Mayer P, Montplaisir JY, Lavigne GJ. A significant increase in breathing amplitude precedes sleep bruxism. Chest. 2008;134(2):332–7.
121. Janković SM, Sokić DV, Vojvodić NM, Ristić AJ, Kostić VS. Multiple rhythmic movement disorders in a teenage boy with excellent response to clonazepam. Mov Disord. 2008;23(5):767–8.
122. Merlino G, Serafini A, Dolso P, Canesin R, Valente M, Gigli GL. Association of body rolling, leg rolling, and rhythmic feet movements in a young adult: a video-polysomnographic study performed before and after one night of clonazepam. Mov Disord. 2008;23(4):602–7.
123. Vetrugno R, Montagna P. Sleep-to-wake transition movement disorders. Sleep Med. 2011;12(Suppl 2):S11–6.
124. Monderer RS, Wu WP, Thorpy MJ. Nocturnal leg cramps. Curr Neurol Neurosci Rep. 2010;10(1):53–9.
125. Butler JV, Mulkerrin EC, O'Keeffe ST. Nocturnal leg cramps in older people. Postgrad Med J. 2002;78(924):596–8.

126. Frauscher B, Brandauer E, Gschliesser V, Falkenstetter T, Furtner MT, Ulmer H, et al. A descriptive analysis of neck myoclonus during routine polysomnography. Sleep. 2010;33(8):1091–6.
127. Frauscher B, Gabelia D, Mitterling T, Biermayr M, Bregler D, Ehrmann L, et al. Motor events during healthy sleep: a quantitative polysomnographic study. Sleep. 2014;37(4):763–73, 773A–B.
128. Wehrle R, Bartels A, Wetter TC. Facio-mandibular myoclonus specific during REM sleep. Sleep Med. 2009;10(1):149–51.
129. Yoshida T, Kato T, Awaya T, Shibata M, Saito K, Ide M, et al. A boy with unilateral neck myoclonus of cortical origin independently on both sides. Seizure. 2015;24:131–3.
130. Pérez-Carbonell L, Silva C, Gaig C, Carreño M, Santamaria J, Iranzo A. Periodic neck myoclonus during sleep. Sleep Med. 2017;38:71–2.
131. Provini F, Vetrugno R, Meletti S, Plazzi G, Solieri L, Lugaresi E, et al. Motor pattern of periodic limb movements during sleep. Neurology. 2001;57(2):300–4.

Part II

Sleep in Neurologic Conditions

Sleep in Seizure Disorders

11

Silvia Miano

Introduction

The importance of sleep in epilepsy is well known since ancient times. Aristotele wrote that sleep is similar to epilepsy and epilepsy to sleep, and sleep deprivation was a known trigger for seizures [1]. At the end of the nineteenth century, Gower described patients with seizures exclusively during sleep [2], while the effects of sleep on epileptiform discharges (IEDs) were described for the first time by Gibbs and Gibbs in 1947 [3, 4]. Nowadays, it is well known that IEDs are activated by non-rapid eye movement (NREM) but decrease significantly and are more localized during REM sleep [5]. Mirroring the behaviour of IEDs, nocturnal seizures occur mostly during stages N1 and N2 NREM and very rarely during REM sleep [6]. The majority of generalized seizures occur during sleep-wake transitional states, while the majority of frontal lobe seizures occur during sleep. Only 20% of temporal lobe seizures occur during sleep. Although the atonia during REM sleep may inhibit the motor activation of a seizure, it is not clear how REM sleep really inhibits the ictal event or, in the case of temporal lobe seizures, increases the rate of secondary generalization [5]. In addition, the occurrence of partial seizures induces a suppression of REM sleep, lasting for one night, while status epilepticus induces a stronger suppression, lasting for days [5]. Sleep may help to discriminate real seizures from psychogenic ones, since the latter do not occur in sleep [6].

When correlated with dim light melatonin onset (DLMO), temporal lobe seizures occurred most frequently 6 hours before DLMO and frontal lobe seizures mainly in the 6–12-hour window after the DLMO, suggesting that seizures are synchronized to the circadian rhythm [7]. Information on the circadian nature of seizures can help physicians to choose the right therapies (administered at bedtime), including a future possibility of light therapy [6].

S. Miano (✉)
Neurocenter of Southern Switzerland, Civic Hospital, Lugano, Switzerland
e-mail: Silvia.miano@eoc.ch

L. M. DelRosso, R. Ferri (eds.), *Sleep Neurology*,
https://doi.org/10.1007/978-3-030-54359-4_11

In this chapter, a description of sleep-related epilepsy, along with co-morbidity and differential diagnosis with sleep disorders, is presented. The impairment of cognitive capabilities and the role of sleep-epileptic network are also additionally discussed.

Sleep-Related Epilepsy

Specific epilepsies occur primarily during sleep and others primarily upon awakening. The American Academy of Sleep Medicine created the term sleep-related epilepsy (SRE) to define those epilepsies in which more than 70% of seizures occur during sleep [8]. Epilepsies with seizures occurring upon awakening from sleep are primary generalized seizures upon awakening and juvenile myoclonic epilepsy (JME), while SRE are sleep-related hypermotor epilepsy, benign focal epilepsy of childhood with centrotemporal spikes, Panayiotopoulos syndrome, tonic seizures of Lennox-Gastaut syndrome, electrical status epilepticus/continuous spike-waves during sleep and Landau-Kleffner syndrome.

Sleep-Related Hypermotor Epilepsy

Nocturnal frontal lobe epilepsy (NFLE) involves the attractive world of sleep [9]. It was discovered in Bologna in 1981 when Lugaresi and Cirignotta described five patients with frequent and clustering sleep episodes, characterized by bizarre movements and/or dystonic-tonic posturing of the limbs, strongly suspected to be epileptic, considering the optimal response to carbamazepine [10].

NFLE have been recently renamed as sleep-related hypermotor epilepsy (SHE), in accordance with a consensus conference of sleep and epileptology experts, who defined the diagnostic criteria [11]. The term "nocturnal" was considered misleading because it implies a circadian rather than a state specificity of the occurrence of seizures during sleep. The typical seizures may arise from other cerebral regions such as temporal or parietal regions. Finally, the original name did not specify the typical clinical hypermotor pattern of seizures [11]. According to the consensus conference, SHE is a rare disease, with an estimated prevalence of 1.8/100,000 individuals, without a gender predominance, and with a peak onset during childhood and adolescence. Seizures are abrupt in onset and offset, typically brief (2 minutes), and have a highly stereotyped hypermotor pattern, usually accompanied to vegetative signs, vocalization, emotional facial expression and asymmetric tonic/dystonic seizures with or without head/eye deviation. Awareness of seizure is common. More rarely, protracted ambulatory behaviour known as epileptic nocturnal wandering (ENW) has been described (lasting more than 2 minutes).

Patients have usually several episodes per night, but this is not regular. Patients may also complain of non-restorative sleep and excessive daytime sleepiness.

The majority of patients are of normal intelligence. However, intellectual disability and behavioural disorders have been reported. Carbamazepine is effective at low doses in two-thirds of patients. Identified aetiologies are heterogeneous and include structural anomalies, acquired injuries and genetic causes. When genetic, sporadic is the most frequent form. In some patients with drug-resistant SHE, the aetiology may involve a surgically treatable lesion, in particular type II focal cortical dysplasia (TFCD), which does not exclude a genetic origin. Nobili et al. found a strong association between nocturnal epilepsy and TFCD in drug-resistant epileptic patients, ascribed to the particular firing pattern of TFCD during NREM sleep [12]. During NREM sleep, the interictal activity is mainly organized in pseudoperiodically recurrent short bursts of fast discharges, often spreading to surrounding non-lesion-related regions and subsequently developing into a seizure [13].

A minority of familial cases has a known genetic mutation. Scheffer et al. described a large Australian family with autosomal dominant NFLE (named ADNFLE, now renamed ADSHE). CHRNA4 was the first epilepsy gene discovered, coding the alpha4 subunit of the neuronal nicotinic acetylcholine receptor (nAChR) [14]. Until now, other mutations in genes (CHRNA2 and CHRNB2) coding for other subunits (alpha2 and beta2) of the nAChR have been identified [15]. The dysfunction of nAChRs is probably at the origin of the additional major neurological or psychiatric symptoms observed in many patients, such as intellectual disability, regression, depression, psychosis, aggression and personality disorder [16, 17]. Furthermore, severe ADSHE has been associated with mutations of the sodium-activated potassium channel encoded by KCNT1, which is also mutated in a severe epileptic encephalopathy with migrating focal seizures of infancy (MFSI). Similarly, in some families with ADSHE, mutations have been found in DEPDC5, another gene implicated in MFSI, encoding a repressor of the mammalian target of rapamycin (mTOR) pathway, a key regulator of cell growth.

Seizures may be preceded by a sudden arousal or a distinct aura. A warning sensation (consisting of fear, especially associated with epigastric discomfort or déjà vu) and auditory aura seem to be more suggestive of a temporal onset [18, 19]. Patients with nocturnal insulo-opercular epilepsy often reported viscero-sensitive (laryngeal and throat sensations, breathing discomfort, unpleasant or rising epigastric sensations) and somatosensory (unpleasant or electrical paresthesiae, diffused or restricted to a small cutaneous area) manifestations and auditory hallucinations. Visual hallucinations are indicative of occipital involvement [20]. Infrequent feeling of levitation has been described in a patient with cortical dysplasia in the right praecuneus region [21].

Considering that epileptic foci are located in deeper areas of the brain like the orbitofrontal or mesial structures, difficult to detect with scalp EEG, stereo-EEG studies on patients who underwent surgery have provided much more information on the anatomo-electro-clinical correlations of SHE. Patients with an asymmetric tonic or dystonic posturing showed an early activation of the supplementary motor area and involvement of the posterior mesial and cingulated frontal cortex. Patients with hyperkinetic ictal behaviour showed the involvement of mesial-dorsolateral, orbitopolar, opercular or larger lobar cortical regions. The epileptic manifestations characterized by fear and prolonged organized motor behaviours, i.e., ENW, involve the activation of anterior cingulate, orbitopolar and temporal regions. In other words, a network including frontal, and possibly extrafrontal, limbic structures is responsible for these complex epileptic manifestations [15, 19]. The increasing complexity of ictal motor behaviours reflects a different duration and propagation of the discharge within the frontal lobe [15]. Apart from the ictal semiology, a long delay (10–20 s) between the electrical and the clinical onset of motor seizure suggests an extrafrontal origin of SHE [19]. Recently a stereo-EEG study revealed that the mean electrographic seizure duration was shorter (38.5 sec vs. 61.8 sec), the mean elapsed time from EEG onset to the first video detectable movement was lower (4.3 sec vs. 9.5 sec), the delay between the first movement and the onset of hypermotor manifestation was shorter (2.2 sec vs. 11.4 sec) and the duration of clinical manifestation was shorter (32.3 sec vs. 52 sec) in frontal than in extrafrontal SHE [22]. Once the hypermotor manifestation began, no differences in seizure phenotype were observed, supporting the hypothesis of a seizure induced-frontal release of central pattern generators (CPGs) [19, 22].

Even though many of SHE's core features have been clarified, some critical issues remain: the semiological overlap between SHE and sleep disorders is challenging even for sleep and epilepsy experts (see Fig. 11.1). The behavioural patterns of NREM arousal parasomnias, REM behaviour disorders and SHE have some similarities. Ictal motor sequences in SHE contain either epileptic features like dystonic posturing and choreic or ballistic movements, either more parasomnic behaviours like repetitive rocking or rolling, deambulation and even pseudoperiodic patterns [15], reflecting the release of the abovementioned CPGs [23]. CPGs are innate motor patterns present in all organisms and localized in the spinal cord and mesencephalon pons and bulb, essential for survival [24]. In adults, these motor sequences are normally under the control of the mature neopallium but may re-emerge during transient loss of neocortical control, such as during an epileptic seizure, cerebral anoxia or parasomnia [23]. Many of these behaviours should be automatically present when the neomammalian cortex is still "in progress". Indeed

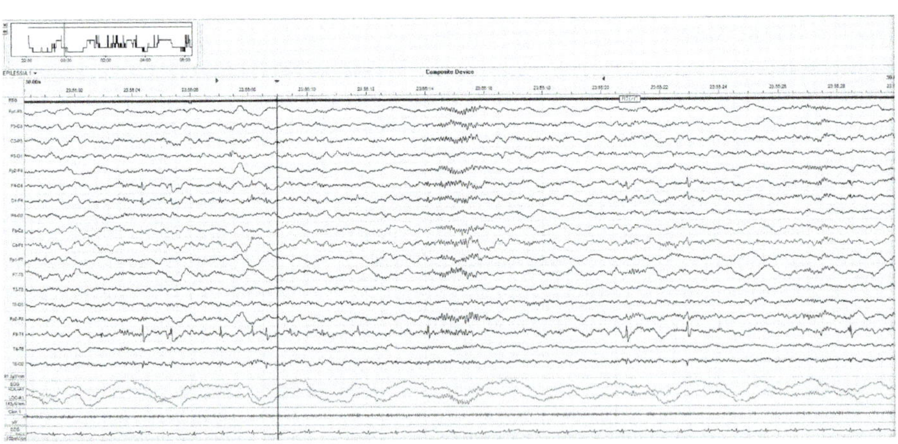

Fig. 11.1 Paroxysmal arousal from N3: dissociative state with slow frontal rhythm activity, mixed activity posteriorly and dystonic posture

alimentary, locomotory and grasping behaviours occur "spontaneously" in foetuses and newborns [25]. This hypothesis is supported by the widely acknowledged existence of different frontal-subcortical circuits reciprocally connecting frontal lobe, striatum, globus pallidus/substantia nigra and thalamus in humans [8, 26]. Another critical topic challenging the differential diagnosis is the high prevalence of arousal disorders in the personal and family histories of patients with SHE, suggesting a common impairment of the pathway controlling physiologic arousal for both conditions [15]. Stereo-EEG case studies revealed a specific dissociative pattern during a NREM sleep parasomnia, with increased sleep-like delta waves over fronto-parietal associative networks, spindles in the hippocampus and wake-like patterns over motor cortex and limbic structures. A GABAergic and nociceptive dysfunction has been hypothesized, similar to SHE, and a similar increase of sleep instability during slow wave sleep responsible for triggering the episodes was found [27]. Other findings showed a significant overlap between NFLE and parasomnias; muscle and movement artefact, rhythmic non-epileptiform θ or δ activity (arousal patterns) over the frontal regions and diffuse attenuation in EEG amplitude (a manifestation of seizure onset or state change) are common findings in both conditions, and did not have a real discriminatory value [28] (Fig. 11.1).

Two instruments have been developed to help clinicians to discriminate parasomnia from SHE: the Frontal Lobe Epilepsy and Parasomnias (FLEP) scale [29] and the Structured Interview for NFLE [30]. The FLEP scale was found to be associated with a real risk of misdiagnosis in some patients, especially subjects presenting ENWs, which were misinterpreted as arousal parasomnias, lowering sensitivity; meanwhile, in patients with REM behaviour disorder, the scale gave misleading epileptic diagnosis, lowering its specificity [31].

The Clinical Spectrum of Benign Epilepsy During Sleep

Based on the current knowledge, some authors postulated the concept of "system epilepsies", responsible of the clinical spectrum of benign epilepsy during sleep [32].

Panayiotopoulos syndrome (PS) is a common idiopathic childhood-specific epilepsy, characterized by seizures, often prolonged, with predominantly autonomic symptoms and shifting and/or multiple foci, often with occipital predominance on EEG [32]. Onset is from age 2 to 11 years. The hallmark of PS is ictal autonomic emesis, associated to other autonomic manifestations, such as pallor, sphincteric incontinence, hypersalivation, cyanosis, mydriasis or miosis, coughing, abnormalities of intestinal motility, breathing, cardiac irregularities and syncopal-like manifestations [32]. PS is a model of childhood autonomic epilepsy [33] (Fig. 11.2).

Benign epilepsy with centrotemporal spikes (BECTS), also called self-limited epilepsy with centrotemporal spikes or childhood epilepsy with centrotemporal spikes (CECTS), is the most common type of focal epilepsy in children, presumably genetic. Age of onset ranges from 1 to 14 years, and the prevalence is about 15–20% of epilepsy in children younger than 15 years of age [34]. BECTS is characterized by infrequent hemifacial sensorimotor seizures during sleep, which may

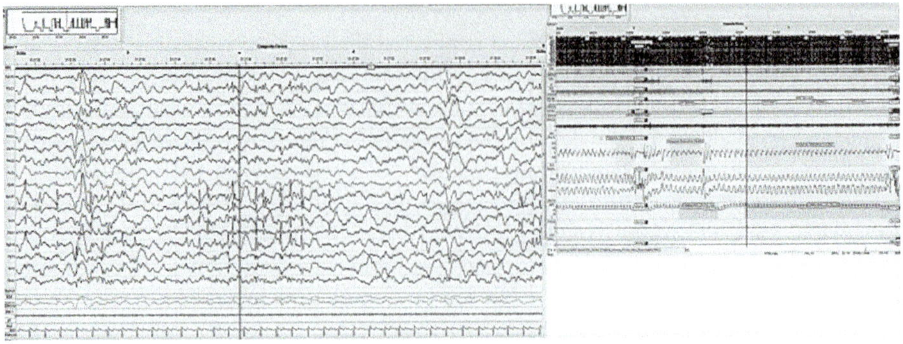

Fig. 11.2 A case of Panayiotopoulos syndrome, with excessive sweating during sleep, and dyslexia. Right centrotemporal sharp waves during sleep with a good response and reduction of spiking with sulthiame, disappearing of sweating and improvement of reading

secondarily generalize in half of the cases or rarely induce a status epilepticus, usually reflecting non-lesion cortical excitability from Rolandic regions. These seizures are often associated with oro-pharyngo-laryngeal symptoms, such as making strange noises, speech arrest and salivation. The seizures predominantly occur during sleep, often in the early morning hours; their frequency is low, typically 2–5 total seizures, but also quite variable, ranging from a single lifetime episode to multiple seizures per day [35]. Interestingly, children with BECTS may present with autonomic seizures referable to PS, while others may alternately have autonomic and Rolandic seizures [32]. Drowsiness and sleep increase the rate of discharges, and an extreme discrepancy between rarity of seizures and the abundant activity of the EEG foci is common [35]. The onset of focal seizures is frequently preceded in early childhood by various developmental deficits, including speech dyspraxia and impairments in language, literacy, attention and behaviour. These neuropsychological deficits cluster in families of BECTS patients, who do not have epilepsy themselves [35].

The prognosis is usually excellent, notwithstanding the abovementioned neuropsychological deficits. Neuroimaging studies have shown subtle cortical abnormalities in the frontal-temporal, perisylvian and parietal region [36]. It seems that neuroimaging abnormalities precede BECTS onset and evolve over time as a marker of a more complex cerebral maturation problem, responsible for both seizures and learning disabilities [35]. These data suggest an association between BECTS and systemic brain disorganization with functional defects in Rolandic areas leading to neurocognitive impairments [37].

The atypical evolution of BECTS can result in:

1. Atypical benign childhood focal epilepsy (ABCFE)
2. Status epilepticus of BECTS (SEBECTS)

3. Landau-Kleffner syndrome (LKS)
4. Epileptic encephalopathy with continuous spikes and waves during sleep (CSWS)

All of the above are considered part of a single spectrum of disorders [34]. Notwithstanding epileptic encephalopathies being within the spectrum of CSWS syndromes, ABFEC and SEBECTS have a favourable outcome when right therapeutic measures are taken. Prognosis of LKS and CSWSS syndrome instead is not so good in terms of full recovery [32].

Atypical features in BECTS comprise many clinical and electrophysiological findings: seizures only occur in the daytime, are associated to Todd's paralysis or even present as status epilepticus. Besides hemifacial contraction, the seizures may present with dysfunctions of the lip, tongue and pharynx, including speech arrest, dysarthria, excessive drooling, oromotor dyspraxia and swallowing difficulties (pseudo-opercular syndrome), further including generalized tonic-clonic seizures, atypical absence, myoclonic seizures, atonic seizures and negative myoclonus at a later stage (pseudo Lennox-Gastaut). The associated seizures tend to be resistant to many antiepileptic drugs; however, they disappear before adolescence, along with the neuropsychological deficits. SEBECTS refers to status epilepticus that can be convulsive or non-convulsive and either generalized or focal. Although some patients remain mentally retarded even after the remission of the seizures, ultimate neurocognitive outcome appears good when the disorder is treated [34] (Fig. 11.3). The EEG usually shows a marked increase and a bilateral synchronization of epileptiform discharges in the Rolandic area or otherwise ripples (80–250 Hz), superimposed on Rolandic spikes [38], which may become continuous during NREM sleep or may evolve in CSWS. Recently, two types of HFO have been distinguished: (a) ripples with slower frequency (80–160 Hz) as physiological activity and (b) fast ripples with higher frequency (200–500 Hz) as a pathological activity, associated with epilepsy [39]. Spike dipole was directed anteriorly in patients with good

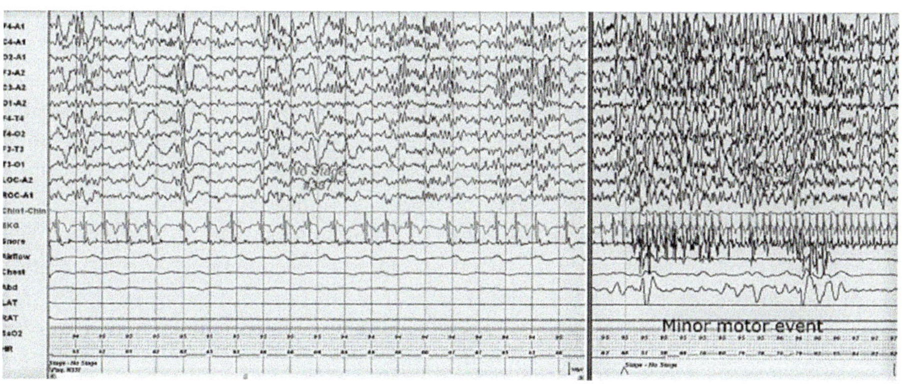

Fig. 11.3 Example of atypical benign epilepsy with centrotemporal spikes, attention deficit disorders and obstructive sleep apnoea, with remission of attention disorders and improvement of epileptiform activities after treatment with ethosuximide and adenotonsillectomy

outcome (seizure control), whereas dipole orientation headed posteriorly was observed in patients with poor seizure control or neurocognitive deficits [40].

LKS, known as acquired epileptic aphasia, is an epileptic encephalopathy characterized by various types of seizures and insidious or sudden-onset acquired aphasia with verbal auditory agnosia. More than half of these children have problems with receptive language, auditory processing, auditory working memory and verbal memory, as well as learning difficulties and attentional and behavioural problems. Benign EEG patterns, like focal epileptiform discharges, may evolve into CSWS sleep in about half of children with LKS. Adequate and early medical intervention may avoid language and cognitive impairments. Acquired opercular epilepsy with oromotor dysfunction is a condition that is difficult to categorize because the clinical features look similar to SEBECTS and LKS [34].

Epileptic encephalopathy with CSWS is a rare childhood epilepsy syndrome and may represent between 0.2% and 2% of the epilepsies. Tassinari et al. first introduced these terms in 1977 [41]. CSWS is the extreme end of atypical BECTS. The onset of seizures varies, but the seizures tend to peak at about 5 years of age, before evolving into epileptic encephalopathy within 1–2 years. Seizure types are not specific. Linguistic, neurocognitive decline and neuropsychiatric features are commonly associated with this condition. It appears that the longer the duration of CSWS, the poorer the outcome is [34]. The EEG pattern at time of diagnosis shows almost continuous slow (typically 1.5–3 Hz) spikes and waves, seen in slow sleep. Some authors provide percentages of slow wave sleep that must be occupied by continuous spikes and waves, e.g., >50% or >85%. The Commission on Classification and Terminology of the International League Against Epilepsy (1989) does not require these criteria [42]. A recent review found that approximately 68% of the cases were diagnosed with CSWS epilepsy-aphasia spectrum when they had spike-wave index >50% clearly activated during sleep [43]. The abnormal EEG activity interferes with sleep-related physiological functions and possibly neuroplasticity processes mediating higher cortical functions, such as learning and memory consolidation [44]. CSWS can concur with other syndromes like Rett syndrome, tuberous sclerosis complex and autism spectrum disorder [43]. The underlying aetiology is unknown although brain malformations, immune disorders and genetic factors have been reported. Immunity disorders with evidence of onconeuronal antibodies have been reported [43]. The SCN2A gene encodes subunits of voltage-gated sodium channel, whereas KCNB1, KCNQ2 and KCNA2 genes encode for subunits of the potassium channel which is highly expressed in brain neurons. The GRIN2A gene encodes *N*-methyl-*D*-aspartate (NMDA) glutamate receptor α2 subunit. The NMDA receptor is a glutamate-activated ion channel permeable to sodium, potassium and calcium and found at excitatory synapses throughout the brain. CNKSR2 gene encodes connector enhancer of KSR2 which is a synaptic protein involved in Ras signalling-mediated neuronal proliferation, migration and differentiation. The SLC6A1 gene encodes voltage-dependent gamma-aminobutyric acid (GABA) transporter 1 (GAT-1), one of the main GABA transporters in the central nervous system. Several mutations (in terms of copy number variations) of these transporters have been found in CSWS epilepsy-aphasia spectrum. Therefore, channelopathy may play a major role in pathogenesis of CSWS/epilepsy-aphasia spectrum [43, 45]. The recommendation is to avoid the classic AEDs (phenobarbital, phenytoin and carbamazepine) and some of the new AEDs, such as oxcarbazepine, lamotrigine, topiramate and levetiracetam and to start treatment with ethosuximide, benzodiazepines or sulthiame. In refractory cases, a ketogenic diet, corticosteroids such as methylprednisolone pulse therapy or dexamethasone, intravenous immunoglobulin and even surgical interventions are suggested [32, 34].

Sleep Macrostructure

Epilepsy per se and/or the seizures themselves may promote sleep disruption and significantly affect the quality, quantity and the architecture of sleep [46]. Increased sleep fragmentation and higher percentage of wakefulness and light sleep, with a decrease in SWS and REM sleep, are common polysomnographic findings in individuals with epilepsy [47, 48]. When seizures occur during the nocturnal period, in addition to a decrease in the duration of REM sleep and increase in REM sleep latency, sleep efficiency and total sleep time drop, causing sleep fragmentation [49]. Generally, a good outcome and response to treatment is accompanied with a significant amelioration of sleep quality, as it has been demonstrated in children with childhood absence epilepsy [50].

Few data are available on sleep macrostructure alterations related to specific epileptic syndromes.

In a group of patients with SHE, a significant increase in wake after sleep onset, SWS duration and REM latency were found, whereas REM sleep duration was significantly lower [51].

A systematic review and meta-analysis of the polysomnographic aspect of sleep in children with epilepsy with centrotemporal spikes showed a longer sleep latency as the strongest outcome associated to the epilepsy [52]. Some studies report that children affected by symptomatic partial epilepsy show permanent modifications in sleep architecture, including a significant decrease of slow wave sleep [53, 54, 55].

Sleep Microstructure

During NREM sleep, phasic EEG events, such as K-complexes, vertex waves, delta wave bursts, and short-lasting arousals, show a peculiar time arrangement, indicated as "cyclic alternating pattern" or CAP. CAP has been described as consisting of transient complexes (phase A) that periodically interrupt the tonic theta/delta activities of NREM sleep (phase B). Functionally, CAP is thought to translate a condition of sustained arousal instability oscillating between a greater arousal level (phase A) and a lesser arousal level (phase B). CAP oscillations are maximally expressed over the frontal areas [56] and may imply that epileptogenic foci could be more prone to be activated by these oscillatory mechanisms. CAP has been analysed in several forms of epilepsy. In genetic generalized epilepsy, a higher CAP rate is reported, along with spikes occurring during phase A1, and similar findings have been reported in lesional temporal lobe epilepsy. In patients with SHE, a significant increase of CAP time, CAP rate, CAP cycles and mean duration of a CAP sequence has been reported, with a significant increase in the number of all CAP phase A subtypes and, in particular, of A1 subtype. The majority of the motor epileptic events are associated with the occurrence of phase A [51, 57], although the minor motor events, coupled to arousal and CAP A phases, have not been considered true epileptic manifestations [58].

In children with BECTS, CAP analysis showed reduced CAP instability due to the presence of spikes which replaced phase A1 and suggested that the centrotemporal spikes may disrupt the physiological synchronization mechanism [59], in contrast with children with drug-resistant epilepsy, in whom an increase of CAP rate was found [55, 60].

Sleep Disorders, Co-morbidity, Differential Diagnosis and Consequences

Approximately one-third to one-half of patients with epilepsy will have a sleep-related complaint. Co-morbid sleep disorders may aggravate the epileptic state, facilitating spikes as well as nocturnal seizures, increasing daytime sleepiness and reducing quality of life [61]. Studies regarding epilepsy and sleep disturbances frequently lack descriptive and statistical analysis for specific sleep disorders (such as insomnia, parasomnia and sleep apnoea) and rather focus on sleep complaints and sleep quality more generally [62]. The most common sleep-wake complaint among adults with epilepsy is sleep maintenance insomnia, occurring with a prevalence of up to 79%, whereas the second sleep-wake complaint is excessive sleepiness [62]. Complaint of insomnia or sleepiness in adults with epilepsy warrants consideration of co-morbid depression, anxiety and suicidal ideation [61, 63]. Insomnia severity and poor sleep quality may be significantly related to the number of antiepileptic medications, and amelioration of insomnia is accompanied by remission of [61]. Specific evidence for the usefulness of cognitive behavioural therapy for insomnia (CBTi) in epilepsy is currently lacking. CBTi can be recommended as an initial treatment strategy although sleep restriction should be considered cautionary in epileptic population [61].

Despite the increase in studies regarding sleep disorders in children with epilepsy over the last decade, none evaluated insomnia as the primary endpoint, and few assessed risk factors and treatment outcomes related to sleep disturbances [62]. Insomnia symptoms were more common in children with epilepsy and developmental delay [62], mostly related to bedtime resistance and co-sleeping. Higher frequency of co-sleeping and room sharing in epileptic children could be explained by parental concern, especially relating to nocturnal seizures [62]. Lower melatonin levels have been found in patients with refractory epilepsy compared with healthy controls [64], and melatonin has been demonstrated to be an optional treatment for insomnia [65]. In a randomized placebo-controlled study, melatonin improved sleep with non-significant improvement in IEDs and seizure frequency [66]. However, negative effects have also been noted, including EEG abnormalities in patients with temporal lobe epilepsy and increased seizure activity in neurologically disabled children.

After the discussion of 26 papers reporting an association between melatonin and epilepsy or seizures, a recent review suggests that in view of conflicting results from earlier work, more large-scale, double-blind, randomized, placebo-controlled clinical trials were needed [67].

Recently, an anecdotal report demonstrated the efficacy of pitolisant, a drug used to treat refractory diurnal sleepiness, to suppress generalized IEDs for 4 hours, in 6 patients out of 14 photosensitive adults [68].

Sleep-related breathing disorders may trigger paroxysmal events during sleep and may exacerbate pre-existing seizures. Sleepiness, a common complaint of patients with epilepsy that is frequently attributed to antiepileptic drugs, may be linked to undiagnosed sleep disorders such as restless legs syndrome or OSA [69]. Patients with epilepsy should be questioned about the presence of other sleep disorders, such as OSA, restless legs syndrome (RLS), insomnia and disorders of hypersomnia, such as narcolepsy. Although it has not been validated in epilepsy patients, the Epworth Sleepiness Scale (ESS) can be used in the evaluation of EDS in epilepsy patients for the presence of these sleep disorders [4].

Co-morbid Sleep-Disordered Breathing in Epilepsy

Both obstructive sleep apnoea (OSA) and central sleep apnoea (CSA) have been reported in epilepsy patients, but CSA has received less attention [61]. At least mild severity OSA (AHI ≥5) was found in nearly 30% of patients, while moderate-to-severe OSA (AHI ≥15) was seen in 10% of patients. Men were three times more likely to have OSA [70]. The prevalence of OSA did not significantly differ according to seizure type (focal versus generalized seizures), epilepsy control (refractory versus controlled epilepsy) or number of antiepileptic drugs [70]. The sleepiness and cognitive dysfunction caused by OSA are usually ignored in epilepsy [65]. Patients with epilepsy and co-morbid OSA were more likely to experience seizures during sleep than patients without OSA; the onset of OSA symptoms may temporally coincide with a change in seizure control – either an increase in seizure frequency or a new onset of seizures or status epilepticus [61]. PSG should be considered when a patient exhibits refractory sleep-related epilepsy [65]. According to a recent meta-analysis, epilepsy patients treated with continuous-positive airway pressure (C-PAP) were >5 times more likely to have a significant reduction in seizure frequency and daytime sleepiness compared to untreated patients [70]. It has been recently described that a 50-year-old man with focal epilepsy controlled by AEDs during daytime underwent PSG for suspected OSA and daytime somnolence. The video PSG revealed many breathing events appearing immediately after falling asleep, suggesting a pattern of so-called sleep-onset central apnoea hypopneas (SOCA). Most of the breathing events ended with brief (2–8 s), stereotyped paroxysmal motor episodes characterized by a ventral flexion of the head, eyes opening, dystonic upper limb extension and mild abduction. Bilateral frontal synchronous bursts of spikes or 2-Hz spike-wave complexes started around 1 s before the onset of each event. Normalization of the respiratory pattern was obtained on adaptive servo-ventilation. The motor events also disappeared, and a striking improvement of sleep was achieved [71]. Interestingly, successful epilepsy surgeries improve OSA symptoms, apnoea hypopnea index and related sleep parameters [65]. Treatment with a compound containing carbonic anhydrase inhibitor properties such as topiramate, zonisamide and acetazolamide may improve both CSA and OSA and should be taken into consideration when a sleep respiratory disorder is co-morbid to epilepsy [65]. Treatment of excessive daytime sleepiness with modafinil is warranted in adults with epilepsy and OSA, which may have benefit also in seizure control, reducing the deleterious effect of drowsiness [65].

The relation between OSA and epilepsy is complex, especially in developmental age: OSA and epilepsy may be in co-morbidity, or recurrent hypoxia and sleep fragmentations may increase the risk to develop epilepsy, and finally OSA may be a misleading diagnosis for epilepsy, and finally OSA as a misleading diagnosis for epilepsy. The prevalence of OSA is nearly 40% in children with refractory epilepsy [72, 73]. Children with neurodevelopmental disability and epilepsy, with or without specific syndromes (such as Angelman syndrome), have higher prevalence of co-morbid OSA (more than 30% of cases) and numerous periodic limb movements during sleep (more than 40% of cases) [74, 75]. As for adults, the precocious diagnosis and treatment of OSA (mostly adenotonsillectomy) may reduce seizure frequency [76].

Few reports showed activation of epilepsy by OSA in children without a previous history of epilepsy. A preliminary study showed IEDs during sleep in 18 of 127 (14.2%) children with OSA. IEDs during sleep are mostly represented by CTS, occipital in the younger and frontal on the older. Children with IEDs and OSA were older and had a longer OSAS duration of disease [77]. Considering that PSG recordings revealed an epileptiform activity during sleep in 1.45% of otherwise healthy children [78], OSA may represent an additional risk factor to develop epilepsy. Subjects with IEDs had a high occurrence of perinatal injuries, suggesting the possibility that a primary brain insult may predispose to both OSAS and the paroxysmal EEG activity [77]. Furthermore CAP analysis showed findings similar to BECTS: a lower CAP rate and A1 index during SWS and a lower A2 index during total NREM sleep, similar to CAP [77]. The same results were replicated in a population of 298 children with OSA or snoring: 48 (16.1%) children were found to have IEDs, whereas 3 were also found to have nocturnal seizures (2 females diagnosed with Rolandic epilepsy and a male diagnosed with SHE). After 6 months, the IEDs had disappeared in those subjects who reported improvement of OSA, while it appeared or persisted in those without changes in OSA parameters [69]. These findings suggest that the risk of developing IEDs, which may or may not be associated with nocturnal seizures, increases if sleep-disordered breathing persists. A similar finding was found in a cohort of Spanish children who underwent nap or standard PSG for suspected OSA: 6 of 25 met the criteria for sleep-disordered breathing and also had PA/IEDs and/or seizure [60]. Four of them (16% of all the samples) met the criteria for OSAS and epilepsy, with IEDs and seizures during sleep. In one patient, the video PSG showed many IEDs with minor motor events, mostly during SWS, and a diagnosis of SHE was established. The magnetic resonance imaging (MRI) showed signs of focal cortical dysplasia involving the right Sylvian fissure. A further two cases met the criteria for BECTS, with partial motor seizures during sleep. The last case received a diagnosis of focal symptomatic epilepsy, and the MRI performed after the PSG study revealed the presence of a malacic area on the left subcortical white matter. These results suggest that children with SBD and neurological conditions may have an additional risk of developing IEDs and/or epilepsy [60]. Another study found higher apnoea hypopnea index in children with BECTS compared to normal control, as well as a trend towards lower oxygen saturation and longer duration of apnoea [79].

A case report of an obese child with severe OSA in whom nocturnal frontal lobe seizures started within a week after he began nasal C-PAP therapy corroborated the hypothesis about a complex relationship between OSA and epilepsy. In this case, treatment caused a rebound of SWS, in particular, of slow wave activity, in terms of significant increase of A1 component of CAP, triggering motor seizure, which was masked by the sleep fragmentation caused by respiratory events. MRI disclosed signs of focal cortical dysplasia involving the right inferior frontal circumvolution. The diagnosis was NFLE, and antiepileptic treatment was started with topiramate (3 mg/kg per day). Topiramate induced a decrease of CAP rate and of arousal index (already decreased by ventilation therapy), suggesting that the control of the seizures was mediated by decreasing SWS-CAP sleep instability [80].

Some symptoms of seizures (e.g., snoring and/or apnoea; vocalizations; coughing; tachypnea; tachycardia; bruxism; salivation; and head, body or limb movements during sleep) can mimic OSA, thus leading clinicians to mistake symptoms of epileptic seizures for obstructive sleep respiratory events [60, 81], especially in patients with SHE from frontal, temporal or insular origin [65]. PSG with extended EEG and video is the gold standard tool to evaluate patients with apnoea due to suspected seizures [69]. Epilepsy in children can be difficult to diagnose, when the only manifestation is central apnoea: in a case report, seizure was documented as the cause of apnoea and desaturation in an otherwise healthy 24-month-old child. The aetiology of the spells was partial seizures originating in the right anterior mid-temporal region. In this case, the absence of associated bradycardia to central apnoea is a red flag for seizure [81]. Choking, air hunger and agitation, from SWS associated to desaturation, were the only symptoms of nocturnal seizure of SHE in a 4-year-old girl with a long history for episodes of choking, four to five times per night [82]. A similar case diagnosis was described in a 3-year-old girl referred for snoring and apnoea, who underwent a cardiorespiratory PSG revealing the presence of atypical obstructive sleep apnoea with stereotyped artefact. A following sleep EEG revealed a motor seizure with ictal discharge from frontal regions. For this reason, she performed a full PSG recording which showed several ictal obstructive apnoea before the onset of motor component of the seizures [83]. Parrino et al. demonstrated that an unattended cardiorespiratory PSG showing repeated ictal polygraphic patterns may indicate the presence of seizures during the night [84]. A video polysomnography should be performed to confirm the ictal nature of such patterns. After being referred for OSA and parasomnias at night, a paediatric Spanish cohort of 24 out of 190 children were diagnosed with SHE, because of repeated recorded motor seizures demonstrated by the attended PSG recording [85].

Co-morbid Restless Legs Syndrome (RLS) in Epilepsy

RLS co-morbidity appears to be frequent, yet highly variable, in different epileptic patient cohorts, and diagnosis was based on symptoms, ranging from 10% to 30% of population investigated [61]. Very rarely, aura symptoms of focal seizures may mimic RLS symptoms, as reported in a single case of right posterior frontal focal cortical dysplasia that presented with nocturnal unilateral left leg paresthesias and cramps that preceded generalized tonic clonic seizures [86].

Co-morbid Parasomnia in Epilepsy

The NREM parasomnia has been discussed in the differential diagnosis and common features of SHE. Co-morbidity with REM behaviour disorder (RBD) has been poorly described in epilepsy [61]. In one large Turkish outpatient epilepsy-clinic cohort [87], 12% of epilepsy patients had probable RBD in response to a survey-based instrument, compared to only 6% of controls. RBD was relatively frequently seen (12.5%), especially in elderly epilepsy patients when systematically investigated for concurrent parasomnias, with RBD prevailing in cryptogenic epilepsy [88]. Conversely, 26.6% of patients with idiopathic RBD had interictal epileptiform activity on EEG [89]. Rarely, autoimmune encephalitis, especially when paraneoplastic, may concomitantly give rise to RBD and focal epilepsy, accompanied by other syndromic cognitive, motor or autonomic features typical of limbic encephalitis. RBD may be treated with clonazepam or melatonin [61].

Association of Sleep with Sudden Unexpected Death in Epilepsy (SUDEP)

SUDEP is a rare but tragic phenomenon, defined as death in people with epilepsy occurring in the absence of a known structural cause of death. The frequency of SUDEP is 1.2 per 1000 patient-years in people with epilepsy overall [90].

SUDEP risk factors have been identified, including a major association with uncontrolled tonic-clonic seizures, as well as medication non-compliance, male sex, epilepsy onset before 16 years and duration of epilepsy >15 years [90]. Patients at particularly high risk are those with developmental and epileptic encephalopathies due to sodium channel (SCN) gene mutations, such as Dravet syndrome. A recent systematic review and meta-analysis found that 69.3% of all reported SUDEP cases occurred during sleep and 30.7% of them occurred during wakefulness [91]. Patients were 6.3 times more likely to die in the prone position during sleep than during wakefulness. These circumstances of SUDEP are remarkably similar to those of sudden infant death syndrome (SIDS) and suggest that sleep might be an important risk factor for SUDEP [91]. Nocturnal seizures are associated with more prominent autonomic dysfunction than are diurnal seizures. Generalized tonic-clonic seizures (GTCS) are the most important risk factor for SUDEP, as they have been observed in a vast majority of witnessed and monitored SUDEP cases. The low incidence of SUDEP in people with NFLE might reflect the low occurrence of GTCS [91]. In the MORTEMUS study, GTCS preceded all SUDEP cases. An expert panel discovered a consistent pattern of tachypnea (18–50 breaths per min) immediately after seizure termination, which was followed by a transient or terminal cardiorespiratory arrest within 3 min [92]. Postictal generalized EEG suppression (PGES) is associated commonly with GTCS, with impaired postictal arousal and with immobility, which may compromise the brainstem autoresuscitation mechanism [91].

The Effect of Antiepileptic Drugs (AEDs) on Sleep

The effect of treatment is summarized on Table 11.1 [4]. The effect on sleep architecture may be mediated by the remission of epilepsy [6]. Ketogenic diet, successful epilepsy surgery and vagus nerve stimulator improved nocturnal sleep. The latter also improved daytime sleepiness, and it may increase the rise of OSA and central apnoeic patterns during sleep [93].

Table 11.1 The effect of antiepileptic treatment of sleep and sleep disorders

Sleep parameter/ disorder	Reduction	Increase
Sleep onset	Benzodiazepines, phenytoin Gabapentin, phenobarbital	Felbamate (rarely levetiracetam, lamotrigine, topiramate, zonisamide)
Stage 2 NREM		Phenobarbital, benzodiazepines (with increase of spindle activity)
Slow wave sleep	Benzodiazepines, ethosuximide Levetiracetam	Gabapentin, pregabalin Carbamazepine
REM sleep	Benzodiazepines, phenytoin Phenobarbital, carbamazepine	Gabapentin, ethosuximide
Arousals	Phenytoin, gabapentin, carbamazepine	Phenobarbital
Excessive daytime sleepiness	Vagus nerve stimulation	Phenobarbital, acid valproic, levetiracetam
Sleep apnoea	Topiramate, zonisamide (both decrease weight), acetazolamide (for central apnoea)	Benzodiazepines, pregabalin (increase weight), valproate (increase weight) Vagus nerve stimulation
Restless legs syndrome	Lamotrigine	Carbamazepine, gabapentin, pregabalin

Medication selection can be tailored towards sleep complaints. Patients with difficulties sleeping at night may benefit from a higher dose of a sedating agent, and individuals with hypersomnia may benefit from alerting AEDs during the day [4]. If seizures tend to occur during the transition from wakefulness to sleep, immediate-release formulations can be more effective than extended-release ones, which would be more efficacious in seizures happening later at night or during sleep-wakefulness transition [94].

Sleep and System Epilepsy: The Cognitive Impact

ADHD is the most prevalent neuropsychiatric co-morbidity in epileptic patients, especially in the refractory cases, presented in up to 60–70% of patients, worsening the psychosocial prognosis [95]. It is less difficult to explain cognitive deficits associated with symptomatic epilepsies when they are the result of recognizable cerebral lesions rather than of epilepsy per se or inducing encephalopathic epilepsy. More difficult to explain is why the so-called idiopathic benign epilepsies are associated with cognitive impairment starting at the onset of or during the natural course of epilepsy [96].

Over the last years, several new aspects have been elaborated contributing to understand better the activation of interictal and ictal epileptic phenomena by sleep: the interrelationship between the sleep-wake circuitry and the different epileptic networks, and the microstructure of sleep associated with epileptic activation, identified within the system of CAP [39].

In summary, IED seems to follow either sigma or delta power, depending on the type of epilepsy. IED rate (ISR) was correlated with sigma power in childhood epilepsy syndromes such as BECTS or CSWS whereas followed delta power in adults with structural temporal or frontal epilepsy (inducing decreasing or increasing of CAP instability, respectively). Recently, the classical dichotomy of partial and generalized epilepsy has been replaced by the unifying network concept. The so-called "generalized" epilepsies involve a bilaterally represented large cerebral system, namely, the thalamocortical system. "Focal" epilepsies involve more or less wide, sometimes bilateral regional circuitries [39]. Activation during sleep is related to the network properties of the particular epileptic syndromes, with interference with cognitive process: (1) via the thalamocortical system as main route and (2) other ways like (a) frontal epilepsy by epileptic transformation of the cholinergic arousal system in NREM sleep and (b) activation of temporal-limbic epilepsy by NREM and REM sleep via hippocampal changes during sleep [39].

In the idiopathic generalized epilepsy (IGE), several studies proved that during interictal spike-wave discharges, a transient cognitive impairment is detectable [97]. The working mode of the system by which the spike pattern develops is proved to be the same that works during NREM sleep opposed to waking and REM state [39]. There is a close relationship between vigilance level and expression of spike-wave paroxysms. Spontaneous paroxysms are promoted by transitory decreases of vigilance level during awake state, after awakening, after lunch, in evening sleepiness, during a boring task and after sleep deprivation [39]. Both in deep NREM sleep and in absences the cortical activity is reduced in certain (mainly frontal) areas. The functional neuroimaging, neurophysiologic and clinical data point to the thalamocortical network as a common substrate of NREM sleep and IGE. Spike-wave discharges and sleep spindles can also be interpreted by the different degrees of GABAergic inhibition in the thalamus. According to the slogan of Steriade – "sleep and epilepsy are bedfellows" – spike-wave discharges of IGE represent the epileptic exaggeration of the bursting mode of the thalamocortical system [39]. Association of generalized spike-wave pattern in IGE with sleep instability in NREM sleep can be measured by CAP rate, usually increased, proportional with sleep instability [98].

Heterogeneous cognitive deficits have been described since the 1990s in typical BECTS, affecting both non-verbal cognitive functions (visual, executive, fine motor execution, attention, memory and speed processing) and verbal functions, during the active phase of the epilepsy. BECTS is associated with a high frequency of learning disorders (10–40%) and academic underachievement [32]. In a prospective study of 44 children with BECTS, the atypical group had significant lower full-scale IQ and verbal IQ [99]. Some patients, especially those in whom the epileptic process is localized around the perisylvian cortex, present with features of autistic spectrum disorder, but unlike primary autism, there is no loss of social interaction [100]. Functional magnetic resonance imaging (fMRI) data support a functional deficit of the default mode network (DMN). This dysfunction is most apparent in the precuneus, a key region of the DMN. In particular, children with BECTS show reduced activation of the DMN during the rest condition and a deactivation during cognitive effort. In addition, reduced functional connectivity was demonstrated between the sensorimotor network and the left inferior frontal gyrus (Broca's area) [32]. This functional decoupling might be a clue to understand language impairment in typical RE and is in line with the identified neuropsychological profile of anterior language dysfunction [101]. The nature and severity of interictal cognitive symptoms are closely related to localization within the network and amount of epileptic interictal discharges during sleep [39]. The brain tissue responsible for the disorder is localized around the Sylvian fissure and strongly correlated with speech function and cognition. Several authors have reported a higher incidence of nocturnal epileptiform EEG discharges in children with specific language impairment, without epilepsy. The benefit of antiepileptic drug treatment remains questionable in this case [32].

In PS, a slightly different network is implicated. Autonomic symptoms are usually generated by activation or inhibition of parts of the central autonomic network that involves the insular cortex, medial prefrontal cortex, amygdala, hypothalamus and ventrolateral medulla. Thus, irrespective of the localization of their onset, ictal discharges may activate the lower threshold autonomic centres [32].

In benign epilepsy during sleep, the amount and persistence time of interictal discharges seem to correlate with the degree of cognitive deficits, and there is a correlation between the degree of spiking during sleep and the degree of cognitive deficits across syndromes. The epileptic discharges are accompanied with a slow wave component, associated with cognitive deficits rather than frequent seizures. This characteristic situation has been called "cognitive epilepsy". The slow wave components of the discharges may protect against conventional sustained, depolarization-based seizures but on the other hand interfere with normal cortical functioning. This assumption is strongly supported by several new results showing that sleep has a use-dependent homeostatic function, connected with sleep slow wave activity, needed for the plastic functions, impaired when abundant IEDs interfere. CSWS affects auditory discrimination and may have a long-lasting impact on cognitive function, whereas in typical children with BECTS with a lower degree of IEDs, plastic brain reorganization or the preservation of networks may prevent such difficulty [102] or reduce the impact of cognitive deficits. Slow wave downscaling measuring according to amount of slow wave activity, amplitude of slow waves, slope of waves and amount of multipeak waves during night sleep are impaired in CSWS patients [102, 103]. After remission of CSWS, the slope decreased significantly overnight. Analysis of slow waves might serve as a prognostic factor regarding cognitive outcome [104].

Recently, the characteristic feature of NREM sleep in Lennox-Gastaut syndrome (LGS), namely, the runs of generalized paroxysmal fast activity (GPFA), has been demonstrated to be a distortion of physiological leading to giant pathological spindles [39]. In Doose syndrome, where GPFA is not present, mental deterioration is less frequent and not as severe as in LGS. GPFA may have a special worsening effect on memory consolidation and can be considered a malignization marker of primary and secondary generalized epilepsies (also induced by benzodiazepine drugs and barbiturates) [39].

Temporo-limbic network epilepsy (TLNE) is the most frequent epilepsy type in adulthood. The main substrate of TLNE is held to be the hippocampus; however, more widespread temporal structural damage plays a role. Within NREM sleep, the activation of temporal spiking was found to be the highest during SWS. The characteristic cognitive deficit conjoining TLNE is disturbance in declarative memory, due to hippocampal dysfunction which is in a certain extent side specific related to verbal memory in the dominant side and to visuospatial memory in subdominant side. There are only some lines of evidence for the role of IEDS during NREM sleep in the memory deficit interfering with hippocampal-cortical dialog during NREM sleep [39].

The role of the frontal lobe during memory process has gained attention in the last years: several studies showed that specific areas within the frontal cortex are involved in long-term memory, contrary to the traditional view that the frontal lobe role is limited to working memory [105]. A systematic neuropsychological study on a representative sample of patients affected by SHE showed neuropsychological deficits in more than half of cases (53.33%), associated with mental disability in 11.7% or a concomitant cognitive decline in 15% of them. A discrepancy strongly emerged between verbal and non-verbal IQ, irrespective of lateralization of seizure foci. A significant worse total IQ mean score was found in patients carrying mutations in the known genes for SHE compared to patients without mutations, independently of the specific gene involved [105]. Even among patients with normal intelligence, deficits involving memory, visuospatial abilities and selected executive functions (phonemic fluency, inhibitory control and working memory), with preserved shifting abilities and planning [105], were found. The finding of memory deficits can be readily explained by both the possible origin of hypermotor seizures from extrafrontal (temporal) networks and the main involvement of frontal areas [105]. In the SHE, similar to parasomnia, the ACh mutation within the frontal system is mainly affected. The number of microarousal during NREM sleep is increased [51], but with opposite function compared to the IGE networks, since the cholinergic arousal system inhibits the thalamic reticular activity [39], inducing a frontal-temporal altered networking. One study reveals a different network structure and inter-area connectivity characterized by a higher connectivity in the occipitoparietal regions and lower values in the frontal areas [106]. Furthermore, an alteration of homeostatic process has been hypothesized in patients with SHE. It has been found that ictal events occur almost exclusively in SWS (72%), with a predominance in the first sleep cycle and decreasing in frequency together with the exponential decay of SWS across the night. In the first sleep period, the repetitive interruptions due to epileptic facilitation produce a sleep fragmentation with a significant increase of REM latency and a homeostatic rebound of deep sleep in the second part of the night, resulting in an unexpected enhancement of SWS. No other untreated sleep disorders, including insomnia, share such a long REM latency with high amount of SWS [51].

Finally, in humans, only 1% of seizures occurred in REM sleep, and the rate of seizures, focal and generalized IEDs and focal high-frequency oscillations (HFOs) was all lowest in REM sleep. Cortical desynchronization in REM sleep suppresses seizure, IEDs and HFO distribution. Cortical desynchronization makes it harder for spontaneously occurring aberrant activity to organize into an IED and then harder still for IEDs to organize into seizures [107]. REM sleep has also been shown to selectively and persistently rebound after successful surgery of epilepsy. The absence of orexinergic activity in REM sleep, which is characterized by diffuse cortical desynchronization, may be overall protective against IEDs and seizures [107]. On the contrary, the presence of orexinergic activity in wakefulness, SWS and state transitions, which are characterized by a relatively greater degree of cortical synchronization, is overall less protective against IEDs and seizures [107].

In Table 11.2, the different states of vigilance implicated in sleep-related epilepsy, the NREM sleep instability related and the suggested appropriate treatment are summarized.

Table 11.2 The relationship between sleep, system epilepsy and treatment

	Alteration of vigilance state	NREM sleep instability	Antiepileptic drug	Effect of antiepileptic drug on sleep stages
Idiopathic generalized epilepsy	Transition from wakefulness to sleep (wake to N1 sleep stage)	Increase of CAP rate, especially A phases during stage N1, N2	Acid valproic, ethosuximide, lamotrigine, levetiracetam	Increase of REM sleep?
Benign focal epilepsy during sleep	Excessive stabilization of stage N2, transitions N2/N3	Decrease of CAP rate and of A phases during N2	Ethosuximide, sulthiame, acetazolamide, benzodiazepines	Reduction of slow wave sleep, increase of spindle activity
Continuous spike-wave during sleep/ Landau-Kleffner syndrome	Excessive stabilization of stage N2, transitions N2/N3	Decrease of CAP rate and of A phases during stage N2, if IED rate allows the analysis	Ethosuximide, sulthiame, acetazolamide, benzodiazepines	Reduction of slow wave sleep, increase of spindle activity
Sleep hypermotor epilepsy (including temporal lobe epilepsy)	Excessive instability of N3, transition N3/REM	Increase of CAP rate and of A phase during N3	Carbamazepine, zonisamide	Increase of slow wave sleep

References

1. Degen R, Degen HE. Sleep and sleep deprivation in epileptology. Epilepsy Res Suppl. 1991;2:235–60.
2. Gowers W. Epilepsy and other chronic convulsive diseases. London: William Wood; 1885.
3. Bazil CW, Malow BA, Sammaritano MR. Sleep and epilepsy: the clinical spectrum. 1st ed. Amsterdam/New York: Elsevier; 2002.
4. Kataria L, Vaughn BV. Sleep and epilepsy. Sleep Med Clin. 2016 Mar;11(1):25–38.
5. Bazil CW. Seizure modulation by sleep and sleep state. Brain Res. 2018. https://doi.org/10.1016/j.brainres.2018.05.003. Epub ahead of print.
6. Jain SV, Kothare SV. Sleep and epilepsy. Semin Pediatr Neurol. 2015;22(2):86–92.
7. Hofstra WA, Gordijn MC, van der Palen J, et al. Timing of temporal and frontal seizures in relation to the circadian phase: a prospective pilot study. Epilepsy Res. 2011;94:158–62.
8. Iber C, Ancoli-Israel S, Chesson A, Quan SF, editors. The AASM manual for the scoring of sleep and associated events: rules, terminology and technical specifications. 1st ed. Westchester: American Academy of Sleep Medicine; 2007.
9. Tinuper P, Bisulli F, Provini F, Montagna P, Lugaresi E. Nocturnal frontal lobe epilepsy: new pathophysiological interpretations. Sleep Med. 2011;12(Suppl 2):S39–42.
10. Lugaresi E, Cirignotta F. Hypnogenic paroxysmal dystonia: epileptic seizures or a new syndrome? Sleep. 1981;4:129–38.
11. Tinuper P, et al. Definition and diagnostic criteria of sleep-related hypermotor epilepsy. Neurology. 2016;86(19):1834–42.
12. Nobili L, Cardinale F, Magliola U, Cicolin A, Didato G, Bramerio M, et al. Taylor's focal cortical dysplasia increases the risk of sleep-related epilepsy. Epilepsia. 2009;50(12):2599–604.

13. Francione S, Nobili L, Cardinale F, Citterio A, Galli C, Tassi L. Intralesional stereo-EEG activity in Taylor's focal cortical dysplasia. Epileptic Disord. 2003;5(Suppl 2):S105–14.
14. Scheffer IE, Bhatia KP, Lopes-Cendes I, Fish DR, Marsden CD, Andermann F, et al. Autosomal dominant frontal epilepsy misdiagnosed as sleep disorder. Lancet. 1994;343(8896):515–7.
15. Tinuper P, Bisulli F. From nocturnal frontal lobe epilepsy to sleep-related hypermotor epilepsy: a 35-year diagnostic challenge. Seizure. 2017;44:87–92.
16. Derry CP, Heron SE, Phillips F, Howell S, MacMahon J, Phillips HA, Duncan JS, Mulley JC, Berkovic SF, Scheffer IE. Severe autosomal dominant nocturnal frontal lobe epilepsy associated with psychiatric disorders and intellectual disability. Epilepsia. 2008;49(12):2125–9.
17. Steinlein OK. Genetics of idiopathic epilepsies. Handb Clin Neurol. 2012;107:145–51.
18. Ferri L, et al. Auditory aura in nocturnal frontal lobe epilepsy: a red flag to suspect an extra-frontal epileptogenic zone. Sleep Med. 2014;15(11):1417–23.
19. Proserpio P, Cossu M, Francione S, Gozzo F, Lo Russo G, Mai R, Moscato A, Schiariti M, Sartori I, Tassi L, Nobili L. Epileptic motor behaviors during sleep: anatomo-electro-clinical features. Sleep Med. 2011a;12(Suppl 2):S33–8.
20. Proserpio P, et al. Insular-opercular seizures manifesting with sleep-related paroxysmal motor behaviors: a stereo-EEG study. Epilepsia. 2011b;52(10):1781–91.
21. Gibbs SA, Figorilli M, Casaceli G, Proserpio P, Nobili L. Sleep related hypermotor seizures with a right parietal onset. J Clin Sleep Med. 2015;11(8):953–5.
22. Gibbs SA, Proserpio P, Francione S, Mai R, Cossu M, Tassi L, Nobili L. Seizure duration and latency of hypermotor manifestations distinguish frontal from extrafrontal onset in sleep-related hypermotor epilepsy. Epilepsia. 2018;59:e130–4.
23. Tassinari CA, Cantalupo G, Högl B, Cortelli P, Tassi L, Francione S, Nobili L, Meletti S, Rubboli G, Gardella E. Neuroethological approach to frontolimbic epileptic seizures and parasomnias: the same central pattern generators for the same behaviors. Rev Neurol (Paris). 2009;165(10):762–8.
24. Grillner S, Wallen P. Central pattern generators for locomotion, with special reference to vertebrates. Ann Rev Neurosci. 1985;8:233–61.
25. Tassinari CA, et al. Central pattern generators for a common semiology in fronto-limbic seizures and in parasomnias. A neuroethologic approach. Neurol Sci. 2005;26(Suppl 3):s225–32.
26. Gardella E, Rubboli G, Francione S, Tassi L, Lo Russo G, Grillner S, Tassinari CA. Seizure-related automatic locomotion triggered by intracerebral electrical stimulation. Epileptic Disord. 2008;10(4):247–52.
27. Castelnovo A, Lopez R, Proserpio P, Nobili L, Dauvilliers Y. NREM sleep parasomnias as disorders of sleep-state dissociation. Nat Rev Neurol. 2018;14(8):470–81.
28. Derry CP, Harvey AS, Walker MC, Duncan JS, Berkovic SF. NREM arousal parasomnias and their distinction from nocturnal frontal lobe epilepsy: a video EEG analysis. Sleep. 2009;32(12):1637–44.
29. Derry CP, Davey M, Johns M, Kron K, Glencross D, Marini C, et al. Distinguishing sleep disorders from seizures: diagnosing bumps in the night. Arch Neurol. 2006;63:705–9.
30. Bisulli F, Vignatelli L, Naldi I, Pittau F, Provini F, Plazzi G, et al. Diagnostic accuracy of a structured interview for nocturnal frontal lobe epilepsy (SINFLE): a proposal for developing diagnostic criteria. Sleep Med. 2012;13:81–7.
31. Manni R, Terzaghi M, Repetto A. The FLEP scale in diagnosing nocturnal frontal lobe epilepsy, NREM and REM parasomnias: data from a tertiary sleep and epilepsy unit. Epilepsia. 2008;49:1581–5.
32. Pal DK, et al. Idiopathic focal epilepsies: the "lost tribe". Epileptic Disord. 2016;18(3):252–88.
33. Panayiotopoulos CP, Michael M, Sanders S, Valeta T, Koutroumanidis M. Benign childhood focal epilepsies: assessment of established and newly recognized syndromes. Brain. 2008;131(9):2264–86.
34. Lee YJ, Hwang SK, Kwon S. The clinical spectrum of benign epilepsy with centro-temporal spikes: a challenge in categorization and predictability. J Epilepsy Res. 2017;7(1):1–6.

35. Parisi P, Paolino MC, Raucci U, Ferretti A, Villa MP, Trenite DK. "Atypical forms" of benign epilepsy with centrotemporal spikes (BECTS): how to diagnose and guide these children. A practical/scientific approach. Epilepsy Behav. 2017;75:165–9.
36. Overvliet GM, Besseling RM, Jansen JF, van der Kruijs SJ, Vles JS, Hofman PA, et al. Early onset of cortical thinning in children with rolandic epilepsy. Neuroimage Clin. 2013;2:434–9.
37. Xiao F, Lei D, An D, Li L, Chen S, Chen F, et al. Functional brain connectome and sensorimotor networks in rolandic epilepsy. Epilepsy Res. 2015;113:113–25.
38. van Klink NE, van't Klooster MA, Leijten FS, Jacobs J, Braun KP, Zijlmans M. Ripples on rolandic spikes: a marker of epilepsy severity. Epilepsia. 2016;57(7):1179–89.
39. Halász P. How sleep activates epileptic networks? Epilepsy Res Treat. 2013;2013:425697. https://doi.org/10.1155/2013/425697. Epub 2013 Sep 12.
40. Kim H, Yoo IH, Lim BC, Hwang H, Chae JH, Choi J, et al. Averaged EEG spike dipole analysis may predict atypical outcome in benign childhood epilepsy with centro-temporal spikes (BCECTS). Brain Dev Jpn Soc Child Neurol. 2016;38(10):903e8.
41. Tassinari CA, Dravet CR. Encephalopathy related to electrical status epilepticus during slow sleep. Electroencephalogr Clin Neurophysiol. 1977;43:529e30.
42. Commission on Classification and Terminology of the International League Against Epilepsy. Proposal for revised classification of epilepsies and epileptic syndromes. Epilepsia. 1989;30(4):389e99.
43. Kessi M, Peng J, Yang L, Xiong J, Duan H, Pang N, Yin F. Genetic etiologies of the electrical status epilepticus during slow wave sleep: systematic review. BMC Genet. 2018;19(1):40.
44. Striano P, Capovilla G. Epileptic encephalopathy with continuous spikes and waves during sleep. Curr Neurol Neurosci Rep. 2013;13(7):360.
45. de Kovel CGF, Syrbe S, Brilstra EH, Verbeek N, Kerr B, Dubbs H, Bayat A, Desai S, Naidu S, Srivastava S, et al. Neurodevelopmental disorders caused by De novo variants in KCNB1 genotypes and phenotypes. JAMA Neurol. 2017;74(10):1228–36.
46. Kothare SV, Kaleyias J. Sleep and epilepsy in children and adolescents. Sleep Med. 2010;11:674–85.
47. Touchon J, Baldy-Moulinier M, Billiard M, Besset A, Cadilhac J. Sleep organization and epilepsy. Epilepsy Res Suppl. 1991;2:73–81.
48. Nunes ML, Ferri R, Arzimanoglou A, Curzi L, Appel CC, Costa da Costa J. Sleep organization in children with partial refractory epilepsy. J Child Neurol. 2003;18:763–6.
49. Wang YQ, Zhang MQ, Li R, Qu WM, Huang ZL. The mutual interaction between sleep and epilepsy on the neurobiological basis and therapy. Curr Neuropharmacol. 2018;16(1):5–16.
50. Dinopoulos A, Tsirouda MA, Bonakis A, Pons R, Pavlopoulou ID, Tsoumakas K. Sleep architecture and epileptic characteristics of drug naïve patients in childhood absence epilepsy spectrum. A prospective study. Seizure. 2018;59:99–107.
51. Parrino L, De Paolis F, Milioli G, Gioi G, Grassi A, Riccardi S, Colizzi E, Terzano MG. Distinctive polysomnographic traits in nocturnal frontal lobe epilepsy. Epilepsia. 2012;53(7):1178–84.
52. Halal CDS, Horta BL, Nunes ML. Polysomnographic aspects of sleep architecture on self-limited epilepsy with centrotemporal spikes: a systematic review and meta-analysis. Sleep Sci. 2017;10(4):161–7.
53. Deray MJ, Epstein MA. Sleep abnormalities in children with intractable epilepsy. Sleep Res. 1991;20:98.
54. Bruni O, Cortesi F, Giannotti F, Curatolo P. Sleep disorders in tuberous sclerosis: a polysomnographic study. Brain and Development. 1995;17:52–6.
55. Pereira AM, Bruni O, Ferri R, Palmini A, Nunes ML. The impact of epilepsy on sleep architecture during childhood. Epilepsia. 2012;53(9):1519–25.
56. Ferri R, Bruni O, Miano S, Terzano MG. Topographic mapping of the spectral components of the cyclic alternating pattern (CAP). Sleep Med. 2005;6:29–36.
57. Terzano MG, et al. Cyclic alternating pattern as a provocative factor in nocturnal paroxysmal dystonia. Epilepsia. 1997;38(9):1015–25.
58. Terzaghi M, Sartori I, Mai R, Tassi L, Francione S, Cardinale F, Castana L, Cossu M, LoRusso G, Manni R, Nobili L. Coupling of minor motor events and epileptiform discharges with arousal fluctuations in NFLE. Epilepsia. 2008;49(4):670–6.

59. Bruni O, Novelli L, Luchetti A, Zarowski M, Meloni M, Cecili M, Villa M, Ferri R. Reduced NREM sleep instability in benign childhood epilepsy with centro-temporal spikes. Clin Neurophysiol. 2010;121(5):665–71.
60. Miano S, Bachiller C, Gutiérrez M, Salcedo A, Villa MP, Peraita-Adrados R. Paroxysmal activity and seizures associated with sleep breathing disorder in children: a possible overlap between diurnal and nocturnal symptoms. Seizure. 2010;19(9):547–52.
61. Latreille V, St Louis EK, Pavlova M. Co-morbid sleep disorders and epilepsy: a narrative review and case examples. Epilepsy Res. 2018;145:185–97.
62. Macêdo PJOM, Oliveira PS, Foldvary-Schaefer N, Gomes MDM. Insomnia in people with epilepsy: a review of insomnia prevalence, risk factors and associations with epilepsy-related factors. Epilepsy Res. 2017;135:158–67.
63. Grigg-Damberger MM, Ralls F. Sleep disorders in adults with epilepsy: past, present, and future directions. Curr Opin Pulm Med. 2014;20(6):542–9.
64. Paprocka J, Dec R, Jamroz E, et al. Melatonin and childhood refractory epilepsy—a pilot study. Med Sci Monit. 2010;16:CR389–96.
65. Liu F, Wang X. Diagnosis and treatment of epilepsy and sleep apnea comorbidity. Expert Rev Neurother. 2017;17(5):475–85.
66. Jain SV, Horn PS, Simakajornboon N, et al. Melatonin improves sleep in children with epilepsy: randomized, double-blind cross-over study. Sleep Med. 2015;16:637.
67. Jain S, Besag FM. Does melatonin affect epileptic seizures? Drug Saf. 2013;36(4):207–15.
68. Kasteleijn-Nolst Trenité D, Parain D, Genton P, Masnou P, Schwartz JC, Hirsch E. Efficacy of the histamine 3 receptor (H3R) antagonist pitolisant (formerly known as tiprolisant; BF2.649) in epilepsy: dose-dependent effects in the human photosensitivity model. Epilepsy Behav. 2013;28(1):66–70.
69. Miano S, Tabarrini A, Vitelli O, Mazzotta A, Del Pozzo M, Rabasco J, Barreto M, Parisi P, Ferretti A, Villa MP. The cooccurrence of interictal discharges and seizures in pediatric sleep-disordered breathing. Epilepsy Behav. 2013;29(3):508–12.
70. Lin Z, Si Q, Xiaoyi Z. Obstructive sleep apnoea in patients with epilepsy: a metanalysis. Sleep Breath. 2017;21:263–70.
71. Khachatryan SG, Prosperetti C, Rossinelli A, Pedrazzi P, Agazzi P, Ratti PL, Manconi M. Sleep-onset central apneas as triggers of severe nocturnal seizures. Sleep Med. 2015;16(8):1017–9.
72. Jain SV, Simakajornboon S, Shapiro SM, Morton LD, Leszczyszyn DJ, Simakajornboon N. Obstructive sleep apnea in children with epilepsy: prospective pilot trial. Acta Neurol Scand. 2012;125(1):e3–6.
73. Kaleyias J, Cruz M, Goraya JS, Valencia I, Khurana DS, Legido A, Kothare SV. Spectrum of polysomnographic abnormalities in children with epilepsy. Pediatr Neurol. 2008;39(3):170–6.
74. Miano S, Bruni O, Elia M, Musumeci SA, Verrillo E, Ferri R. Sleep breathing and periodic leg movement pattern in Angelman syndrome: a polysomnographic study. Clin Neurophysiol. 2005;116(11):2685–92.
75. Miano S, Bruni O, Aricò D, Elia M, Ferri R. Polysomnographic assessment of sleep disturbances in children with developmental disabilities and seizures. Neurol Sci. 2010b;31(5):575–83.
76. Segal E, Vendrame M, Gregas M, Loddenkemper T, Kothare SV. Effect of treatment of obstructive sleep apnea on seizure outcomes in children with epilepsy. Pediatr Neurol. 2012;46:359–62.
77. Miano S, Paolino MC, Peraita-Adrados R, Montesano M, Barberi S, Villa MP. Prevalence of EEG paroxysmal activity in a population of children with obstructive sleep apnea syndrome. Sleep. 2009;32(4):522–9.
78. Sans Capdevila O, Dayyat E, Kheirandish-Gozal L, Gozal D. Prevalence of epileptiform activity in healthy children during sleep. Sleep Med. 2008;9:303–9.
79. Gogou M, Haidopoulou K, Eboriadou M, Pavlou E. Sleep disturbances in children with rolandic epilepsy. Neuropediatrics. 2017;48(1):30–5.
80. Miano S, Pelliccia A, Evangelisti M, Pagani J, Villa MP. Role of continuous positive airway pressure therapy on the pathogenesis of sleep-related frontal lobe epilepsy in a child with obstructive sleep apnea syndrome. J Child Neurol. 2008;23(1):124–8.
81. Andrade EO, Arain A, Malow BA. Partial epilepsy presenting as apneic seizures without posturing. Pediatr Neurol. 2006;35:359–62.

82. Seo WH, Park M, Eun SH, Rhie S, Song DJ, Chae KY. My child cannot breathe while sleeping: a report of three cases and review. BMC Pediatr. 2017;17(1):169.
83. Vitelli O, Miano S, Tabarrini A, Mazzotta AR, Supino MC, Forlani M, Villa MP. Epilepsy and sleep-disordered breathing as false friends: a case report. J Child Neurol. 2014;29(10):NP114–7.
84. Parrino L, Milioli G, Grassi A, et al. Atypical patterns in portable monitoring for sleep apnoea: features of nocturnal epilepsy? Eur Respir J. 2013;41:368–75.
85. Miano S, Peraita-Adrados R. Nocturnal frontal lobe epilepsy is often misdiagnosed as sleep disorders in children: a case series. Rev Neurol. 2013;56(5):257–67.
86. Arico I, Condurso R, Granata F, Nobili L, Bruni O, Silvestri R. Nocturnal frontal lobe epilepsy presenting with restless leg syndrome-like symptoms. Neurol Sci. 2011;32:313–5.
87. Ismayilova V, Demir AU, Tezer FI. Subjective sleep disturbance in epilepsy patients at an outpatient clinic: a questionnaire-based study on prevalence. Epilepsy Res. 2015;115:119–25.
88. Manni R, Terzaghi M, Zambrelli E. REM sleep behavior disorder in elderly subjects with epilepsy: frequency and clinical aspects of the comorbidity. Epilepsy Res. 2007;77:128–33.
89. Manni R, Terzaghi M, Zambrelli E, Pacchetti C. Interictal, potentially misleading, epileptiform EEG abnormalities in REM sleep behavior disorder. Sleep. 2006;29:934–7.
90. Myers KA, Bello-Espinosa LE, Symonds JD, Zuberi SM, Clegg R, Sadleir LG, Buchhalter J, Scheffer IE. Heart rate variability in epilepsy: a potential biomarker of sudden unexpected death in epilepsy risk. Epilepsia. 2018;59(7):1372–80. https://doi.org/10.1111/epi.14438.
91. Ali A, Wu S, Issa NP, Rose S, Towle VL, Warnke P, Tao JX. Association of sleep with sudden unexpected death in epilepsy. Epilepsy Behav. 2017;76:1–6.
92. Ryvlin P, Nashef L, Lhatoo SD, Bateman LM, Bird J, Bleasel A, et al. Incidence and mechanisms of cardiorespiratory arrests in epilepsy monitoring units (MORTEMUS): a retrospective study. Lancet Neurol. 2013;12:966–77.
93. Parhizgar F, Nugent K, Raj R. Obstructive sleep apnea and respiratory complications associated with vagus nerve stimulators. J Clin Sleep Med. 2011;7:401–7.
94. Carreño M, Fernández S. Sleep-related epilepsy. Curr Treat Options Neurol. 2016;18(5):23.
95. Ott D, R C, Guthrie D, et al. Measures of psychopathology in children with complex partial seizures and primary generalized epilepsy with absence. J Am Acad Child Adolesc Psychiatry. 2001;40:907–14.
96. Parisi P, Bruni O, Pia Villa M, Verrotti A, Miano S, Luchetti A, Curatolo P. The relationship between sleep and epilepsy: the effect on cognitive functioning in children. Dev Med Child Neurol. 2010;52(9):805–10.
97. Aldenkamp AP, Arends J, Verspeek S, Berting M. The cognitive impact of epileptiform EEG-discharges; relationship with type of cognitive task. Child Neuropsychol. 2004;10(4):297–305.
98. Terzano MG, Parrino L, Anelli S, Halasz P. Modulation of generalized spike-and-wave discharges during sleep by cyclic alternating pattern. Epilepsia. 1989;30(6):772–81.
99. Metz-Lutz MN, Filippini M. Neuropsychological findings in rolandic epilepsy and Landau-Kleffner syndrome. Epilepsia. 2006;47(2):71–5.
100. Deonna T, Roulet-Perez E. Early-onset acquired epileptic aphasia (Landau-Kleffner syndrome, LKS) and regressive autistic disorders with epileptic EEG abnormalities: the continuing debate. Brain and Development. 2010;32(9):746–52.
101. Besseling RM, Jansen JF, Overvliet GM, et al. Reduced functional integration of the sensorimotor and language network in rolandic epilepsy. Neuroimage Clin. 2013;2:239–46.
102. Melissa F, Antonella B, Melania G, Antonella P, Angelo R, Assunta MM, et al. Comparing cortical auditory processing in children with typical and atypical benign epilepsy with centrotemporal spikes: electrophysiologic evidence of the role of non-rapid eye movement sleep abnormalities. Epilepsia. 2015;56:726–34.
103. Tassinari CA, Rubboli G, Volpi L, et al. Encephalopathy with electrical status epilepticus during slow sleep or ESES syndrome including the acquired aphasia. Clin Neurophysiol. 2000;111(supplement 2):S94–S102.

104. Bölsterli BK, Gardella E, Pavlidis E, Wehrle FM, Tassinari CA, Huber R, Rubboli G. Remission of encephalopathy with status epilepticus (ESES) during sleep renormalizes regulation of slow wave sleep. Epilepsia. 2017;58(11):1892–901.
105. Licchetta L, et al. Profile of neuropsychological impairment in Sleep-related Hypermotor Epilepsy. Sleep Med. 2018;48:8–15.
106. Busonera G, Cogoni M, Puligheddu M, Ferri R, Milioli G, Parrino L, Marrosu F, Zanetti GEEG. Spectral coherence analysis in nocturnal epilepsy. IEEE Trans Biomed Eng. 2018;65:2713.
107. Marcus C. Orexin and epilepsy: potential role of REM sleep. Sleep. 2017;40(3).

Sleep in Headaches

12

Oliviero Bruni

Introduction

The reciprocal relationship between headache and sleep has been documented in medical literature for over a century, and clinical texts highlight the importance of sleep as a headache precipitant. The precise nature and magnitude of the headache/sleep association and underlying mechanisms remain, however, poorly understood [1].

Epidemiology, clinical characteristics, and comorbid disorders of migraine as well as the sleep patterns and the clinical manifestations of sleep disorders are related to age.

Different clinical evidence supports the existence of mutual relationships between sleep and pain, in general. It is well known that noxious stimuli and painful disorders interfere with sleep, however; sleep disturbances also affect pain perception. In fact, sleep deprivation may enhance the response to pain stimuli; moreover, sleep restriction is one of the most common triggers of migraine attacks [2]. Furthermore sleep represents a well-documented behavioral state related to the occurrence of some headache syndromes, and, on the other hand, sleep disorders are observed among all headache subgroups.

Both sleep disturbances and headache disorders are widespread health problems during childhood: migraine and tension headaches alone occur in approximately 12% of the pediatric population, and 25% of children have experienced at least one type of sleep problem [3].

In the adults, a specific sleep disorder was found in 55% of subjects with onset of headache during the night [4], and a direct correlation between sleep disturbance and headache severity has been demonstrated [5].

O. Bruni (✉)
Department of Developmental and Social Psychology, S. Andrea Hospital – Sapienza University, Rome, Italy
e-mail: oliviero.bruni@uniroma1.it

L. M. DelRosso, R. Ferri (eds.), *Sleep Neurology*,
https://doi.org/10.1007/978-3-030-54359-4_12

Sleep disorders are the most frequent comorbid disorders in children with migraine, followed by anxiety disorders and depression; further, 66% of migraine children with sleep disorders had enduring headache [6]. In younger children with immature language and cognitive abilities, migraine descriptors can be missed, or sleep disturbances may not be recognized as causative factors for migraine [7], and therefore some manifestations related to sleep can be missed or misdiagnosed.

Recent research suggests that infant colic may also fit into the category of childhood periodic syndromes like benign paroxysmal torticollis, benign paroxysmal vertigo, abdominal migraine, and cyclic vomiting syndrome [8]. These conditions are thought to represent early-life expression of migraine genes that later in life are expressed as migraine headache.

Pathophysiological Considerations

The pathophysiological links between headache and sleep are mediated by different aspects: common anatomical pathways and genetic mechanisms, same neurotransmitters, and similar chronobiological patterns; finally they might have a bidirectional impact (Table 12.1).

Several neuroanatomical structures are involved in the control of the sleep-wake cycle and in the modulation of pain through the action of the serotonergic and dopaminergic systems: the trigeminal nucleus caudalis in the pons and midbrain, the hypothalamus, the pineal gland, the antinociceptive system represented by the rostral ventromedial medulla oblongata, the serotonergic raphe nuclei, the noradrenergic locus coeruleus, and the periaqueductal gray (PAG) matter [9].

The serotonergic and dopaminergic system are involved in the regulation of sleep and in the headache pathogenesis.

Table 12.1 Relations between sleep and headache

Headache/sleep association
Intrinsic origin (modulation through the same neurotransmitters)
Extrinsic origin (i.e., fibromyalgia syndrome)
Reinforcement (bad sleep hygiene)
Sleep-related headache (during or after sleep)
Sleep stage relationship: REM sleep (migraine, cluster), slow-wave sleep (migraine)
Sleep as trigger or relieving factor for headache
Headache as cause of sleep disruption (e.g., attacks occurring during sleep)
Comorbidity with sleep disorders (parasomnias, restless legs syndrome)
Headache/sleep comorbidities
Insomnia
Restless legs syndrome and PLMS
Parasomnias
Narcolepsy
Sleep deprivation

Serotonergic activation is needed in order to maintain the behavioral sleep and has a role in the inhibition of REM sleep and in triggering the head pain [10], while a dopaminergic receptor hypersensitivity or a dopaminergic imbalance is involved in the migraine pathogenesis and could account for awakenings and for the decrease in activity and cortical activation in migraine subjects [11]. An imbalance of the dopaminergic system is responsible for some premonitory symptoms of migraine, such as nausea, yawning, and dizziness [11]. Furthermore dopamine is involved in sleep disorders such as restless legs syndrome (RLS) and periodic leg movements of sleep (PLMS).

The relation of head pain attacks with events linked to chronobiological timing (seasonal, menstrual, and circadian) suggested a hypothalamic involvement and also a role of melatonin that resulted in decreased migraine subjects suggesting a chronobiological dysfunction in migraineurs with insomnia [12, 13]. Migraine attacks showed a circadian periodicity with a peak during the first hours of waking (between 4 and 9 AM), a menstrual periodicity with a peak after the onset of menses, and a weak seasonal periodicity [14].

Headache might be a symptom of sleep disturbance and a side effect of sleep- or wake-modulating treatments [15, 16].

Sleep deprivation and sleep schedule changes are common headache triggers for both migraine and tension-type headache, and even in children, one of the most common self-perceived triggers of head pain was the lack of sleep [17]. On the other hand, sleep seems efficacious to relieve head pain or terminate the headache attack [18, 19]. Frequency of falling asleep during attacks is significantly more common in patients <8 years of age than in older children, and in these children, there is higher resolution of attacks with sleep [20].

Likewise in adults the most frequent precipitating factors were lack of sleep and fatigue but also food and/or drinks, menstruation, heat/cold/weather. Additionally sleep disturbances can be premonitory symptoms similar to mood changes and food cravings, which can occur many hours before the migraine attack [21].

From a physiological point of view, we could hypothesize that sleep could trigger an autonomic reset that determined the head pain relief [22].

Furthermore, treating headache improves sleep (5-HT, serotoninergic drugs), and, conversely, treating sleep improves headache (CPAP treatment, sleep hygiene, dopaminergic agents for restless legs syndrome, stimulant for narcolepsy, fibromyalgia treatment) [12].

The Specific Links Between Headache and Sleep

The most common sleep complaints in children with headache are represented by reduced sleep duration, difficult bedtime settling, longer sleep latency, night awakenings, and nocturnal symptoms like nightmares, parasomnias, or restless sleep and daytime sleepiness [23–25].

Family history is a strong predictor of the presence of headache or migraine. Children with headache reported a higher prevalence of sleep disturbances in

Table 12.2 Mutual interactions between headache and sleep

(a) Time of occurrence (headache occurs during sleep, after sleep, and in relationship with sleep stages)
(b) Quantitative associations (excess, lack, bad quality, or short duration of sleep may trigger headache)
(c) Reciprocal link with pain: noxious stimuli and painful disorders interfere with sleep, and sleep disturbances affect pain perception

infancy, of colic, and an elevated level of familiarity for migraine or for sleep disturbances, showing the presence of a genetic link and of a common neurobiological substrate that might act from the beginning of life [26].

The interactions between sleep and headache are mediated by different quantitative and qualitative factors (Table 12.2).

Sleep Disturbances in Subjects with Headache

Infants with sleep problems have an increased probability for the development or persistence of headache. Sleep disorders in the first months of life were found to be present in 78% of children with enduring headache vs. 25% of children showing headache remission [27]. Furthermore it has been shown that sleep disruption at age 3 predicted headaches at 6 years [28].

Surveys in large pediatric populations have confirmed the strong association between headache and different sleep disorders such as parasomnias, insomnia, sleep breathing disorders, and daytime sleepiness [26]. Frequency and duration of migraine attacks predicted specific sleep disturbances such as sleep anxiety, parasomnias, and bedtime resistance [29].

In further support of this association, it has been demonstrated that the lack of sleep (69.6%) was the second disorder most commonly reported by children, preceded only by stress (75.5% of patients). Accordingly, in a large epidemiologic study on childhood migraine, we reported that "a bad sleep" was the primary predisposing or causative factor for headache attacks followed by emotional distress [30].

Following this line of research, infantile colic, which represents one of the earliest manifestations of pain and crying in healthy infants, has been identified as an early precursor of migraine, and there is an increased prevalence of infantile colic in migraine children [31, 32]. A positive history of colic was reported in 38.4% of subjects with migraine, significantly higher than in controls (26.9%) and in subjects with tension-type headache (TTH) (25.2%) [26] confirming the specificity of the association of colic only with migraine [28]. In one report, an infant with colic experienced improvement after starting antimigraine (cyproheptadine) therapy, reinforcing the relationships between the two conditions [33].

Headache sufferers reported bad sleep more frequently than controls and slept significantly shorter with a longer sleep latency [34].

Seventy-five adult chronic headache patients compared to 50 healthy reported more difficulty at sleep onset, more awakenings, more nocturnal symptoms (hypnagogic startles, restless legs syndrome, pain, respiratory problems, sweating, bruxism), and more daytime symptoms like non-refreshing sleep, fatigue, and daytime somnolence [35]. In a large epidemiological study, about 15% of adults reported severe headaches in the past 3 months, and they were more likely to have insomnia, excessive sleepiness, recurrent pain, and depression or anxiety symptoms during the preceding 12 months [36].

A recent study explored headache-related factors that predict sleep disturbance and compared sleep complaints with other forms of headache-related disability among youth with migraines. Patients 10–18 years old with migraine or probable migraine and without daily sleep complaints completed a 90-day Internet-based headache and sleep diary. Headache intensity and timing of headache onset were predictive of sleep disturbances. Sleep disturbances correlated positively and significantly with daily headache disability scores. These findings demonstrated that specific headache factors predict sleep disturbances among youth with primary headaches [37].

Specific Sleep Disorders in Subjects with Headache

The most common sleep disturbances in children with headache are represented by insufficient sleep, difficulties falling asleep, anxiety related to sleep, restless sleep, night waking, nightmares, daytime sleepiness, and parasomnias [38–40].

An old report in 48 children with headache confirmed the association of primary headache with night wakings (41.7%), difficulty falling asleep (20.8%), pavor nocturnus and nightmares (14.6%), enuresis (8.3%), and somnambulism (6.3%) [41].

Our pioneer study on a large group of pediatric patients with headache highlighted the strong association between sleep disorders and both migraine and tension headache with an increased prevalence of nocturnal symptoms, such as sleep breathing disorders, restless sleep, and parasomnias and of daytime sleepiness in the migraine group [26].

While no significant differences were found between the migraine and headache groups, in respect to controls, children with migraine and TTH presented shorter sleep duration and longer sleep latency; a higher prevalence of difficulty falling asleep and night awakenings and of sleep talking and bruxism; and reports of frightening dreams, whereas no differences were observed for sleepwalking, bed-wetting, or sleep terrors. Sleep breathing problems were more frequent in subjects with migraine, while restless sleep and daytime sleepiness occurred more frequently in subjects with migraine and tension headache vs. controls.

These findings were confirmed by several subsequent studies [29, 42, 43]. Another study correlated the duration and frequency of migraine with sleep disturbances showing that the frequency of migraine predicted parasomnias, whereas duration of migraine predicted sleep anxiety and bedtime resistance [29].

A more recent report confirmed the presence of excessive daytime sleepiness, narcolepsy, and insomnia in children with headaches while failed to corroborate the higher prevalence of other symptoms like sleep apnea, restlessness, and parasomnias [42].

Insomnia

Several studies in adults highlighted the strict relation between headache and insomnia. In 50 insomniacs, 48% complained of headache, mostly migraine without aura (37.5%) or episodic tension-type headache (50%) [44].

In the USA, adults with headache were more likely than those without headache to report insomnia symptoms like difficulty initiating sleep, maintaining sleep, and waking up early and daytime fatigue [45].

In general, subjects with insomnia were significantly more likely, compared to those without insomnia, to suffer from headache and migraine and insomnia; furthermore, among headache-free subjects at baseline, there was a significantly increased risk of headache 11 years later [46].

A Chinese study showed that the prevalence of DSM-IV insomnia was higher in women with headache than in those without headache (19.9% vs. 5.3%; $p < 0.01$). Furthermore, women with insomnia had a 4.0-fold increased risk of having headache at least once per week [47].

Non-rapid Eye Movement Sleep Parasomnias

NREM parasomnias have been frequently reported to be associated with headache in children. Sleepwalking has been firstly associated with migraine, mainly in adults: Barabas et al. [48] found history of at least two episodes of somnambulism in 30% of migraineurs vs. 4.8% of those with non-migraine headaches, 5% of those with learning disabilities/neurologic impairment, and 6.6% of epileptics; Pradalier et al. [49] found sleepwalking in 21.9% of migraine subjects vs. 6.6% of controls and Giroud et al. [50] in 29.4% of migraine subjects vs. 5.4% of non-migraine headache subjects. Other disorders of arousal have been found to be associated with migraine: Dexter [51] found an incidence of 71% of pavor nocturnus (vs. 11% of controls), of 55% of somnambulism (vs. 16% of controls), and of 41% of enuresis (vs. 16% of controls).

Sleep terror history in childhood was found to be a predictive factor for the development of migraine in adolescence [52].

It has been hypothesized that somnambulism and migraine represented a different age-related expression of a dysfunctional pathway of serotonin. The hypercapnic acidosis, secondary to sleep-disordered breathing, could stimulate the serotonergic neurons, resulting in increased excitability of motoneurons leading to the appearance of somnambulism. This evidence is supported by the fact that sleepwalking and headache can be precipitated by sleep-disordered breathing [53, 54].

A form of headache that could be linked to a disorder of arousal is the cluster headache (CH) with attacks during sleep, in which patients jump out of the bed before being fully awake, but it is still not clear if the disorder of arousal precedes or is the consequence of the pain attack [55].

Sleep-Disordered Breathing

Different studies demonstrated a link between sleep-disordered breathing and headaches. The prevalence of headache in patients with OSAS ranged from 32.9% to 58.5% [56–58].

The clinical presentation is a morning tension-type headache, mainly frontal, frontotemporal, or temporal (38.9%) in location, tightening or pressing (78.9%), and mild to moderate (84.2%) [44]. Guilleminault et al. [59] reported a 36% incidence of morning headaches in 50 patients with sleep apnea. Other studies reported the connection between chronic headache and OSAS when studied with polysomnography (PSG) [60, 61], and this association has been reported also for cluster headache [62]. Chronic headache was seven times more common among subjects with OSAS than in the general population, and treatment of sleep apnea (nCPAP) leads to headache improvement.

In a population-based cohort study, the incidence of migraine was significantly higher in the sleep-related breathing disorder (SBD) cohort than in the control cohort [63].

On the contrary, other studies reported that OSA's prevalence is not higher than in general population and do not find a statistically significant relationship between headache and AHI or minimal oxygen saturation [64, 65].

It has been hypothesized that CH could be triggered by sleep-disordered breathing and the treatment of the sleep apnea solved or greatly improved the head pain [66].

Although several studies have been conducted to explain the pathogenesis of migraine in adult subjects with obstructive sleep apnea syndrome (OSAS), few data are available on the relationship between sleep apnea and headache or migraine in children.

A varied range of symptoms and signs are associated with OSAS in the pediatric population. In children and adolescents with OSAS, the most common clinical manifestation reported is snoring but also obesity, excessive daytime sleepiness, heavy habitual snoring, and neuropsychological disturbances [57].

Guilleminault et al. first describing 50 OSA patients reported that 18 of them suffered from frontal or diffuse morning headache [67]. In adults, 33% of OSAS patients had headache [57] but this is not a specific symptom of sleep apnea [56]. Therefore it is plausible that migraine attacks are secondary to sleep disruption rather than to sleep apnea. Children with migraine reported a higher prevalence of sleep-disordered breathing (56.6%) in respect to nonspecific headache (54%) and chronic migraine (27%) [68].

The association between headache and OSAS is probably based on a combination of factors: hypercarbia, hypoxemia, altered cerebral blood flow, increased intracranial pressure, alterations in sympathetic nerve activity and increases in blood pressure secondary to multiple arousals, and brainstem dysfunction. However, it has been hypothesized that migraine attacks could be secondary to sleep disruption rather than to sleep apnea by itself [69].

Future studies may address the question if sleep apnea is the primary event leading to headache or if sleep disruption is the main pathogenetic factor for morning headache.

Restless Legs Syndrome and Periodic Limb Movements (PLMS)

Restless legs syndrome (RLS) is characterized by an urge to move the legs, accompanied by unpleasant leg sensations, occurring at night, worsened by rest, and improved by movements. RLS is reported to be associated with migraine with a prevalence ranging from 8.7% to 39.0%, whereas migraine prevalence in RLS ranges from 15.1% to 62.6% [70, 71].

Another study suggested that RLS might affect migraine clinical presentation, being associated with chronic and highly disabling migraine [72].

Restless legs syndrome prevalence in migraine was higher than in controls (16.9% vs. 8.7%) and more severe [73]. Moreover, poorer sleep quality was independently associated with RLS occurrence and RLS severity in migraine patients.

The hypotheses to explain the coexistence of RLS and migraine are based on a common genetic basis and brain structures involved, as well as cerebral dopamine imbalance and altered iron metabolism. Dopamine antagonists are effective in migraine attacks, and usually iron deficiency is associated with an increase in extracellular dopamine levels and in the changes in dopamine levels with circadian rhythm that could explain the appearance of RLS symptoms at night and the relationship with migraine because of a dopaminergic hyperfunction reached during the day after lower trough levels [74]. On the other hand, sleep deprivation or sleep instabilities caused by RLS could trigger migraine [75].

The common pathophysiological substrate for migraine and RLS involved [76] a disturbance of iron metabolism and a dopaminergic dysfunction [77].

Dopaminergic premonitory symptoms like yawning, irritability, and mood changes, as well as nausea and vomiting, may be caused by dopamine and are present in 47.6% of patients with RLS and migraine but only in 13.1% in those without RLS [78, 79].

Only 1 pediatric study on children and adolescents aged 5–18 years focused on RLS and headaches: 24 patients with migraine (21.6%), 4 (5%) headache-free controls, and 9 (8.3%) healthy children met the diagnostic criteria for definite RLS. This study showed an approximately fourfold higher frequency of RLS in pediatric patients with migraine compared with headache-free controls [80].

In the only published study on PLMS and migraine in a pediatric group of patients, there was an increase of periodic limb movement index (PLMI) in 26.47%

of children with migraine that presented higher frequency, intensity, duration, and life impairment of migraine and lower efficacy of prophylactic and acute pharmacologic treatment, with respect to children with migraine without aura and normal PLMI [81].

Sleep Bruxism (SB)

SB, a sleep-related movement disorder characterized by teeth grinding and clenching, is frequently associated with orofacial pain and headaches. Children with SB may report approximately three times as many headaches than non-SB subjects with an odds ratio of 4.3; on the other hand, children with migraine showed a high prevalence (29%) of SB [82].

Narcolepsy

Few studies have been conducted to highlight the association between headache and narcolepsy.

Tension headache was found to be more prevalent in narcolepsy vs. controls (60.3% vs. 40.7%) [83], and migraine prevalence showed a twofold to fourfold increase in patients with narcolepsy, and the onset of narcolepsy symptoms was 12.3 years before the onset of migraine symptoms. The increased prevalence of migraine was independent of the pharmacologic treatment of narcolepsy and of severity of narcolepsy symptoms [84].

On the contrary, the German Migraine and Headache Society Study Group reported a similar prevalence of migraine in narcoleptics (21.9%) and controls (19.8%) [85].

The only study on the relation between migraine in children and narcolepsy demonstrated that migraine in children less than 18 years of age was associated with a 5.30-fold increased risk of narcolepsy. This association was still present after controlling for the effects of potential confounders such as demographic characteristics, comorbidities, and concurrent medication uptake. Furthermore, the elevated risk in migraine children was consistently detected in migraine with aura and migraine without aura subtypes, indicating that the study results were robust [86].

The relationship between narcolepsy and migraine might be mediated by the orexinergic neurons of the posterior hypothalamus that are involved both in inhibition of analgesia and in narcolepsy.

Circadian Rhythm Disorders

It is evident that sleep disorders may precede the appearance of migraines and could act as trigger factors (TFs). A recent study showed that stress was the most frequently reported TF (75.5%), followed by the lack of sleep (69.6%), warm climate (68.6%), and video games (64.7%) [87].

These data were confirmed by another study on 749 children and adolescents with different pain disorders: weather conditions (33%), illness (30.7%), physical exertion (21.9%), and lack of sleep (16.2%) were the most frequent self-perceived triggers for pain noted by the respondents [17]. Similarly another research showed that the most frequent causative factors of the headache attack in children and adolescents were represented by "bad sleep" and emotional distress [30].

The sleep-wake pattern of migraine analyzed with actigraphy in children showed that during the interictal period, sleep parameters of children suffering from migraine did not differ from those of controls, but in the night preceding the migraine attack, there was a decrease in nocturnal motor activity, indicating a decrease in cortical activation during the sleep period preceding migraine attacks [88].

Objective Studies in Headaches

Some sleep disorders may present with headache and may be misdiagnosed: about 50% of patients with migraine showed underlying sleep disorders like periodic limb movement of sleep, obstructive sleep apnea syndrome, and fibromyalgia syndrome [89].

Sleep architecture in headache patients is generally preserved, and no specific alterations have been found in the adult population. Some migraine attacks seem to be associated with a large amount of deep sleep, and some others are linked to REM stages [90, 91]; (b) cluster headache is triggered by REM and NREM sleep stage II [92]; and (c) chronic paroxysmal hemicrania is associated with a reduction of total sleep time and of REM phase, with an increase of awakenings during REM [93].

Even if considered linked to a disorder of arousal, the onset of the nocturnal attacks in cluster headache was either from REM sleep [90] or NREM sleep [91] and characterized by a marked fragmentation of sleep. The same fragmentation has been demonstrated in chronic paroxysmal hemicrania, characterized by frequent nocturnal arousals, that occurs mainly from REM sleep [93].

Another interesting finding is the decrease of cortical activation in adult migraineurs that showed, the day before the crisis, a decreased number of arousals and lower REM density and alpha power [94]. Accordingly, a decreased EEG complexity was observed in the first two NREM cycles in patients with spontaneous nocturnal attacks [95].

Only 1 polysomnographic study has been published on 90 children with migraine (60), chronic migraine (11), tension headache (6), and nonspecific headache (13). Sleep-disordered breathing was more frequent among children with migraine (56.6%) and nonspecific headache (54%) vs. chronic migraine (27%), while tension headache was associated with bruxism (50%).

Severe migraine and chronic migraine were associated with shorter sleep time, longer sleep latency, and shorter rapid eye movement and slow-wave sleep [68].

Actigraphic studies show that children and adolescents with headache had a poorer sleep quality than controls, with excessive daytime sleepiness, less time spent in quiet motionless sleep, and waking significantly earlier in the morning [96].

A previous study showed that during the interictal period, sleep parameters of children suffering from migraine did not differ from those of controls, but in the night preceding the migraine attack, there was a decrease in nocturnal motor activity, indicating a decrease in cortical activation during the sleep period preceding migraine attacks [88].

Hypnic Headache Syndrome

First described by Raskin in 1988, this is a rare recurrent, benign headache disorder occurring exclusively during sleep and in older subjects [97]. Few cases younger than age 40 years have been reported, including five pediatric cases [98]. The common symptom is regular awakening from nocturnal sleep caused by headache attacks lasting for 30–60 min. Polysomnographic studies of this syndrome showed a more consistent association with REM sleep, but later studies showed that the majority of hypnic headache attacks arise from non-REM sleep, mainly sleep stage N2 [99].

The apparent cyclicity, the similar duration, and response to lithium suggest a pathophysiologic connection with cluster headache linked to a hypothalamic dysfunction [100].

Treatment for Both Migraine and Sleep

Nonpharmacologic Treatment

The use of nonpharmacologic preventive measures in children with migraine includes lifestyle adjustments (dietary changes, sleep hygiene), reassurance, stress management, biofeedback, relaxation techniques, and other behavioral therapies [101].

Educating patients about headache and its management, identifying and managing triggers (via diaries), modifying lifestyles, and understanding the importance of adopting and adhering to interventions (either pharmacologic or nonpharmacologic) are relevant to all persons with headache.

Diet

Certain foods are anecdotally considered potential headache triggers, but many of these do not have strong evidence. The foods most likely to be triggers include monosodium glutamate (MSG), aspartame, nitrates, alcohol (particularly red wine), and caffeine in excess, with less evidence for tyramine and chocolate [102].

Behavioral Interventions

Behavioral interventions have been recommended as first-line options for headache prevention and are often used to reduce frequency and severity of headaches and headache-related disability instructing the patient to manage their physiological and psychological response to stressors. Cognitive-behavioral therapy (CBT) is effective for both migraine and TTH, with studies showing that 40–50% of individuals experience 50% or greater reduction in headache frequency [103].

Biofeedback

Numerous researches reported that biofeedback is effective for both migraine and TTH, with studies showing on average 47–70% of individuals experiencing 50% or greater reduction in headache frequency [104].

Complementary and Alternative Medicine (CAM) Interventions

Data on the efficacy of CAM treatments is limited by small sample sizes, poor study designs, and weak results. Outside of acupuncture, there is limited empirical evidence supporting the efficacy of these treatments [105].

Natural Treatments

Natural remedies have long been used in folk medicine to treat migraine and other headache disorders; some of these treatments are potentially beneficial, but there are limited evidence from controlled studies. An extract of the rhizome of the butterbur plant 100 mg or 150 mg daily seems to be effective in reducing the frequency of migraine headaches. Also vitamin B2 in doses of 200 mg twice daily and coenzyme Q10 are effective for the prevention of migraine [106].

Nonpharmacologic Treatment in Children and Adolescents

A recent study applied a nonpharmacologic treatment for migraine in 32 preschool children, and 60 older school-age children were instructed to follow (a) sleep hygiene, (b) proper diet (refraining from food additives, with elimination of smoked lunch meats, smoked cheese, yellow cheese, chocolate and foods containing chocolate, pizza, and foods containing monosodium glutamate), and (c) no direct sun exposure. The improvement of migraine was significantly higher in preschoolers demonstrating that they were more sensitive than older children to nonpharmacologic treatment [107].

In a pioneering study, the simple application of sleep hygiene rules in 70 children and adolescents with migraine determined a reduction in the mean duration and frequency of migraine attacks, while the severity of the pain did not change. After 6 months of follow-up, the sleep hygiene group reported lower mean headache duration and a reduced frequency of migraine attacks than the control group, suggesting that better sleep quality led to altered migraine patterns [108].

Although this study represents an indirect measure of the effects that sleep disturbances can have on migraine, it supports the direction of the relationship (i.e., sleep disturbance can exacerbate migraine).

A recent attempt to evaluate the feasibility and preliminary effectiveness of Internet CBT for adolescents with headache showed that it was well received by the youth and parents, but sleep patterns do not improve in response to specialized headache treatment alone or with adjunctive Internet CBT. Although specific interventions targeting sleep disturbance have been evaluated in adults with chronic pain with promising results for both sleep and pain outcomes, similar work has not yet been conducted in adolescents with headache [109].

Pharmacologic Treatment

Based on the similarities in pathophysiology, it is not surprising that the drugs used for prophylaxis or treatment of migraine can improve sleep and vice versa. The effect is mediated by the action on pain threshold or by modifying/improving sleep [110].

Antihistaminics

The histaminergic system might have a role as a potent modulator of meningeal nociceptors' activity in migraine. Activation of inhibitory H3 receptors has been suggested for migraine prophylaxis, and both H3R and H4R ligands may theoretically have prophylactic properties, but the lack of specificity and undesired side effects discouraged the potential exploratory studies [111].

Cyproheptadine, an antagonist at the 5-HT2, histamine H1, and muscarinic cholinergic receptors, is widely used in the prophylactic treatment of migraine in children. The total dose ranges from 12 to 36 mg per day (given two to three times per day or at bedtime). Common adverse events are sedation and weight gain, while dry mouth, nausea, lightheadedness, ankle edema, aching legs, and diarrhea are less common. Cyproheptadine may inhibit growth in children and reverse the effects of SSRIs. A single class II study showed cyproheptadine (4 mg per day) was as effective as propranolol (80 mg per day) in reducing migraine frequency and severity [112].

We can assume, however, that the antihistaminics could act in migraine indirectly through the improvement of sleep, and this effect could decrease the pain in migraine children.

Melatonin

Melatonin is involved in the biological regulation of circadian rhythms, sleep, mood, and aging and in various headache syndromes. Contribution of melatonin in the pathophysiology of headache may be related to its anti-inflammatory effects (i.e., free radical scavenging and reduction of pro-inflammatory cytokines upregulation), an increase of nitric oxide synthase activity, inhibition of dopamine and glutamate release, membrane stabilization, potentiation of gamma aminobutyric acid (GABA)- and opiate-induced analgesia, protection from glutamate neurotoxicity, neurovascular regulation and serotoninergic modulation, and the similarity of its chemical structure to that of indomethacin [13, 113–115].

Different studies showed the effectiveness of melatonin in patients with headache or migraine, and circadian rhythm disorders related also to the demonstration of a decrease of melatonin levels in these individuals [116, 117]. The efficacy of melatonin in these cases could be related to regularization of the sleep-wake pattern through its chronobiological and "sleep hygiene" effect [118].

Despite several studies demonstrating a decrease of melatonin levels in adults with migraine, a recent report showed no significant difference in urinary

6-sulfatoxymelatonin between the migraine children and control group, indicating that nocturnal production of melatonin is not reduced in children with migraine [119].

In adults, melatonin treatment decreased headache in 78.6% of 328 patients with circadian rhythm sleep disorders and headache [120], while in children with primary headache, melatonin 3 mg twice daily reduced the number (by more than 50%), intensity, and duration of headache attacks, showing a better efficacy in migraine vs. tension headache form [121].

A study of melatonin in a single dose of 0.3 mg/kg for 3 months indicated that melatonin might be considered as an effective and safe drug in the prophylaxis of migraine in children showing a significant decrease in the frequency of attacks and in the severity and duration of headache [122].

However, all these studies were nonrandomized and conducted in small samples; therefore, there is still no definitive consensus about the therapeutic use of melatonin for headache in children.

Serotonergic Drugs

It might be hypothesized that a congenital imbalance of serotonergic and dopaminergic pathways might predispose individuals to sleep disorders and to headache [123]. This common substrate might act since the early period of life leading to sleep disorders during infancy (i.e., colic, insomnia) and to the development of migraine or headache later in life [7].

Serotonin (5-HT) decrease may lower the threshold of pain perception but also might disrupt the sleep structure and predispose to headache [124, 125]. Further, a reduction in brain synthesis of 5-HT intensifies photophobia and other migrainous symptoms [126].

A double-blind crossover study of 27 migraine children aged 6–12 years treated with 5-hydroxytryptophan (5-HTP) (5 mg/kg body weight) vs. placebo showed that both L-5-HTP and placebo led to a significant reduction of the migraine index and frequency of migraine attacks with no differences on final efficacy [127]. On the other hand, another report on 48 elementary and junior high school students with primary headache associated with sleep disorders showed that treatment with L-5-HTP in these patients determined the improvement of both conditions, headache and sleep disorders, in particular frequent awakenings and some parasomnias [41].

Tricyclic antidepressants are used for migraine prevention; however, only one tricyclic antidepressant (amitriptyline) has proven efficacy in migraine. A randomized, double-blind, placebo-controlled trial of amitriptyline (1 mg per kg of body weight per day), topiramate (2 mg per kg per day), and placebo in children and adolescents 8–17 years of age with migraine showed no significant between-group differences in the primary outcome (50% reduction of attacks), which occurred in 52% of the patients in the amitriptyline group, 55% of those in the topiramate group, and 61% of those in the placebo group. There were also no significant between-group differences in headache-related disability, headache days, or the percentage of patients who completed the 24-week treatment period. Amitriptyline adverse events were fatigue (30%) and dry mouth (25%), and three patients had serious adverse events of altered mood [128].

Conclusion

The high comorbidity between the migraine or headache and disturbed sleep indicates that a common pathophysiological substrate is present and its knowledge could help in the management of the headache syndromes.

Several studies in the last years have demonstrated that the link between sleep and migraine is more complex and that sleep disturbances represent comorbid, predisposing, predictive, or even prognostic factors for headache development or endurance.

Therefore, we should assume that screening for sleep disorders with the use of proper tests including PSG and referral to a sleep clinic, when appropriate, could be very helpful. Comorbid sleep conditions should always be screened in children with migraine in order to improve patient management and to choose the most appropriate treatment.

References

1. Kelman L, Rains JC. Headache and sleep: examination of sleep patterns and complaints in a large clinical sample of migraineurs. Headache. 2005;45(7):904–10.
2. Mitsikostas DD, Viskos A, Papadopoulos D. Sleep and headache: the clinical relationship. Headache. 2010;50(7):1233–45.
3. Abu-Arefeh I, Russell G. Prevalence of headache and migraine in school children. BMJ. 1994;309(6957):765–9.
4. Paiva T, Farinha A, Martins A, Batista A, Guilleminault C. Chronic headaches and sleep disorders. Arch Intern Med. 1997;157(15):1701–5.
5. Boardman HF, Thomas E, Millson DS, Croft PR. The natural history of headache: predictors of onset and recovery. Cephalalgia. 2006;26(9):1080–8.
6. Guidetti V, Galli F, Fabrizi P, Giannantoni AS, Napoli L, Bruni O, et al. Headache and psychiatric comorbidity: clinical aspects and outcome in an 8-year follow-up study. Cephalalgia. 1998;18(7):455–62.
7. Guidetti V, Dosi C, Bruni O. The relationship between sleep and headache in children: implications for treatment. Cephalalgia. 2014;34(10):767–76.
8. Gelfand AA. Migraine and childhood periodic syndromes in children and adolescents. Curr Opin Neurol. 2013;26(3):262–8.
9. Evers S. Sleep and headache: the biological basis. Headache. 2010;50(7):1246–51.
10. Rains JC, Poceta SJ, Penzien DB. Sleep and headaches. Curr Neurol Neurosci Rep. 2008;8:167–75.
11. Peroutka SJ. Dopamine and migraine. Neurology. 1997;49:650–6.
12. Dodick DW, Eross EJ, Parish JM, et al. Clinical, anatomical, and physiologic relationship between sleep and headache. Headache. 2003;43:282–92.
13. Peres MFP. Melatonin, the pineal gland and their implications for headache disorders. Cephalalgia. 2005;25(6):403–11.
14. Fox AW, Davis RL. Migraine chronobiology. Headache. 1998;38:436–41.
15. Holland PR. Headache and sleep: shared pathophysiological mechanisms. Cephalalgia. 2014;34(10):725–44.
16. Nesbitt AD, Leschziner GD, Peatfield RC. Headache, drugs and sleep. Cephalalgia. 2014;34(10):756–66.
17. Roth-Isigkeit A, Thyen U, Stöven H, Schwarzenberger J, Schmucker P. Pain among children and adolescents: restrictions in daily living and triggering factors. Pediatrics. 2005;115(2):e152–62.

18. Blau JN. Resolution of migraine attacks: sleep and the recovery phase. J Neurol Neurosurg Psychiatry. 1982;45(3):223–6.
19. Wilkinson M, Williams K, Leyton M. Observations on the treatment of an acute attack of migraine. Res Clin Stud Headache. 1978;6:141–6.
20. Aaltonen K, Hämäläinen ML, Hoppu K. Migraine attacks and sleep in children. Cephalalgia. 2000;20(6):580–4.
21. Chabriat H, Danchot J, Michel P, Joire JE, Henry P. Precipitating factors of headache. Prospective study in a national control-matched survey in migraineurs and nonmigraineurs. Headache. 1999;39:335–8.
22. Sahota PK, Dexter JD. Sleep and headache syndromes: a clinical review. Headache. 1990;30(2):80–4.
23. Dosi C, Riccioni A, Della Corte M, Novelli L, Ferri R, Bruni O. Comorbidities of sleep disorders in childhood and adolescence: focus on migraine. Nat Sci Sleep. 2013;5:77–85.
24. Engstrøm M, Hagen K, Bjørk M, Gravdahl GB, Sand T. Sleep-related and non-sleep-related migraine: interictal sleep quality, arousals and pain thresholds. J Headache Pain. 2013;14:68.
25. Jennum P, Jensen R. Sleep and headache. Sleep Med Rev. 2002;6(6):471–9.
26. Bruni O, Fabrizi P, Ottaviano S, Cortesi F, Giannotti F, Guidetti V. Prevalence of sleep disorders in childhood and adolescence with headache: a case-control study. Cephalalgia. 1997;17(4):492–8.
27. Balottin U, Termine C, Nicoli F, Quadrelli M, Ferrari-Ginevra O, Lanzi G. Idiopathic headache in children under six years of age: a follow-up study. Headache. 2005;45(6):705–15.
28. Aromaa M, Sillanpää ML, Rautava P, Helenius H. Childhood headache at school entry: a controlled clinical study. Neurology. 1998;50(6):1729–36.
29. Miller VA, Palermo TM, Powers SW, Scher MS, Hershey AD. Migraine headaches and sleep disturbances in children. Headache. 2003;43(4):362–8.
30. Bruni O, Russo PM, Ferri R, Novelli L, Galli F, Guidetti V. Relationships between headache and sleep in a non-clinical population of children and adolescents. Sleep Med. 2008;9(5):542–8.
31. Jan MM, Al-Buhairi AR. Is infantile colic a migraine-related phenomenon? Clin Pediatr (Phila). 2001;40(5):295–7.
32. Katerji MA, Painter MJ. Infantile migraine presenting as colic. J Child Neurol. 1994;9(3):336–7.
33. Romanello S, Spiri D, Marcuzzi E, Zanin A, Boizeau P, Riviere S, et al. Association between childhood migraine and history of infantile colic. JAMA. 2013;309(15):1607–12.
34. Spierings EL, van Hoof J. Fatigue and sleep in chronic headache sufferers: an age- and sex-controlled questionnaire study. Headache. 1997;37(9):549–52.
35. Paiva T, Martins P, Batista A, Esperanca P, Martins I. Sleep disturbances in chronic headache patients: a comparison with healthy controls. Headache Q. 1994;5:135–41.
36. Strine TW, Chapman DP, Balluz LS. Population-based U.S. study of severe headaches in adults: psychological distress and comorbidities. Headache. 2006;46(2):223–32.
37. Heyer GL, et al. Specific headache factors predict sleep disturbances among youth with migraine. Pediatr Neurol. 2014;51(4):489–93.
38. Dosi C, Figura M, Ferri R, Bruni O. Sleep and headache. Semin Pediatr Neurol. 2015;22(2):105–12.
39. Esposito M, Roccella M, Parisi L, Gallai B, Carotenuto M. Hypersomnia in children affected by migraine without aura: a questionnaire-based case-control study. Neuropsychiatr Dis Treat. 2013;9:289–94.
40. Isik U, Ersu RH, Ay P, Save D, Arman AR, Karakoc F, et al. Prevalence of headache and its association with sleep disorders in children. Pediatr Neurol. 2007;36(3):146–51.
41. De Giorgis G, Miletto R, Iannuccelli M, Camuffo M, Scerni S. Headache in association with sleep disorders in children: a psychodiagnostic evaluation and controlled clinical study – L-5HTP versus placebo. Drugs Exp Clin Res. 1987;13(7):425–33.
42. Luc ME, Gupta A, Birnberg JM, Reddick D, Kohrman MH. Characterization of symptoms of sleep disorders in children with headache. Pediatr Neurol. 2006;34(1):7–12.
43. Carotenuto M, Guidetti V, Ruju F, Galli F, Tagliente FR, Pascotto A. Headache disorders as risk factors for sleep disturbances in school aged children. J Headache Pain. 2005;6(4):268–70.

44. Alberti A, Mazzotta G, Gallinella E, et al. Headache characteristics in obstructive sleep apnea syndrome and insomnia. Acta Neurol Scand. 2005;111(5):309–16.
45. Lateef T, Swanson S, Cui L, Nelson K, Nakamura E, Merikangas K. Headaches and sleep problems among adults in the United States: findings from the national comorbidity survey replication study. Cephalalgia. 2010;31:648–53.
46. Odegård SS, Sand T, Engstrøm M, Stovner LJ, Zwart JA, Hagen K. The long-term effect of insomnia on primary headaches: a prospective population based cohort study (HUNT-2 and HUNT-3). Headache. 2011;51:570–80.
47. Yeung WF, Chung KF, Wong CY. Relationship between insomnia and headache in community-based middle-aged Hong Kong Chinese women. J Headache Pain. 2010;11:87–95.
48. Barabas G, Ferrari M, Matthews WS. Childhood migraine and somnambulism. Neurology. 1983;33(7):948–9.
49. Pradalier A, Giroud M, Dry J. Somnambulism, migraine and propranolol. Headache. 1987;27(3):143–5.
50. Giroud M, d'Athis P, Guard O, Dumas R. Migraine and somnambulism. A survey of 122 migraine patients. Rev Neurol (Paris). 1986;142(1):42–6.
51. Dexter JD. The relationship between disorders of arousal from sleep and migraine. Headache. 1986;26:322.
52. Fialho LMN, Pinho RS, Lin J, Minett TSC, de Souza Vitalle MS, Fisberg M, et al. Sleep terrors antecedent is common in adolescents with migraine. Arq Neuropsiquiatr. 2013;71(2):83–6.
53. Guilleminault C, Palombini L, Pelayo R, Chervin RD. Sleepwalking and sleep terrors in prepubertal children: what triggers them? Pediatrics. 2003;111(1):e17–25.
54. Juszczak GR, Swiergiel AH. Serotonergic hypothesis of sleepwalking. Med Hypotheses. 2005;64(1):28–32.
55. Wolfe H. Headache and other head pain. 2nd ed. New York: Oxford University Press; 1963.
56. Aldrich MS, Chauncey JB. Are morning headaches part of obstructive sleep apnea syndrome? Arch Intern Med. 1990;150(6):1265–7.
57. Neau JP, Paquereau J, Bailbe M, Meurice JC, Ingrand P, Gil R. Relationship between sleep apnoea syndrome, snoring and headaches. Cephalalgia. 2002;22(5):333–9.
58. Sand T, Hagen K, Schrader H. Sleep apnoea and chronic headache. Cephalalgia. 2003;23(2):90–5.
59. Guilleminault C. Clinical overview of the sleep apnea syndromes. In: Guilleminault C, Dement W, editors. Sleep apnea syndromes. New York: Alan R. Liss; 1978. p. 1–12.
60. Dexter J. Headache as a presenting complaint of the sleep apnea syndrome. Headache. 1984;24:171.
61. Mathew NT, Glaze D, Frost JJR. Sleep apnea and other sleep abnormalities in primary headache disorders. In: Rose C, editor. Migraine: proceedings of the 5th international migraine symposium. London: Karger; 1985. p. 40–9.
62. Kudrow L, McGinty D, Phillips E, Stevenson M. Sleep apnea in cluster headache. Cephalalgia. 1984;4:33–8.
63. Harnod T, Wang YC, Kao CH. Association of migraine and sleep-related breathing disorder: a population-based cohort study. Medicine. 2015;94(36):e1506.
64. Jensen R, Olsborg C, Salvesen R, Torbergsen T, Bekkelund SI. Is obstructive sleep apnea syndrome associated with headache? Acta Neurol Scand. 2004;109:180–4.
65. Kallweit U, Hidalgo H, Uhl V, Sandor PS. Continuous positive airway pressure therapy is effective for migraines in sleep apnea syndrome. Neurology. 2011;76:1189–91.
66. Zalleck SN, Chervin RD. Improvement in cluster headache after treatment for obstructive sleep apnea. Sleep Med. 2000;1(2):135–8.
67. Guilleminault C. Obstructive sleep apnea. The clinical syndrome and historical perspective. Med Clin North Am. 1985;69(6):1187–203.
68. Vendrame M, Kaleyias J, Valencia I, Legido A, Kothare SV. Polysomnographic findings in children with headaches. Pediatr Neurol. 2008;39(1):6–11.
69. Tran DP, Spierings ELH. Headache and insomnia: their relation reviewed. Cranio. 2013;31(3):165–70.

70. Schürks M, Winter A, Berger K, Kurth T. Migraine and restless legs syndrome: a systematic review. Cephalalgia. 2014;34(10):777–94.
71. Chen P-K, Fuh J-L, Chen S-P, Wang S-J. Association between restless legs syndrome and migraine. J Neurol Neurosurg Psychiatry. 2010;81(5):524–8.
72. Lucchesi C, Bonanni E, Maestri M, Siciliano G, Murri L, Gori S. Evidence of increased restless legs syndrome occurrence in chronic and highly disabling migraine. Funct Neurol. 2012;27:91–4.
73. van Oosterhout WP, van Someren EJ, Louter MA, Schoonman GG, Lammers GJ, Rijsman RM, Ferrari MD, Terwindt GM. Restless legs syndrome in migraine patients: prevalence and severity. Eur J Neurol. 2016; https://doi.org/10.1111/ene.12993.drom.
74. Marmura MJ. Use of dopamine antagonists in treatment of migraine. Curr Treat Options Neurol. 2012;14:27–35.
75. Fernández-Matarrubia M, Cuadrado ML, Sánchez-Barros CM, et al. Prevalence of migraine in patients with restless legs syndrome: a case-control study. Headache. 2014;54(8):1337–46.
76. Cannon PR, Larner AJ. Migraine and restless legs syndrome: is there an association? J Headache Pain. 2011;12(4):405–9.
77. Akerman S, Goadsby PJ. Dopamine and migraine: biology and clinical implications. Cephalalgia. 2007;27(11):1308–14.
78. d'Onofrio F, Bussone G, Cologno D, Petretta V, Buzzi MG, Tedeschi G, et al. Restless legs syndrome and primary headaches: a clinical study. Neurol Sci. 2008;29(Suppl 1):S169–72.
79. Cologno D, Cicarelli G, Petretta V, d'Onofrio F, Bussone G. High prevalence of dopaminergic premonitory symptoms in migraine patients with restless legs syndrome: a pathogenetic link? Neurol Sci. 2008;29(Suppl 1):S166–8.
80. Seidel S, Bock A, Schlegel W, Kilic A, Wagner G, Gelbmann G, et al. Increased RLS prevalence in children and adolescents with migraine: a case-control study. Cephalalgia. 2012;32(9):693–9.
81. Esposito M, Parisi P, Miano S, Carotenuto M. Migraine and periodic limb movement disorders in sleep in children: a preliminary case-control study. J Headache Pain. 2013;14:57.
82. Carra MC, Bruni O, Huynh N. Topical review: sleep bruxism, headaches, and sleep-disordered breathing in children and adolescents. J Orofac Pain. 2012;26(4):267–76.
83. Dahmen N, Kasten M, Wieczorek S, Gencik M, Epplen JT, Ullrich B. Increased frequency of migraine in narcoleptic patients: a confirmatory study. Cephalalgia. 2003;23(1):14–9.
84. Lecendreux M, Poli F, Oudiette D, Benazzouz F, Donjacour CEHM, Franceschini C, et al. Tolerance and efficacy of sodium oxybate in childhood narcolepsy with cataplexy: a retrospective study. Sleep. 2012;35(5):709–11.
85. Khil L, Pfaffenrath V, Straube A, et al. Incidence of migraine and tension-type headache in three different populations at risk within the German DMKG headache study. Cephalalgia. 2012;32(4):328–36.
86. Yang C-P, Hsieh M-L, Chiang J-H, Chang H-Y, Hsieh VC-R. Migraine and risk of narcolepsy in children: a nationwide longitudinal study. PLoS One. 2017;12(12):e0189231. https://doi.org/10.1371/journal.pone.0189231.
87. Neut D, Fily A, Cuvellier JC, et al. Prevalence of triggers in paediatric migraine: a questionnaire study in 102 children and adolescents. J Headache Pain. 2012;13:61–5.
88. Bruni O, Russo PM, Violani C, et al. Sleep and migraine: an actigraphic study. Cephalalgia. 2004;24:134–9.
89. Paiva T, Batista A, Martins P, Martins A. The relationship between headaches and sleep disturbances. Headache. 1995;35:590–6.
90. Dexter J, Weitzman ED. The relationship of nocturnal headaches to sleep stage patterns. Neurology. 1970;20:513–8.
91. Dexter J. The relationship between stage III+IVREM sleep and arousals with migraine. Headache. 1970;19:364–9.
92. Pfaffenrath V, Pöllmann W, Rüther E, Lund R, Hajak G. Onset of nocturnal attacks of chronic cluster headache in relation to sleep stages. Acta Neurol Scand. 1986;73:403–7.
93. Kayed K, Godtlibsen O, Sjaastad O. Chronic paroxysmal hemicrania IV: "REM sleep locked" nocturnal attacks. Sleep. 1978;1:91–5.

94. Göder R, Fritzer G, Kapsokalyvas A, et al. Polysomnographic findings in nights preceding a migraine attack. Cephalalgia. 2001;21(1):31–7.
95. Strenge H, Fritzer G, Göder R, et al. Non-linear electroencephalogram dynamics in patients with spontaneous nocturnal migraine attacks. Neurosci Lett. 2001;309(2):105–8.
96. Bursztein C, Steinberg T, Sadeh A. Sleep, sleepiness, and behavior problems in children with headache. J Child Neurol. 2006;21:1012–9.
97. Raskin NH. The hypnic headache syndrome. Headache. 1988;28:534–6.
98. Liang JF, Wang SJ. Hypnic headache: a review of clinical features, therapeutic options and outcomes. Cephalalgia. 2014;34(10):795.
99. Holle D, Naegel S, Obermann M. Pathophysiology of hypnic headache. Cephalalgia. 2014;34(10):806–12.
100. Holle D, Naegel S, Obermann M. Hypnic headache. Cephalalgia. 2013;33(16):1349–57.
101. Sándor PS, Afra J. Nonpharmacologic treatment of migraine. Curr Pain Headache Rep. 2005;9(3):202–5.
102. Martin PR. Behavioral management of migraine headache triggers: learning to cope with triggers. Curr Pain Headache Rep. 2010;14:221–7.
103. Nicholson RA, Buse DC, Andrasik F, et al. Curr Treat Options Neurol. 2011;13:28.
104. Nestoriuc Y, Martin A, Rief W, Andrasik F. Biofeedback treatment for headache disorders: a comprehensive efficacy review. Appl Psychophysiol Biofeedback. 2008;33:125–40.
105. Whitmarsh T, Buse DC. Biobehavioral, complementary, and alternative treatments for headache. In Handbook of headache. Martelletti P, Steiner TJ (Eds). New York: Springer 2011: pp. 667–686.
106. Evans RW, Taylor FR. “Natural” or alternative medications for migraine prevention. Headache. 2006;46:1012–8.
107. Eidlitz-Markus T, Haimi-Cohen Y, Steier D, Zeharia A. Effectiveness of nonpharmacologic treatment for migraine in young children. Headache. 2010;50(2):219–23.
108. Bruni O, Galli F, Guidetti V. Nonpharmacologic treatments for migraine and tension-type headache: Sleep hygiene and migraine in children and adolescents. Cephalalgia. 1999;19(S25):58–60.
109. Law EF, Beals-Erickson SE, Noel M, Claar R, Palermo TM. Pilot randomized controlled trial of internet-delivered cognitive-behavioral treatment for pediatric headache. Headache. 2015;55(10):1410–25.
110. El-Chammas K, Keyes J, Thompson N, Vijayakumar J, Becher D, Jackson JL. Pharmacologic treatment of pediatric headaches: a meta-analysis. JAMA Pediatr. 2013;167(3):250–8.
111. Alstadhaug KB. Histamine in migraine and brain. Headache. 2014;54(2):246–59.
112. Rao BS, Das DG, Taraknath VR, Sarma Y. A double blind controlled study of propranolol and cyproheptadine in migraine prophylaxis. Neurol India. 2000;48(3):223–6.
113. Reiter RJ, Tan DX, Galano A. Melatonin: exceeding expectations. Physiology (Bethesda). 2014;29(5):325–33.
114. Srinivasan V, Pandi-Perumal SR, Spence DW, Moscovitch A, Trakht I, Brown GM, et al. Potential use of melatonergic drugs in analgesia: mechanisms of action. Brain Res Bull. 2010;81(4–5):362–71.
115. Wilhelmsen M, Amirian I, Reiter RJ, Rosenberg J, Gögenur I. Analgesic effects of melatonin: a review of current evidence from experimental and clinical studies. J Pineal Res. 2011;51(3):270–7.
116. Chazot G, Claustrat B, Brun J, Jordan D, Sassolas G, Schott B. A chronobiological study of melatonin, cortisol growth hormone and prolactin secretion in cluster headache. Cephalalgia. 1984;4(4):213–20.
117. Claustrat B, Loisy C, Brun J, Beorchia S, Arnaud JL, Chazot G. Nocturnal plasma melatonin levels in migraine: a preliminary report. Headache. 1989;29(4):242–5.
118. Bruni O, Alonso-Alconada D, Besag F, Biran V, Braam W, Cortese S, et al. Current role of melatonin in pediatric neurology: clinical recommendations. Eur J Paediatr Neurol. 2015;19(2):122–33.

119. Abou-Khadra MK, Kishk NA, Shaker OG, Hassan A. Urinary 6-Sulphatoxymelatonin levels and sleep disorders in children with migraine. J Child Neurol. 2013;29(7):947–51.
120. Rovers J, Smits M, Duffy JF. Headache and sleep: also assess circadian rhythm sleep disorders. Headache. 2014;54(1):175–7.
121. Miano S, Parisi P, Pelliccia A, Luchetti A, Paolino MC, Villa MP. Melatonin to prevent migraine or tension-type headache in children. Neurol Sci. 2008;29(4):285–7.
122. Fallah R, Shoroki FF, Ferdosian F. Safety and efficacy of melatonin in pediatric migraine prophylaxis. Curr Drug Saf. 2015;10(2):132–5.
123. Mascia A, Afra J, Schoenen J. Dopamine and migraine: a review of pharmacological, biochemical, neurophysiological, and therapeutic data. Cephalalgia. 1998;18(4):174–82.
124. Supornsilpchai W, Sanguanrangsirikul S, Maneesri S, Srikiatkhachorn A. Serotonin depletion, cortical spreading depression, and trigeminal nociception. Headache. 2006;46(1):34–9.
125. Monti JM. Serotonin control of sleep-wake behavior. Sleep Med Rev. 2011;15(4):269–81.
126. Drummond PD. Tryptophan depletion increases nausea, headache and photophobia in migraine sufferers. Cephalalgia. 2006;26(10):1225–33.
127. Santucci M, Cortelli P, Rossi PG, Baruzzi A, Sacquegna T. L-5-hydroxytryptophan versus placebo in childhood migraine prophylaxis: a double-blind crossover study. Cephalalgia. 1986;6(3):155–7.
128. Powers SW, Coffey CS, Chamberlin LA, Ecklund DJ, Klingner EA, Yankey JW, et al. Trial of amitriptyline, topiramate, and placebo for pediatric migraine. N Engl J Med. 2017;376:115–24.

Sleep in Cerebral Palsy

13

Michelle Sobremonte-King

Introduction

Cerebral palsy (CP) is a group of nonprogressive developmental disorders that affects the development of movement and posture [1]. It is the most common motor disability of childhood occurring at a rate of 1.5–4 per 1000 live births [2]. Children afflicted by CP often have muscle spasms, musculoskeletal pain, contractures, epilepsy, antiepileptic medications, and behavioral abnormalities that predispose them to a wide range of sleep disorders (Table 13.1) [3]. The handful of studies looking at the prevalence of sleep disorders in children with CP use different criteria but collectively demonstrate a higher prevalence as compared to the typically developing pediatric population, with a prevalence ranging from 19% to 44% [3–6]. Sleep disorders in patients with CP can have a significant impact on a child's physical, emotional, and cognitive development with far-reaching consequences that also affect caregivers and families who in turn also become sleep deprived [7]. Despite the substantial health consequences of sleep disorders in CP patients and their families, there is a lack of both awareness and applied clinical research geared to CP-related sleep disorders [8]. In a study involving 286 children with mild to profound intellectual delay (including patients with CP), only 19% of the parents of children with a current sleep problem received any advice about their child's sleep disorder from a healthcare professional [9]. As the multidisciplinary care and longevity of children with CP improve, recognizing and addressing sleep disorders in these patients becomes an important component in improving and maintaining their quality of life.

M. Sobremonte-King (✉)
Department of Pediatric Pulmonary and Sleep Medicine, The Breathing Institute, Children's Hospital Colorado, Aurora, CO, USA
e-mail: michelle.sobremonte-king@seattlechildrens.org

L. M. DelRosso, R. Ferri (eds.), *Sleep Neurology*,
https://doi.org/10.1007/978-3-030-54359-4_13

Table 13.1 Common sleep disorders in cerebral palsy

Sleep disorder	Specific disorder	Possible etiologies	Management
Insomnia	Primary insomnia (inadequate sleep hygiene and behavioral insomnia of childhood) Secondary insomnia (insomnia due to medical condition)	Pain Abnormal muscle tone Inability to reposition Poor sleep hygiene and environment (i.e., bed-sharing, etc.)	Pain/spasticity control (i.e., baclofen pump) Behavioral intervention (i.e., graduated extinction, faded bedtime, bedtime routines, etc.) Melatonin
Sleep-related breathing disorder	Obstructive sleep apnea Hypoventilation	Maxillofacial abnormalities (i.e., maxillary hypoplasia, palatal hypotonia, glossoptosis, retrognathia, etc.) Abnormal neuromuscular control of upper airway Kyphoscoliosis Gastroesophageal reflux disease	Diagnostic polysomnogram Tonsillectomy and adenoidectomy Tongue base suspension Noninvasive ventilation (CPAP) Secretion management and airway clearance
Circadian rhythm disorder	Free-running circadian disorder Delayed sleep phase	Visual impairment Poor sleep hygiene (i.e., exposure to blue-spectrum light)	Melatonin Light therapy Sleep hygiene
Hypersomnia	Hypersomnia due to a medical condition or drug effect	Epilepsy Seizure medications (i.e., benzodiazepines, phenobarbital, phenytoin, valproic acid, etc.)	Seizure control Avoiding sedating anticonvulsants Avoiding anticonvulsants that disrupt sleep architecture

CPAP continuous positive airway pressure

Spectrum of Sleep Disorders in Children with CP

An integrative review by Lelis et al. reported 21 factors associated with the development of sleep disorders in children with CP [10]. These risk factors were divided into intrinsic and extrinsic risk factors. Intrinsic factors were related to physiological features, and extrinsic factors were related to environmental issues. The succeeding section will summarize (Table 13.2) and detail what we think are the most important factors according to the framework suggested by Lelis et al. Among the sleep disorders described are sleep hyperhidrosis, disorders of arousal, sleep anxiety, difficult morning awakening, bruxism, insomnia, nightmares, parasomnias, difficulties in initiating and maintaining nighttime sleep, sleep-wake transition disorders, and sleep-disordered breathing [5, 10]. In general, these can be grouped into six broad categories: sleep-related breathing disorder, insomnia, circadian rhythm disorders, parasomnias, hypersomnia, and sleep-related movement disorders.

Table 13.2 Factors associated with the development of sleep disorders in patients with cerebral palsy

Intrinsic factors	Extrinsic factors
Structural abnormalities	Medications (antiepileptics)
Airway obstruction	Posturing devices
Obesity	Sleep practices (bed-sharing)
Postural limitations/spasticity	Socioeconomic
Epilepsy	Maternal unemployment
Visual disturbances	
Cognitive impairment	
Pain	

Sleep-Related Breathing Disorders

Sleep-related breathing disorders (SRBD) are a group of respiratory disorders that are observed or become more severe during sleep and include obstructive sleep apnea (OSA), central apnea, and hypoventilation. These disorders occur more frequently in children with CP as compared to normally developing children at a prevalence of 15–18% [5, 11]. In particular, children with CP are at increased risk for developing OSA. This occurs when pharyngeal dilator muscles are unable to maintain the patency of the airway against the subatmospheric pressure generated during inspiration. Pathophysiologic mechanisms for the development of OSA can be broadly divided into anatomic factors that reduce the caliber of the upper airway, such as craniofacial abnormalities, obesity, and adenotonsillar hypertrophy, and factors that increase upper airway collapsibility, such as neurologically based alterations in upper airway muscle tone [12]. The latter are of particular concern in CP as these children have a wide range of structural and functional abnormalities that compromise upper airway muscle tone and predispose to OSA. These include maxillary hypoplasia, palatal hypotonia, glossoptosis, retrognathia, laryngomalacia, and laryngeal dystonia [13–15]. They can also have abnormal neuromuscular control of their upper airways that further complicate structural abnormalities [11]. Untreated OSA is associated with diminished quality of life and neurocognitive detriment and can also result in cardiorespiratory failure and death in severe cases [16–18]. Unfortunately, airway obstruction is underappreciated in children with CP, and the potential for OSA is often only identified from a lengthy history of snoring [10]. In a small retrospective case series of eight children with severe spastic quadriplegic CP presenting with upper airway obstruction, seven of the children required admission to the intensive care unit and two required placement of a tracheostomy. Interestingly, despite the relatively severe symptoms and presentation, only two of these children had previously been referred for sleep studies [15].

Identifying children with sleep disorders can be particularly challenging as they can present with a myriad of subtle symptoms that can be difficult to distinguish from symptoms related to other comorbidities. These symptoms can include behavioral changes, cognitive impairment, failure to thrive, decreased appetite, and developmental delay [19]. As such, a structured evaluation with screening questionnaires can be an effective strategy to evaluate for OSA and other SRDBs in patients with CP. In a study by Garcia et al., the Pediatric Sleep Questionnaire (PSQ) and Gross Motor Function Classification System (GMFCS) were used to evaluate symptoms of OSA in 215 children, 18 of which had CP. The investigators found increased PSQ scores (indicating increased risk of OSA) for children with CP (58%) and CP with epilepsy (67%) as compared to normal controls (27%) [20]. Whether this risk for OSA increases with the severity of CP is unclear with some studies showing a relationship and others showing no relationship [11, 20]. In a prospective study evaluating parent responses to the Sleep Disturbance Scale for Children (SDSC) in 173 patients with CP, Newman et al. reported that 14.5% of patients had a pathologic score for disorders of sleep-related breathing on the SDSC [5]. These studies highlight the utility of a structured approach to screening children with CP for sleep-disordered breathing.

For children with CP, an overnight polysomnogram (PSG) remains the gold standard for the diagnosis of OSA. The PSG can also reveal the underlying presence of seizure disorders and other movement disorders that can complicate SRBDs and compromise quality of sleep (Figs. 13.1 and 13.2). Although other diagnostic techniques such as nocturnal pulse oximetry have promising utility in detecting moderate-to-severe OSA, this method has not been validated in patients with neurodevelopmental disorders [21].

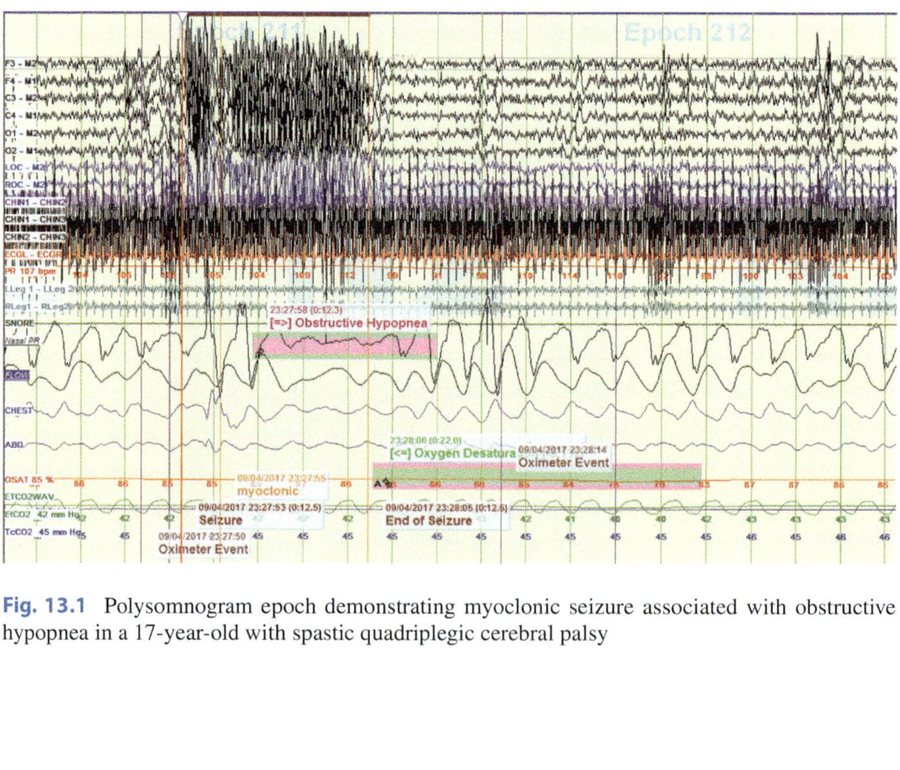

Fig. 13.1 Polysomnogram epoch demonstrating myoclonic seizure associated with obstructive hypopnea in a 17-year-old with spastic quadriplegic cerebral palsy

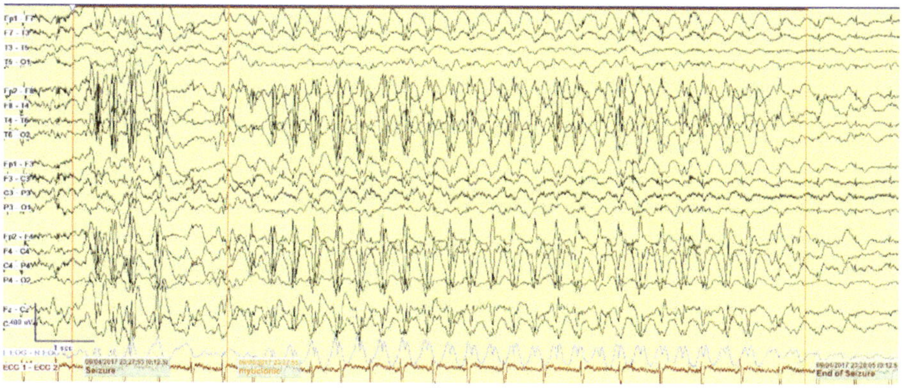

Fig. 13.2 Myoclonic absence seizure, right posterior predominant shown in a 21-channel electroencephalogram (EEG) obtained during polysomnography in a 17-year-old with spastic quadriplegic cerebral palsy

Insomnia

Disorders of initiation and maintenance of sleep are common in children with CP. In a study of 100 children with CP divided into 2 groups (preschool group, 52 patients, and school-aged group, 48 patients), Elsayed et al. found that up to 46.2% of preschool children and 25% of school-aged children had evidence of early insomnia based on results from surveys with sleep questionnaires. Insomnia was also found to be more common in GMFSC grade IV and V [22]. In the study by Newman et al., disorders of initiation and maintenance of sleep were found mostly in patients with spastic quadriplegia and dyskinetic CP [5]. These findings support the concept that children with CP have problems with initiation and maintenance of sleep due predominantly to physical factors that stem from their primary neuromuscular disorder. Factors such as abnormal muscle tone and the inability to alter posture to alleviate pain make it difficult for the child with CP to initiate and maintain sleep. Ironically, caregivers can be distressed by observing these difficulties and decide to co-sleep with their child which in turn has the potential to exacerbate insomnia [5, 10].

Circadian Rhythm Disorders

Establishing and maintaining normal sleep patterns for children with CP can be difficult. As will be evident in subsequent section, children with CP have a multitude of factors that can affect sleep quality and consistency, and addressing all or even some of these factors can be challenging. Perhaps the most significant factor contributing to circadian rhythm disorders in patients with CP is visual deficits. Light is the central regulator of the circadian rhythm in humans via its stimulation of the retinal-hypothalamic tract and suppression of melatonin secretion. Cortical blindness and other visual disturbances either nullify or diminish the ability of these patients to develop a circadian rhythm. In fact, patients with CP often have free-running sleep-wake cycles. For patients who are not blind, neuromuscular immobility can still play a role in children who are exposed to blue-spectrum light and poor sleep hygiene. Careful consideration frequency and timing of blue-spectrum light can be important in children with CP who may not have the same opportunities and abilities to participate in outdoor activities that provide sufficient daytime light exposure [7].

Hypersomnias, Parasomnias, and Sleep-Related Movement Disorders

A study by Newman et al. showed that the principal factor associated with sleep disturbance was epilepsy, which was primarily associated with excessive daytime somnolence [5]. A similar observation was seen in a study of 293 children and adults with learning disabilities where the presence of active epilepsy was the strongest independent predictor of excessive daytime sleepiness [23]. As such, hypersomnia appears to be more of a manifestation of an underlying disease process such as epilepsy, rather than being a primary sleep disorder. Further complicating matters is that some medications used to treat epilepsy can be sedating and cause hypersomnolence. In particular, benzodiazepines which can be used to treat either seizures or muscle spasms can be particularly sedating and can also increase the risk of developing SRBDs [24].

With regard to parasomnias and sleep-related movement disorders, Elsayed et al. observed the following frequencies in a cohort of 100 children with CP: sleepwalking in 30%, sleep bruxism in 38%, and periodic limb movement disorder/restless legs syndrome in 46% [22]. Although attribution of underlying factors to the development of parasomnias and sleep-related movement disorders is difficult to specify, CP children with pain had a higher prevalence of parasomnias [7].

Risk Factors Associated with the Development of Sleep Disorders in Children with CP

Intrinsic Factors

Structural Abnormalities

Children with CP have a wide range of structural and functional abnormalities that predispose them to a variety of sleep disorders. These abnormalities can be grouped anatomically and divided into abnormalities affecting the upper airway or the thoracic cage. Among the reported anatomic abnormalities that lead to dynamic airway obstruction during sleep are maxillary hypoplasia, palatal hypotonia, glossoptosis, retrognathia, and laryngomalacia [13–15]. In addition, there can be abnormal neuromuscular control of their upper airways that further complicates structural abnormalities [11]. For instance, in older children, laryngomalacia is thought to develop due to repetitive collapse of supraglottic structures into the airway due to poor supraglottic muscle tone. This results in reduction of the caliber of the airway and consequent stridor from airway compromise during inspiration [14, 15].

With regard to thoracic abnormalities, kyphoscoliosis is common in children with CP occurring in as much as 39% [14]. It is generally thought that thoracic skeletal abnormalities such as kyphoscoliosis can affect sleep and exacerbate SRBDs by reducing the child's ability to expectorate which contributes to recurrent aspiration, impaired mucociliary clearance, and lower airway obstruction and respiratory failure [14]. Severe kyphoscoliosis has also been associated with hypoventilation and hypoxemia during sleep assuming a supine position [10, 25].

Another anatomic, but also functional, abnormality that can contribute to pulmonary disorders which can lead to sleep disorders is gastroesophageal reflux disease (GERD). Up to 80–90% of children with CP have a chronic gastrointestinal problem of which GERD is the most common [26]. GERD results from reflux of gastric acid into the esophagus through an either anatomically incompetent (hiatal hernia) or functionally incompetent gastroesophageal sphincter [27]. The presence of GERD in patients with CP puts them at significant risk for recurrent aspiration which leads to the development or exacerbation of SRBDs and chronic lung disease [5, 14]. This is particularly important for children with CP since abnormal tone and neuromuscular disorders prevent some children to optimally position themselves to prevent recurrent aspiration. In severe cases, anti-reflux surgery with Nissen fundoplication has been reported to improve symptoms with caregiver reported improvement in sleep in 32% of patients [28].

Epilepsy

The presence of an epilepsy in patients with CP is strongly associated with the presence of a sleep disorder. Epilepsy is present in approximately 20–40% of patients with CP and is most common in children with hemiplegic or quadriplegic CP [29]. Newman et al. conducted one of the first studies to determine the prevalence of sleep disorders in CP patients and found that the presence of active epilepsy was associated with the presence of a sleep disorder (odds ratio [OR] = 17.1) [5]. Another equally large cohort of CP patients demonstrated a similar association where investigators found an abnormal SDSC total score in 34% of children with CP plus active epilepsy as compared to 18% in children with CP and controlled epilepsy and 15% in children with CP without epilepsy [3]. The similar rates of abnormal SDSC scores in patients with controlled epilepsy and normal control underscore the importance of effective antiepileptic treatment in treating sleep disorders in patients with CP. Although the precise nature of the relationship between epilepsy and sleep disorders in patients with neurodevelopmental abnormalities is complex, it is generally viewed that uncontrolled seizure activity can perpetuate a vicious cycle of fragmented sleep architecture that results in sleep deprivation which in turn results in lowering of the seizure threshold [30]. A case-controlled study of 31 patients with drug-resistant epilepsy showed that compared to normal controls, these patients had a significant reduction of time in bed, total sleep time, rapid eye movement (REM) sleep, sleep stage N2, sleep efficiency, and a significant increase in wake after sleep onset [30]. Antiepileptic drugs add to the complexity as they themselves can influence sleep architecture and will be discussed later.

Visual Impairment

Between 20% and 50% of children with CP have cortical visual impairment which significantly increases the risk of developing sleep disorders through dysregulation of the circadian rhythm [8]. Bright light is the major stimulus for the human circadian pacemaker located in the suprachiasmatic nuclei which regulates wakefulness by suppressing melatonin secretion by the pineal gland. Without light, melatonin secretion is disinhibited and the diurnal variation is lost. Consequently, the prevalence of sleep disorders in blind patients is considerably high, and one study showed that up to 83% of patients in a combined pediatric and adult cohort disclosed at least one sleep problem, as compared to 57% in the control population [31]. More specifically, 17% of these patients reported a free-running sleep-wake cycle which is a phenomenon that is described in children with congenital blindness [31, 32].

Cognitive Disability

Cognitive impairment frequently occurs in children with CP and has been increasingly reported to be associated with sleep disorders in children with CP [33]. Some studies demonstrate that the degree of cognitive impairment can be a predictor for the presence of an underlying sleep problem [3, 9]. In contrast, other studies do not find an association between the presence of cognitive impairment making the precise nature of the relationship yet to be fully determined [34, 35].

Pain and Muscle Spasticity

Pain and discomfort during sleep is a common occurrence in children with physical disabilities and has been suggested by some to be the strongest contributing factor for sleep disorders to develop in these children [36, 37]. Pain is particularly common in children with CP and is reported in as high as 67–84% of patients [38]. Many factors such as skin breakdown, pressure ulcers, involuntary movements, abnormal postures, and abnormal muscle tone in the form of spasticity contribute to pain in children with CP and have the ability to decrease the quality of sleep [7]. In a study of 123 children with CP, patients who had pain also had significantly more sleep problems overall, more night awaking, parasomnia, sleep-disordered breathing, and shorter sleep duration [39].

In particular, pain may be associated to the underlying disease-related mechanisms in CP. Lesions in the motor areas of the brain and descending corticospinal tracts result in muscle spasticity, characterized by increased tone, hyperreflexia, clonus, and resistance to stretching. The resulting musculoskeletal pain and primary motor impairment decrease the ability to change body position during the night and may contribute to sleep disturbances [5]. Intervention to decrease pain and discomfort from muscle spasticity has been demonstrated to improve sleep quality in children with CP. A study evaluating the use of an intrathecal baclofen pump in 35 children with bilateral cerebral palsy and severe spasticity improved sleep quality with a reduction in nighttime awakening within 6 months of implantation [40]. As such, pain and abnormal muscle tone are significant factors that interfere with sleep in children with CP and can be addressed with recognition and intervention. Unfortunately, although this area is a major contributor to sleep disorders, it is underappreciated and research activity is sparse [7].

Extrinsic Factors

Antiepileptic Medications

Antiepileptic medications have been described to influence the sleep architecture and the sleep quality of patients who take them [41]. Since many patients with CP with epilepsy will need antiepileptic medications to control their seizure disorder, it is generally viewed that these medications can predispose them to have sleep disorders. However, the interaction is complex since different types of antiepileptics have different effects on sleep, and some CP patients with epilepsy will be on multiple antiepileptic medications. For instance, drugs like phenobarbital, phenytoin, valproic acid, and higher-dose levetiracetam aggravate daytime sleepiness, whereas the antiepileptics topiramate and zonisamide do not [41, 42]. In contrast, drugs like gabapentin, pregabalin, tiagabine, clobazam, and carbamazepine were found to reduce sleep latency and/or actually improve sleep efficiency [42]. Alternatively, other antiepileptic drugs such as benzodiazepines not only affect sleep architecture but also promote somnolence and decrease upper airway muscle tone leading to aggravation of sleep-disordered breathing and OSA [10].

Despite the complexity of these interactions, clinicians should not lose track of the utility and benefits of these medications in controlling seizure activity which, as mentioned, is a stronger factor influencing sleep in patients with CP. It is clear that children with CP and uncontrolled seizures have a higher prevalence of sleep disorders as compared to those who have their epilepsy controlled, regardless of the number of medications they may be on [3].

Positioning and Posturing Devices

Preoccupation with nighttime positioning and the use of uncomfortable devices for improving postural problems in children with CP are common and generally viewed as having the potential to disrupt sleep [10]. Although the body of available data describing this interaction is sparse and varied, the general impression is that posturing devices do not significantly disrupt sleep and may actually improve the quality of sleep. In one study, no significant differences in PSG sleep quality measures were seen in children wearing nighttime posturing equipment as compared to children without [43]. Another study reached the same conclusion using GMFCS levels to evaluate sleep quality [35].

Familial and Cultural Effects

A variety of familial factors including bed-sharing, being a single parent, maternal unemployment, and low socioeconomic status have been associated with an increase in sleep disorders in children with CP [5]. Of note, although several studies have reported an increase in sleep problems associated with the practice of bed-sharing, drawing inferences as to cause and effect can be tricky [44, 45]. Problems during sleep may be the reason for parents to co-sleep with their child, and, expectedly, this setting ideally places them to be able to observe a number of sleep-related events which would have otherwise remained unnoticed [5]. Consequently, parents who co-sleep with their child often report more sleep problems in their children as compared to non-bed-sharing parents [10].

Behavioral Correlates

Behavioral and psychiatric problems are common in children with CP, and both can significantly impact sleep quality and exacerbate underlying sleep disorders [46]. A tool that is often used to evaluate behavioral problems in patients with neurodevelopmental disorders is the Child Behavior Checklist (CBCL), which can group scored syndromes into two "broad band" scales: internalizing problems (anxious/depressed, withdrawn-depressed, and somatic complaints scores) and externalizing problems (rule-breaking and aggressive behavior) [47]. Romeo et al. used the CBCL to estimate the incidence and patterns of behavioral problems and their association with sleep disorders in a prospective single-institution study of 165 children with CP. They found that 24% of children reported an abnormal CBCL total score, with 27% showing internalizing and 9% with externalizing problems. In addition, behavioral problems were often associated with abnormal Sleep Disturbance Scale for Children (SDSC) total scores, as well as total and internalizing CBCL scores [3].

Social Impact of Sleep Disorders on Children with Cerebral Palsy and Their Families

Unrecognized and untreated sleep disorders in patients with CP can have serious and far-reaching consequences that affect both the child and their families. It is well established that poor sleep quality in children with sleep disorders can result in negative effects on behavior and school performance [24, 48, 49]. In addition, the burden of caring for children with CP can be significant for family members especially for those with sleep disorders. Wayte et al. report that nearly 40% of children with CP required parental attention on at least one occasion every night, and 74% of parents reported that their own daytime functioning was affected by their child's sleep disorder [50].

Some parents feel the need to co-sleep due to their child's restlessness and the compulsion of needing to position them to address issues of pain related to abnormal tone. This in turn creates the potential for sleep disorders in the parents, and many studies document that they do not get enough sleep themselves [5, 40]. In a study of 40 children with CP, 40% of mothers were found to have poor sleep quality of whom 44% had a depressed mood. In addition, child and maternal sleep disturbance were significantly correlated, and maternal sleep quality predicted 50% of the variance in maternal depression [50]. Other studies also report that parents of children with sleep problems feel more stressed and irritable and express that their child's sleep problems interfere and negatively impact their own work and social lives [5, 34, 51]. As such, when placed in this broader context, it is clear that sleep disorders in children with CP need to be recognized as of being of wider concern because of the physical and psychological toll that can result for the entire family [7].

Treatment of Sleep Disorders in Children with CP

The management of sleep disorders in children with CP can be particularly challenging due to the multitude and variety of contributing factors that often occur simultaneously. It often necessitates addressing the contributing intrinsic and extrinsic factors, many of which are social and behavioral. As such, Tremblay et al. recommend that the first step in treatment of insomnia in children with CP should always be behavioral interventions. Examples of behavioral interventions include graduated extinction, faded bedtime, parent education, and positive bedtime routines. These can be particularly effective in non-SRDB-related sleep disorders where several studies have demonstrated their efficacy [52, 53]. Modification of factors related to the environment, such as exposure to blue-spectrum light, sound, and wide swings in ambient temperature, has the potential to interfere with sleep is another noninvasive strategy that can be effective in improving sleep quality in patients with CP. For instance, excessive and untimely exposure of children to artificial blue-spectrum light, usually through electronic devices like television and smartphones, can alter the timing, duration, and amount of melatonin synthesis and, therefore, interfere with the circadian rhythm [54]. This becomes a particular issue considering that some parents and caregivers co-sleep with their children. Alternatively, bed-sharing is another modifiable factor that can be addressed to improve sleep quality in children [7].

Pharmacologic treatment may be indicated when behavioral measures and environmental modification are ineffective or do not suffice. However, robust data on pharmacologic treatment strategies are lacking, much more so in children with CP or other neurodevelopmental abnormalities. Chloral hydrate is the only medication currently labeled by the US Food and Drug Administration for the treatment of insomnia in children but is not recommended by the American Academy of Pediatrics due to the risk of hepatotoxicity [55]. The use of other medication appears to be based largely on personal clinical experience and data extrapolated in the treatment of adults [55]. A review by Owens and Mindell describes reports of antidepressants, anticonvulsants, antipsychotics, and alpha-agonists in the treatment of childhood insomnia, but data are mostly derived from small studies and case series and do not describe including patients with CP [55].

Exogenous administration of melatonin has been successfully used in typically developing children to treat circadian rhythm disorders and disorders of initiating and maintaining sleep. Of all the previously described medications, melatonin has received the most attention, and randomized data that includes patients with CP are available. A review of six randomized controlled trials (N = 82) of melatonin in children with intellectual disability suggests that melatonin can be effective in reducing sleep onset latency and increase total sleep time [56]. Wasdell et al. conducted a randomized double-blind placebo-controlled crossover trial to evaluate the efficacy of controlled-release melatonin in the treatment of delayed sleep phase syndrome and impaired sleep maintenance in children with neurodevelopmental disabilities. The study enrolled 51 children ages 2–18 years (26 patients had CP) who did not respond to sleep hygiene intervention. Fifty patients completed the crossover trial and 47 completed the open-label phase. The study reported a significant improvement of total nighttime sleep and sleep latency by approximately 30 minutes as documented by direct observation and by actigraph recordings [57]. Unfortunately, the authors of this study did not provide data or a detailed description of potential toxicities of melatonin during the trial. Potential toxicities that have been described include depression, enuresis, somnolence, and one report of increase in seizure activity [24, 58]. However, a study of melatonin in children with intractable epilepsy did not report an increase in seizure activity [59].

Evidently, some sleep disorders in children with CP are structural and need to be addressed with more invasive measures. OSA is the primary example and is a common cause of sleep-disordered breathing in children with CP. The initial management of OSA in children involves identifying adenotonsillar hypertrophy as a contributing factor and tonsillectomy as the primary treatment [17]. Of note, children with CP are at an increased risk of postsurgical complications, and the need for postoperative monitoring cannot be understated [60].

Tongue base suspension (TBS) is another surgical technique that has been described to potentially benefit patients with OSA and CP in combination with other surgical techniques [61]. A study by Hartzell et al. described PSG data in CP children who received adenotonsillectomy and uvulopalatopharyngioplasty plus TBS vs. adenotonsillectomy and uvulopalatopharyngioplasty alone. Patients who received TBS had significant improvements in their arousal index leading the authors to conclude that children with CP and moderate to severe OSA may potentially benefit from TBS [60].

In children in whom OSA persists despite tonsillectomy or in whom tonsillectomy cannot be done, other measures such as nasopharyngeal airway devices, oxygen supplementation, and continuous positive airway pressure (CPAP) are implemented. A CPAP machine works by delivering an intraluminal pressure that is above the critical closing pressure of the airway that effectively stents the airway open to overcome dynamic obstruction [62]. Unfortunately, compliance with nonsurgical techniques is a common problem, and treatment of OSA in patients with neurodevelopmental abnormalities like CP can be challenging. A guiding principle in managing OSA in patients with CP is recognizing that OSA is a multifactorial disorder and that treatment should be individualized based on the underlying neurologic abnormality and site of obstruction [24].

References

1. Bax M, Goldstein M, Rosenbaum P, Leviton A, Paneth N, Dan B, et al. Proposed definition and classification of cerebral palsy, April 2005. Dev Med Child Neurol. 2005;47(8):571–6.
2. Data & statistics for cerebral palsy [Internet]. Centers for Disease Control and Prevention. Available from: https://www.cdc.gov/ncbddd/cp/data.html#references
3. Romeo DM, Brogna C, Quintiliani M, Baranello G, Pagliano E, Casalino T, et al. Sleep disorders in children with cerebral palsy: neurodevelopmental and behavioral correlates. Sleep Med. 2014;15(2):213–8.

4. Kotagal S, Gibbons VP, Stith JA. Sleep abnormalities in patients with severe cerebral palsy. Dev Med Child Neurol. 1994;36(4):304–11.
5. Newman CJ, O'Regan M, Hensey O. Sleep disorders in children with cerebral palsy. Dev Med Child Neurol. 2006;48(7):564–8.
6. Hayashi M, Inoue Y, Iwakawa Y, Sasaki H. REM sleep abnormalities in severe athetoid cerebral palsy. Brain and Development. 2018;12(5):494–7.
7. Dutt R, Roduta-Roberts M, Brown C. Sleep and children with cerebral palsy: a review of current evidence and environmental non-pharmacological interventions. Children. 2015;2(1):78–88.
8. Jan MMS. Cerebral palsy: comprehensive review and update. Ann Saudi Med. 2006;26:123–32.
9. Didden R, Korzilius H, Van Aperlo B, Van Overloop C, De Vries M. Sleep problems and daytime problem behaviours in children with intellectual disability. J Intellect Disabil Res. 2002;46(7):537–47.
10. Lélis ALPA, Cardoso MVLM, Hall WA. Sleep disorders in children with cerebral palsy: an integrative review. Sleep Med Rev. 2015;30:63–71.
11. Koyuncu E, Turkkani MH, Sarikaya FG, Ozgirgin N. Sleep disordered breathing in children with cerebral palsy. Sleep Med. 2017;30:146–50.
12. King MS, Xanthopoulos MS, Marcus CL. Improving positive airway pressure (PAP) adherence in children. Sleep Med Clin. 2014;9:219–34.
13. Worley G, Witsell DL, Hulka GF. Laryngeal dystonia causing inspiratory stridor in children with cerebral palsy. Laryngoscope. 2003;113(12):2192–5.
14. Fitzgerald DA, Follett J, Van Asperen PP. Assessing and managing lung disease and sleep disordered breathing in children with cerebral palsy. Paediatr Respir Rev. 2009;10:18–24.
15. Wilkinson DJ, Baikie G, Berkowitz RG, Reddihough DS. Awake upper airway obstruction in children with spastic quadriplegic cerebral palsy. J Paediatr Child Health. 2006;42(1–2):44–8.
16. Beebe DW. Cognitive, behavioral, and functional consequences of inadequate sleep in children and adolescents. Pediatr Clin North Am. 2011;58:649–65.
17. Marcus CL, Brooks LJ, Draper KA, Gozal D, Halbower AC, Jones J, et al. Diagnosis and management of childhood obstructive sleep apnea syndrome. Pediatrics. 2012;130(3):e714–55.
18. Kravath RE, Pollak CP, Borowiecki B, Weitzman ED. Obstructive sleep apnea and death associated with surgical correction of velopharyngeal incompetence. J Pediatr [Internet]. 1980;96(4):645–8. Available from: http://www.jpeds.com/article/S0022-3476(80)80730-X/abstract
19. Cohen SR, Lefaivre JF, Burstein FD, Simms C, Kattos AV, Scott PH, et al. Surgical treatment of obstructive sleep apnea in neurologically compromised patients. Plast Reconstr Surg. 1997;99(3):638–46.
20. Garcia J, Wical B, Wical W, Schaffer L, Wical T, Wendorf H, et al. Obstructive sleep apnea in children with cerebral palsy and epilepsy. Dev Med Child Neurol. 2016;58(10):1057–62.
21. Brouillette RT, Morielli A, Leimanis A, Waters KA, Luciano R, Ducharme FM. Nocturnal pulse oximetry as an abbreviated testing modality for pediatric obstructive sleep apnea. Pediatrics [Internet]. 2000;105(2):405–12. Available from: http://pediatrics.aappublications.org/cgi/doi/10.1542/peds.105.2.405
22. Elsayed RM, Hasanein BM, Sayyah HE, El-Auoty MM, Tharwat N, Belal TM. Sleep assessment of children with cerebral palsy: using validated sleep questionnaire. Ann Indian Acad Neurol. 2013;16(1):62–5.
23. Lindblom N, Heiskala H, Kaski M, Leinonen L, Nevanlinna A, Iivanainen M, et al. Neurological impairments and sleep-wake behaviour among the mentally retarded. J Sleep Res. 2001;10(4):309–18.
24. Simard-Tremblay E, Constantin E, Gruber R, Brouillette RT, Shevell M. Sleep in children with cerebral palsy: a review. J Child Neurol. 2011;26:1303–10.
25. Seddon PC, Khan Y. Respiratory problems in children with neurological impairment. Arch Dis Child. 2003;88(1):75–8.
26. Chambers HG. Advances in cerebral palsy. Curr Opin Orthop. 2002;13:424–31.

27. Christie DL, O'Grady LR, Mack DV. Incompetent lower esophageal sphincter and gastroesophageal reflux in recurrent acute pulmonary disease of infancy and childhood. J Pediatr. 1978;93(1):23–7.
28. O'Loughlin EV, Somerville H, Shun A, Gangemi R, La Hei E, Desphande A, et al. Antireflux surgery in children with neurological impairment: caregiver perceptions and complications. J Pediatr Gastroenterol Nutr. 2013;56(1):46–50.
29. Odding E, Roebroeck ME, Stam HJ. The epidemiology of cerebral palsy: incidence, impairments and risk factors. Disabil Rehabil. 2006;28(4):183–91.
30. Pereira AM, Bruni O, Ferri R, Palmini A, Nunes ML. The impact of epilepsy on sleep architecture during childhood. Epilepsia. 2012;53(9):1519–25.
31. Petit D, Paquet JBL, et al. Sleep: an unrecognized actorin child development. Inst la Stat du Quebec. 2010;5(2):1–20.
32. Okawa M, Nanami T, Wada S, Shimizu T, Hishikawa Y, Sasaki H, et al. Four congenitally blind children with circadian sleep-wake rhythm disorder. Sleep. 1987;10(2):101–10.
33. Pruitt DW, Tsai T. Common medical comorbidities associated with cerebral palsy. Phys Med Rehabil Clin N Am. 2009;20:453–67.
34. Richdale A, Francis A, Gavidia-Payne S, Cotton S. Stress, behaviour, and sleep problems in children with an intellectual disability. J Intellect Develop Disabil. 2000;25(2):147–61.
35. Hill CM, Parker RC, Allen P, Paul A, Padoa KA. Sleep quality and respiratory function in children with severe cerebral palsy using night-time postural equipment: a pilot study. Acta Paediatr Int J Paediatr. 2009;98(11):1809–14.
36. Wright M, Tancredi A, Yundt B, Larin H. Sleep issues in children with physical disabilities and their families. Phys Occup Ther Pediatr. 2006;26(3):55–72.
37. Levi R, Hultling C, NASH MS, Seiger A. The Stockholm spinal-cord injury study 1. Medical problems in a regional SCI population. Paraplegia [Internet]. 1995;33(6):308–15. Available from: https://www.nature.com/articles/sc199570.pdf
38. McKearnan KA, Kieckhefer GM, Engel JM, Jensen MP, Labyak S. Pain in children with cerebral palsy: a review. J Neurosci Nurs. 2004;36(5):252–9.
39. Breau LM, Camfield CS. Pain disrupts sleep in children and youth with intellectual and developmental disabilities. Res Dev Disabil. 2011;32(6):2829–40.
40. Ramstad K, Jahnsen R, Lofterod B, Skjeldal OH. Continuous intrathecal baclofen therapy in children with cerebral palsy – when does improvement emerge? Acta Paediatr Int J Paediatr. 2010;99(11):1661–5.
41. Legros B, Bazil CW. Effects of antiepileptic drugs on sleep architecture: a pilot study. Sleep Med. 2003;4(1):51–5.
42. Jain SV, Glauser TA. Effects of epilepsy treatments on sleep architecture and daytime sleepiness: an evidence-based review of objective sleep metrics. Epilepsia. 2014;55:26–37.
43. Shibagaki M, Kiyono S, Takeuchi T. Nocturnal sleep in mentally retarded infants with cerebral palsy. Electroencephalogr Clin Neurophysiol. 1985;61(6):465–71.
44. Lozoff B, Askew GL, Wolf AW. Cosleeping and early childhood sleep problems: effects of ethnicity and socioeconomic status. J Dev Behav Pediatr. 1996;17(1):9–15.
45. Cortesi F, Giannotti F, Sebastiani T, Vagnoni C. Cosleeping and sleep behavior in Italian school-aged children. J Dev Behav Pediatr. 2004;25(1):28–33.
46. McDermott S, Coker AL, Mani S, Krishnaswami S, Nagle RJ, Barnett-Queen LL, et al. A population-based analysis of behavior problems in children with cerebral palsy. J Pediatr Psychol. 1996;21(3):447–63.
47. Achenbach TM, Ruffle TM. The child behavior checklist and related forms for assessing behavioral/emotional problems and competencies. Pediatr Rev. 2000;21(8):265–71.
48. Gozal D. Sleep-disordered breathing and school performance in children. Pediatrics. 1998;102(3):616–20.
49. Marcus CL, Brooks LJ, Draper KA, Gozal D, Halbower AC, Jones J, et al. Diagnosis and management of childhood obstructive sleep apnea syndrome. Pediatrics [Internet]. 2012;130(3):e714–55. Available from: http://www.ncbi.nlm.nih.gov/pubmed/22926176

50. Wayte S, McCaughey E, Holley S, Annaz D, Hill CM. Sleep problems in children with cerebral palsy and their relationship with maternal sleep and depression. Acta Paediatr [Internet]. 2012;101(6):618–23. Available from: http://doi.wiley.com/10.1111/j.1651-2227.2012.02603.x
51. Quine L. Severity of sleep problems in children with severe learning difficulties: description and correlates. J Community Appl Soc Psychol. 1992;2(4):247–68.
52. Mindell JA, Kuhn B, Lewin DS, Meltzer LJ, Sadeh A. Behavioral treatment of bedtime problems and night wakings in infants and young children – an American academy of sleep medicine review. Sleep. 2006;29(10):1263–76.
53. Lancioni GE, O'Reilly MF, Basili G. Review of strategies for treating sleep problems in persons with severe or profound mental retardation or multiple handicaps. Am J Ment Retard. 1999;104(2):170–86.
54. Gooley JJ, Chamberlain K, Smith KA, Khalsa SBS, Rajaratnam SMW, Van Reen E, et al. Exposure to room light before bedtime suppresses melatonin onset and shortens melatonin duration in humans. J Clin Endocrinol Metab. 2011;96(3):E463–72.
55. Owens JA, Mindell JA. Pediatric insomnia. Pediatr Clin N Am. 2011;58(3):555–69.
56. Sajith SG, Clarke D. Melatonin and sleep disorders associated with intellectual disability: a clinical review. J Intellect Disabil Res. 2007;51:2–13.
57. Wasdell MB, Jan JE, Bomben MM, Freeman RD, Rietveld WJ, Tai J, et al. A randomized, placebo-controlled trial of controlled release melatonin treatment of delayed sleep phase syndrome and impaired sleep maintenance in children with neurodevelopmental disabilities. J Pineal Res. 2008;44(1):57–64.
58. Sheldon SH. Pro-convulsant effects of oral melatonin in neurologically disabled children. Lancet. 1998;351(9111):1254.
59. Elkhayat HA, Hassanein SM, Tomoum HY, Abd-Elhamid IA, Asaad T, Elwakkad AS. Melatonin and sleep-related problems in children with intractable epilepsy. Pediatr Neurol. 2010;42(4):249–54.
60. Kosko JR, Derkay CS. Uvulopalatopharyngoplasty: treatment of obstructive sleep apnea in neurologically impaired pediatric patients. Int J Pediatr Otorhinolaryngol. 1995;32(3):241–6.
61. Hartzell LD, Guillory RM, Munson PD, Dunham AK, Bower CM, Richter GT. Tongue base suspension in children with cerebral palsy and obstructive sleep apnea. Int J Pediatr Otorhinolaryngol. 2013;77(4):534–7.
62. Marcus CL, McColley SA, Carroll JL, Loughlin GM, Smith PL, Schwartz AR. Upper airway collapsibility in children with obstructive sleep apnea syndrome. J Appl Physiol. 1994;77(2):918–24.

Sleep and Cerebrovascular Disease

14

Claude Nguyen

Introduction

Stroke is the fifth leading cause of death and the leading cause of disability in the United States [1]. In the quest to identify contributors to cerebrovascular disease, sleep disorders have been found to increase the risk of stroke. Just as important, sleep disorders are a frequent manifestation of stroke and can affect stroke recovery and quality of life. Therefore, it is useful for the clinician to understand the relationship between sleep and stroke, both to empower them to better manage their patients' stroke risk, and to more effectively manage sleep disorders as a vital component of stroke recovery. This chapter aims to describe the relationship between sleep and stroke, exploring the contributions of sleep disorders to stroke risk and covering important clinical issues involving sleep disorders in the stroke patient.

Sleep Disorders and Stroke Risk

Sleep Duration and Its Effects on Stroke Risk Factors

Too little as well as too much sleep can adversely affect stroke risk. Overall, there is an abundance of evidence linking short sleep duration to elevated stroke risk [2]. This observation holds true even in patients with low risk of obstructive sleep apnea (OSA) and normal body mass index [3]. The bulk of this evidence is from animal studies, showing that sleep deprivation leads to impaired metabolism, including decreased

C. Nguyen (✉)
Division of Stroke and Neurocritical Care, California Pacific Medical Center, Sutter Health, San Francisco, CA, USA

L. M. DelRosso, R. Ferri (eds.), *Sleep Neurology*,
https://doi.org/10.1007/978-3-030-54359-4_14

thyroid hormone output and reduced energy balance, in turn leading to decreased cerebral function [4].

Despite this, it should be noted that few studies have used objective testing to measure daytime sleepiness. One study used the multiple sleep latency test (MSLT), finding that stroke patients had more daytime sleepiness [5]; however, the control group was comprised of Parkinson's and Alzheimer's disease patients, thereby limiting the generalizability of their findings.

Conversely, long sleep duration has been less consistently linked to stroke [3], although there has been a lack of laboratory or epidemiologic support to suggest a biological mechanism, suggesting that other factors such as presence of OSA or inflammatory markers could confound this relationship [6].

Laboratory studies involving gene expression have shown that sleep deprivation prior to stroke induces neuroprotective effects [7], altering cellular signaling response by inhibiting inflammatory response, cell cycle pathway, and neuroendocrine regulation. Notably, experiments in rodents have shown upregulation in melanin-concentrating hormone and hypocretin/orexin during the acute phase of stroke [8].

Interestingly, there has been some evidence demonstrating a link between long sleep duration and elevated risk of developing hemorrhagic stroke [9], although some have found this association to affect only women [10]. Regardless, further study in this area is needed.

Sleep duration has been found to have significant effects on traditional stroke risk factors. In terms of age, older persons often have decreased sleep duration and interrupted sleep. Advanced age is also associated with increased REM sleep with resulting decrease in slow-wave sleep [11]. There are other factors associated with increased age which lead to hypoventilation, including reduced ventilation, decreased lung elasticity, depressed respiratory regulation from autonomic dysfunction, or other comorbidities leading to decreased respiratory function. Overall, this results in an increase in oxygen demand, which, in turn, leads to increased stroke risk.

Linkages between sleep duration and hypertension have been well-described. The Sleep Heart Health Study demonstrated that those with less than 7 or more than 8 hours of sleep had a higher risk of having hypertension [4]. Likewise, the Coronary Artery Risk Development in Young Adults Study showed a relation between short sleep duration and elevated systolic as well as diastolic blood pressure [4]. In those with excessive sleep, the root cause for these findings is thought to be a result of increased sympathetic nervous system activity, caused by an increase in excretion of post-sleep urinary norepinephrines. However, studies are inconsistent as to whether short duration among older persons leads to hypertension. To explain the link between short sleep duration and hypertension, during normal sleep, one's blood pressure typically decreases. Short sleep duration has been linked to a lack of this decrease in nocturnal blood pressure, which has been linked to increased cardiovascular mortality [4]. Independent of OSA, those patients with medically refractory hypertension not only had shorter sleep periods compared to people who had normotension or controlled hypertension but also spent less time in rapid eye movement (REM) sleep.

Conversely, prolonged sleep duration has been linked to elevated stroke risk, although it is unclear whether this could be explained by confounding variables such as SDB [6].

Diabetes is another traditional stroke risk factor that can be further impaired by impaired sleep duration. Sleep deprivation has been linked to derangements in metabolism. Both too much sleep and too little sleep have been associated with impaired glucose tolerance, and those with abnormal sleep duration who already have diabetes were shown to have more difficulty with glucose control [4]. Studies have shown that decreased sleep duration is associated with up to a 30% increase in risk of developing diabetes. Akinseye et al. studied a cohort of 26,364 diabetic patients, finding that abnormal sleep duration was associated with increased risk of stroke, and even those with longer sleep duration had elevated stroke risk, although this relationship varied depending on sex and ethnicity [12]. It is believed that sleep deprivation causes increased sympathetic activity, which leads to decreased beta-cell responsiveness.

Obesity is another independent risk factor for stroke that can be exacerbated by changes in sleep. Sleep deprivation has been shown to lead to obesity. Decreased sleep leads to changes in metabolic state, stimulating a decrease in the hormone leptin, which controls satiety [4]. In turn, this stimulates appetite by increasing secretion of ghrelin. In addition, decreased sleep duration may lead to decreased physical activity and energy consumption, leading to increase in weight. Thermoregulatory mechanisms have also been proposed, as reduced sleep time has been associated with reduced body temperature.

Sleep duration can have profound effects on hyperlipidemia. Several studies have shown that women who have either short or long sleep duration tend to have lower levels of serum HDL and increased LDL, total cholesterol, and triglyceride levels; this effect was not observed in men [4]. Sleep deprivation is thought to stimulate appetite especially for fatty foods. Less REM sleep has been found to be associated with an increased desire for fat consumption [4].

Atrial fibrillation is another major risk factor for stroke. In a systematic review, one group found evidence linking frequent awakening and insomnia to atrial fibrillation [13]. Decreased sleep duration has been linked to prolonged P-wave duration and P-wave dispersion, which could portend a diagnosis of atrial fibrillation [4]. Decreased REM sleep has been linked to an increase in the presence of atrial fibrillation as well [4]. Although a direct relationship between sleep duration and atrial fibrillation is unclear, this evidence raises concern that sleep duration should continue to be studied.

Sleep-Disordered Breathing and Risk of Stroke

Sleep-disordered breathing (SDB) includes obstructive sleep apnea (OSA), central sleep apnea (CSA), and Cheyne-Stokes breathing (CSB). Sleep apnea is defined by a cessation or reduction of airflow during sleep. OSA is the disruption of airflow caused by compromise of the upper airway. In contrast, CSA can occur as a result of central nervous system pathology, resulting in a cyclical breathing pattern of increased and depressed respiratory drive. Not only is OSA a significant independent risk factor for stroke, there is correlation between the severity of OSA and stroke risk [14, 15].

SDB is thought to cause stroke by promoting paradoxical emboli [16]: SDB causes hypoxemia and retention of carbon dioxide, which causes an increase in pulmonary vascular resistance. In turn, right ventricular afterload and right atrial pressure are increased, causing transient elevation in right-to-left pressure, promoting shunting of emboli through a patent foramen ovale (PFO).

While cerebral blood flow (CBF) undergoes various changes based on the stage of sleep, in OSA, brain perfusion is altered: CBF is decreased, and intermittent surges in CBF velocity and blood pressure are observed [17]. Some evidence suggests that sympathetic activity is a strong contributor protecting the brain against excessive changes in cerebral perfusion pressure and cerebral blood flow during REM sleep. It is postulated that the increased sympathetic activity works to protect the brain from a decreased blood flow seen in OSA. Various physiologic pathways are involved in this response affecting CBF, including peripheral and central chemoreceptors, PO2 and PCO2 levels, pH level, and regional brain activity [18].

In terms of localization, various studies have shown an association between SDB and brainstem strokes [19]. CSA and CSB were first described in patients with bilateral strokes who have altered consciousness or heart failure; however, more recent studies have shown CSB in unilateral strokes without these associations. Since then, CSA/CSB has been shown to be associated with heart failure.

Stroke patients with OSA have also been found to have more cognitive and functional impairment on comprehensive neuropsychological testing than those without OSA [20]. Specifically, these patients may have deficits in memory, attention, vigilance, executive functioning, visuospatial perception, and motor function. Overall, it is clear that SDB is associated not only with increased risk of stroke but with increased post-stroke morbidity.

Sleep Disorders as a Consequence of Stroke

Insomnia

Up to half of post-stroke patients experience insomnia in the first few months of recovery [19]. Post-stroke insomnia is exacerbated by comorbidities commonly found in stroke patients, including depression, SDB, and pain. The busy hospital environment, set up to closely monitor stroke patients and provide frequent neurological checks by necessity, ironically exacerbates the problem of insomnia. Although rare, insomnia can also be caused by structural injury from stroke, notably in the pontomesencephalic, paramedian thalamic, and left dorsomedial prefrontal cortical regions [19].

Hypersomnia/Excessive Daytime Sleepiness

Hypersomnia, or excessive daytime sleepiness (EDS), was found in over 20% of stroke patients in one study [21], with a decrease in REM sleep being associated with the development of significant EDS. Those with EDS have worse clinical outcomes, since presumably SDB is a cause of EDS. In the Northern Manhattan Study, a large, multiethnic population-based, urban cohort, a modified version of the Epworth Sleepiness Scale, was used to assess for daytime sleepiness. Those with EDS had increased risk of stroke and other vascular events, and notably, the risk was different between ethnic groups, likely due to socioeconomic differences [22].

Hypersomnia has been associated with subcortical and pontomesencephalic strokes [19]. In the subacute recovery period, there can be improvement in hypersomnia, though it is common for fatigue to be a chronic problem. The most severe form of hypersomnia results from paramedian thalamic stroke, which can result in up to 20 hours per day of sleep-like behavior, including impairment in memory, attention, and cognition. Post-stroke insomnia can last months, although symptoms in some cases, particularly bilateral stroke, can last years.

Restless Leg Syndrome/Periodic Limb Movements of Sleep

Restless leg syndrome (RLS) is a disorder causing an increased urge to move, which typically improves with activity and worsens at night with rest. It is frequently accompanied by periodic limb movements of sleep (PLMS), which typically occur during non-REM sleep. Approximately 12–13% of post-stroke patients have RLS, with two-thirds having bilateral symptoms [19]. One study showed a correlation between stroke severity and severity of PLMS [23]; as increased sympathetic activity has been implicated in PLMS, it is thought to be related to cardiovascular complications. In a small cohort of stroke patients, those with RLS had greater neck circumference, worse sleep quality, and higher rates of diabetes compared to those without RLS [24].

Woo et al. found that RLS developed sooner than PLMS in stroke patients, but attributed this to the uncomfortable symptoms of RLS that could prompt an earlier discovery of these symptoms [25]. They found that RLS and PLMS were caused by either lesions in the pons or the corona radiata had either unilateral or bilateral RLS while those with bilateral RLS had larger lesions involving the corona radiata and the basal ganglia. This suggests that the basal ganglia are responsible for bilaterality of symptoms.

REM Sleep Behavior Disorder

During REM sleep behavior disorder (RBD), patients act out their dreams. It is thought to affect approximately 11% of stroke patients [19]. RBD is thought to be linked to lesions in the brainstem [26], potentially caused by degeneration of the glutamatergic neurons in the pons and GABAergic neurons in the medulla. In the Kailuan Study cohort, Ma et al. found that those with probable RBD had higher risk of developing both ischemic and hemorrhagic strokes [27].

Circadian Variation

Previous studies have demonstrated stroke occurrence to exhibit a circadian pattern, with stroke onset peaking during morning hours, independent of stroke subtype or vascular risk factors [28]. As such, sleep's effects on circadian rhythm have been implicated in stroke risk. In non-REM sleep, sympathetic activity decreases and parasympathetic activity increases, leading to increased heart rate, blood pressure, cardiac output, peripheral vascular resistance, and respiratory rate. Conversely, in REM sleep, periodic oscillations in sympathetic and parasympathetic activity occur, resulting in overall reduced parasympathetic and decreased sympathetic tone. This results in an overall surge of sympathetic activity during the morning hours of awakening, which has been postulated to cause the displacement of vulnerable plaques resulting in stroke. Sheppard et al. found that increasing levels of morning blood pressure surge when measured as a continuous variable and not with a predefined threshold could be associated with increased stroke risk [29]. Interestingly, alteration in sleep schedule and resulting disruption of circadian rhythm in night shift workers has been found to be an independent risk factor for stroke and heart disease [30].

Diagnosis of Sleep-Disordered Breathing in Stroke Patients

The American Academy of Sleep Medicine task force group recommends screening for SDB in those with signs and symptoms suggesting an increased risk of moderate or severe OSA [31]. In the latest edition of their acute stroke treatment guidelines, the American Heart Association/American Stroke Association acknowledged the relationship between OSA and stroke, recognizing the increased morbidity and mortality carried by an OSA diagnosis; however, they listed routine screening for OSA in stroke patients as a level III recommendation, suggesting no benefit for screening, as some randomized controlled trials did not show benefit with treatment of moderate to severe OSA in preventing cardiovascular events [32].

The lack of demonstrated benefit of treatment in OSA patients, at least as far as those with stroke are concerned, could be due to inherent differences in these patients. Sacchetti et al. postulate that this could be a result of stroke patients being affected by a combination of OSA and CSA. They propose categorizing stroke patients with SDB into four different groups [18]. The first group consists of those who terminate their apneic spells when they arouse from sleep, who have pure OSA, with upper airway obstruction either preceding or as a result of their stroke. A second group consists of those who alternate between OSA and CSA without arousal from sleep. These patients exhibit decreased oxygen and increased carbon dioxide levels as a result of their OSA, leading to compensatory hyperventilation, which then leads to decreased carbon dioxide levels causing a high loop gain environment that promotes inhibition of central respiratory drive, or CSA. A third group consists of those who suffered impairment of their cyclic alternating patterns of sleep caused directly by the stroke. These patients can exhibit OSA, impaired chemosensation, increased arousability, and a reduction in levels of REM sleep. The last group consists of those whose stroke has caused damage to central structures resulting in pure CSA.

Even though evidence of secondary benefit for treatment of OSA in stroke patients is lacking as noted above, it is clear that treatment of OSA can have other benefits, including improved restfulness and mood. Unfortunately, a survey given to a group of 112 physicians caring for stroke patients found that most do not routinely screen their patients for SDB [33]. The study found that among those who screened for SDB, the most common screening tool was the Epworth Sleepiness Scale (ESS), followed by the Body Sensation Questionnaire (BSQ) and the STOP-BANG. Though STOP-Bang has not been validated in stroke patients, it has been shown in one cohort to have a high sensitivity for OSA [34], which may make it an ideal screening tool. Another study found that the 4-Variable screening tool and the STOP-BAG (STOP-BANG without the neck circumference measurement) were most effective in detecting OSA among stroke patients, although only moderately predictive [35].

Wake-Up Strokes

Stroke is a time-sensitive disease, with every passing minute resulting in the loss of 1.9 million neurons [36]. Acute stroke treatments include intravenous tissue-plasminogen activator (IV t-PA) and endovascular thrombectomy or clot retrieval. Treatment with IV t-PA is limited to a maximum of 4.5 hours from the onset of symptoms, depending on the stroke patient.

Patients with wake-up strokes (WUS), whose symptoms are observed at the time of awakening from sleep, comprise approximately 25% of all strokes [37, 38] and are associated with worse short-term functional outcomes. In the SLEEP TIGHT study, an observational study of minor ischemic stroke, those with WUS are more likely to have severe OSA, more frequent oxygen desaturation events, and higher LDL cholesterol levels, with these findings more prominent among men [39]. As explained above, paradoxical emboli through a PFO are thought to be the culprit in WUS. One cross-sectional study found that those with WUS were much more likely than those with non-WUS to have both OSA and a PFO [40].

WUS patients are a source of frustration for stroke neurologists, as the exact time of stroke onset is not known; as a result, these patients present a unique challenge for treatment. The treating physician often must use the time at which the patients were last known to be normal (often the time that they went to sleep), which often excludes them from eligibility for IV t-PA. Yet, ongoing efforts to expand treatment options in acute stroke have raised the prospects that patients with WUS can be eligible for treatment. There is prospective evidence to suggest that administration of IV t-PA is safe in patients with WUS [41]. In addition, several ongoing studies are focusing on the WUS population, with some employing perfusion neuroimaging to help select patients for thrombolytic therapy [42]. For patients whose stroke is caused by a large vessel occlusion (LVO), recent research suggests that select patients may benefit from endovascular therapy up to 24 hours from stroke onset [43, 44]. However, it is important to note that even though patients with LVO comprise the majority of those afflicted with disability, dependence, and death, they only represent approximately one-third of stroke patients [45].

Stroke Recovery

It is increasingly clear that sleep is integral in its reorganizing and restorative effects on the brain and regulating synaptic plasticity. It is postulated that sleep promotes a general homeostatic rescaling of synaptic function. During sleep, brain regions areas that underwent learning or stimulation were shown to have increased slow-wave activity, while immobilization led to a decrease in slow-wave activity during sleep [46].

One study found stroke patients to have a 0.5% lower baseline and mean oxygen saturation compared to age- and sex-matched controls [47]. Despite this small difference, it is postulated that lower oxygen levels could lead to decreased activity during hours of wakefulness, thereby interfering with rehabilitation regimens.

A systematic review and meta-analysis of studies involving 410 post-stroke sleep polysomnogram characteristics performed ranging from the time of admission to several months after stroke showed that stroke patients have shorter total sleep time, and more time spent in stage 1 sleep, with less time spent in stage 2 sleep [48]. They also found stroke patients to have lower percentage of slow-wave sleep compared to hospitalized controls, with no differences in REM sleep.

Other studies have noted decreased levels of REM sleep in stroke survivors [49], leading to sleep disturbances that in turn impair stroke rehabilitation potential. Impaired sleep architecture in stroke patients leads to non-dipping arterial blood pressure, impaired nocturnal heart rate variability, and baroreflex sensitivity in OSA, which all adversely affect stroke outcomes [50]. Cavalcanti et al. assessed the sleep quality of stroke patients with actigraphy, finding that stroke patients had increased sleep duration, latency, and fragmentation, which also results in decreased activity during waking hours. They found that sleep efficiency had the highest correlation with quality of life. As a result, the group also recommended an emphasis on increased exposure to natural light to aid with circadian synchronization, restriction of daytime sleep to improve nighttime sleep quality.

Modification of sleep stages may improve outcomes in stroke patients. It should be noted that over-the-counter sleep medications, not prescription sleep medications, have been implicated as an independent risk factor for stroke [51]. Acupuncture has been shown to be a promising alternative to medications for the treatment of insomnia, although further study is required [52].

Sleep is often neglected as far as rehabilitation is concerned. However, addressing sleep quality is promising for both short-term and long-term benefits in stroke survivors. It is established that deep sleep is known to promote neural plasticity, learning, and memory consolidation [48]. Furthermore, post-stroke patients are more susceptible to behavioral changes, especially depression [53]. As previously discussed, stroke patients can have sleep disorders as a secondary symptom of their stroke or as a primary sleep disorder. Baglioni et al. proposed a conceptual model describing the relationship between stroke and sleep on stroke outcomes [48]. In their model, stroke causes psychological and physiological changes that adversely affect sleep. In turn, poor sleep affects mental and physiological health, daytime function, and memory, which all contribute to stroke outcome.

It appears as though attention to improving sleep quality would have a potentially high yield toward patient outcomes, even in the absence of firm evidence. Regardless, further studies are needed to review the effects of improving sleep on outcomes of stroke survivors.

References

1. Kochanek KD, Murphy S, Xu J, Arias E. Mortality in the United States, 2016. NCHS Data Brief. 2017;(293):1–8. PMID:29319473.
2. Ge B, Guo X. Short and long sleep durations are both associated with increased risk of stroke: a meta-analysis of observational studies. Int J Stroke. 2015;10(2):177–84.
3. Ruiter Petrov ME, Letter AJ, Howard VJ, Kleindorfer D. Self-reported sleep duration in relation to incident stroke symptoms: nuances by body mass and race from the REGARDS study. J Stroke Cerebrovasc Dis. 2014;23(2):e123–32.
4. Phua CS, Jayaram L, Wijeratne T. Relationship between sleep duration and risk factors for stroke. Front Neurol [Internet]. 2017 [cited 2018 Jan 9];8. Available from: http://journal.frontiersin.org/article/10.3389/fneur.2017.00392/full
5. Bliwise DL, Rye DB, Dihenia B, Gurecki P. Greater daytime sleepiness in subcortical stroke relative to Parkinson's disease and Alzheimer's disease. J Geriatr Psychiatry Neurol [Internet]. 2016 [cited 2018 Apr 18]. Available from: http://journals.sagepub.com/doi/pdf/10.1177/089198870201500202
6. Ramos AR, Gangwisch JE. Is sleep duration a risk factor for stroke? Neurology. 2015;84(11):1066–7. PMID: 25716359.
7. Pace M, Baracchi F, Gao B, Bassetti C. Identification of sleep-modulated pathways involved in neuroprotection from stroke. Sleep. 2015;38(11):1707–18.
8. Pace M, Adamantidis A, Facchin L, Bassetti C. Role of REM sleep, melanin concentrating hormone and orexin/hypocretin systems in the sleep deprivation pre-ischemia. PLoS One. 2017;12(1):e0168430.
9. Kim TJ, Kim CK, Kim Y, Jung S, Jeong H-G, An SJ, et al. Prolonged sleep increases the risk of intracerebral haemorrhage: a nationwide case−control study. Eur J Neurol. 2016;23(6):1036–43.
10. Song Q, Liu X, Zhou W, Wang L, Zheng X, Wang X, et al. Long sleep duration and risk of ischemic stroke and hemorrhagic stroke: the Kailuan prospective study. Sci Rep [Internet]. 2016 [cited 2018 Apr 23];6. Available from: https://www.ncbi.nlm.nih.gov/pmc/articles/PMC5025713/
11. Patyar S, Patyar RR. Correlation between sleep duration and risk of stroke. J Stroke Cerebrovasc Dis. 2015;24(5):905–11.
12. Akinseye OA, Ojike NI, Akinseye LI, Dhandapany PS, Pandi-Perumal SR. Association of sleep duration with stroke in diabetic patients: analysis of the national health interview survey. J Stroke Cerebrovasc Dis. 2016;25(3):650–5.
13. Chokesuwattanaskul R, Thongprayoon C, Sharma K, Congrete S, Tanawuttiwat T, Cheungpasitporn W. Associations of sleep quality with incident atrial fibrillation: a meta-analysis. Intern Med J. 2018;48(8):964–72.
14. Yaggi HK, Concato J, Kernan WN, Lichtman JH, Brass LM, Mohsenin V. Obstructive sleep apnea as a risk factor for stroke and death. N Engl J Med. 2005;353(19):2034–41.
15. Muñoz R, Durán-Cantolla J, Martinez-Vila E. Obstructive sleep apnea–hypopnea and incident stroke: the Sleep Heart Health study. Am J Respir Crit Care Med. 2010;182(10):1332.

16. Guchlerner M, Kardos P, Liss-Koch E, Franke J, Wunderlich N, Bertog S, et al. PFO and right-to-left shunting in patients with obstructive sleep apnea. J Clin Sleep Med. 2012;8(4):375–80.
17. Winklewski PJ, Frydrychowski AF. Cerebral blood flow, sympathetic nerve activity and stroke risk in obstructive sleep apnoea. Is there a direct link? Blood Press. 2013;22(1):27–33.
18. Sacchetti ML, Della Marca G. Are stroke cases affected by sleep disordered breathings all the same? Med Hypotheses. 2014;83(2):217–23.
19. Hermann DM, Bassetti CL. Role of sleep-disordered breathing and sleep-wake disturbances for stroke and stroke recovery. Neurology. 2016;87(13):1407–16.
20. Aaronson JA, van Bennekom CAM, Hofman WF, van Bezeij T, van den Aardweg JG, Groet E, et al. Obstructive sleep apnea is related to impaired cognitive and functional status after stroke. Sleep. 2015;38(9):1431–7.
21. Klobučníková K, Šiarnik P, Čarnická Z, Kollár B, Turčáni P. Causes of excessive daytime sleepiness in patients with acute stroke – a polysomnographic study. J Stroke Cerebrovasc Dis. 2016;25(1):83–6.
22. Boden-Albala B, Roberts ET, Bazil C, Moon Y, Elkind MSV, Rundek T, et al. Daytime sleepiness and risk of stroke and vascular disease: findings from the Northern Manhattan study (NOMAS). Circ Cardiovasc Qual Outcomes. 2012;5(4):500–7.
23. Benbir G, Karadeniz D. Influence of periodic leg movements in sleep on stroke outcome. Sleep Biol Rhythms. 2013;11(3):194–9.
24. Medeiros C a M, de Bruin PFC, Paiva TR, Coutinho WM, Ponte RP, de Bruin VMS. Clinical outcome after acute ischaemic stroke: the influence of restless legs syndrome. Eur J Neurol. 2011;18(1):144–9.
25. Woo HG, Lee D, Hwang KJ, Ahn T-B. Post-stroke restless leg syndrome and periodic limb movements in sleep. Acta Neurol Scand. 2017;135(2):204–10.
26. Tang WK, Hermann DM, Chen YK, Liang HJ, Liu XX, Chu WCW, et al. Brainstem infarcts predict REM sleep behavior disorder in acute ischemic stroke. BMC Neurol. 2014;14:88.
27. Ma C, Pavlova M, Liu Y, Liu Y, Huangfu C, Wu S, et al. Probable REM sleep behavior disorder and risk of stroke: a prospective study. Neurology. 2017;88(19):1849–55.
28. Casetta I, Granieri E, Fallica E, Cecilia O l, Paolino E, Manfredini R. Patient demographic and clinical features and circadian variation in onset of ischemic stroke. Arch Neurol. 2002;59(1):48–53.
29. Sheppard JP, Hodgkinson J, Riley R, Martin U, Bayliss S, McManus RJ. Prognostic significance of the morning blood pressure surge in clinical practice: a systematic review. Am J Hypertens. 2015;28(1):30–41.
30. Brown DL, Feskanich D, Sánchez BN, Rexrode KM, Schernhammer ES, Lisabeth LD. Rotating night shift work and the risk of ischemic stroke. Am J Epidemiol. 2009;169(11):1370–7.
31. Kapur VK, Auckley DH, Chowdhuri S, Kuhlmann DC, Mehra R, Ramar K, et al. Clinical practice guideline for diagnostic testing for adult obstructive sleep apnea: an American Academy of sleep medicine clinical practice guideline. J Clin Sleep Med. 2017;13(3):479–504.
32. Powers WJ, Rabinstein AA, Ackerson T, Adeoye OM, Bambakidis NC, Becker K, et al. 2018 guidelines for the early management of patients with acute ischemic stroke: a guideline for healthcare professionals from the American Heart Association/American Stroke Association. Stroke. 2018;49(3):e46–99;STR.0000000000000158
33. Navalkele DD, Barlinn K, Minagar A, Chernyshev OY. Exploration of screening practices for obstructive sleep apnea in stroke medical community: a pilot study. Pathophysiology. 2016;23(2):105–9.
34. Silva GE, Vana KD, Goodwin JL, Sherrill DL, Quan SF. Identification of patients with sleep disordered breathing: comparing the four-variable screening tool, STOP, STOP-Bang, and epworth sleepiness scales. J Clin Sleep Med. 2011;7(5):467–72.
35. Boulos MI, Wan A, Im J, Elias S, Frankul F, Atalla M, et al. Identifying obstructive sleep apnea after stroke/TIA: evaluating four simple screening tools. Sleep Med. 2016;21:133–9.
36. Saver JL. Time is brain – quantified. Stroke. 2006;37(1):263–6.
37. Kim BJ, Roh JK. Ischemic stroke during sleep: its association with worse early functional outcome. Stroke. 2011;42:1901–6.

38. Koton S, Tanne D, Bornstein NM, Investigators N. Ischemic stroke on awakening: patients' characteristics, outcomes and potential for reperfusion therapy. Neuroepidemiology. 2012;39(3–4):149–53.
39. Koo BB, Bravata DM, Tobias LA, Mackey JS, Miech EJ, Matthias MS, et al. Observational study of obstructive sleep apnea in wake-up stroke: the SLEEP TIGHT study. Cerebrovasc Dis. 2016;41(5–6):233–41.
40. Ciccone A, Proserpio P, Roccatagliata DV, Nichelatti M, Gigli GL, Parati G, et al. Wake-up stroke and TIA due to paradoxical embolism during long obstructive sleep apnoeas: a cross-sectional study. Thorax. 2013;68(1):97–104.
41. Barreto AD, Fanale CV, Alexandrov AV, Gaffney KC, Vahidy FS, Nguyen CB, et al. Prospective, open-label safety study of intravenous recombinant tissue plasminogen activator in wake-up stroke. Ann Neurol. 2016;80(2):211–8.
42. Stern GM, Van Hise N, Urben LM, Korobey MJ, Pitlick JM, Crannage AJ. Thrombolytic therapy in wake-up stroke patients. Clin Neuropharmacol. 2017;40(3):140.
43. Nogueira RG, Jadhav AP, Haussen DC, Bonafe A, Budzik RF, Bhuva P, et al. Thrombectomy 6 to 24 hours after stroke with a mismatch between deficit and infarct. N Engl J Med. 2018;378(1):11–21.
44. Albers GW, Marks MP, Kemp S, Christensen S, Tsai JP, Ortega-Gutierrez S, et al. Thrombectomy for stroke at 6 to 16 hours with selection by perfusion imaging. N Engl J Med. 2018;378(8):708–18.
45. Malhotra K, Gornbein J, Saver JL. Ischemic strokes due to large-vessel occlusions contribute disproportionately to stroke-related dependence and death: a review. Front Neurol [Internet]. 2017 [cited 2018 Apr 23];8. Available from: https://www.ncbi.nlm.nih.gov/pmc/articles/PMC5715197/
46. Sarasso S, Santhanam P, Määtta S, Poryiazova R, Ferrarelli F, Tononi G, et al. Non-fluent aphasia and neural reorganization after speech therapy: insights from human sleep electrophysiology and functional magnetic resonance imaging. Arch Ital Biol. 2010;148(3):271.
47. Roffe C, Frohnhofen H, Sills S, Hodsoll J, Allen MB, Jones PW. Frequency of nocturnal hypoxia in clinically stable patients during stroke rehabilitation. Clin Rehabil. 2010;24(3):267–75.
48. Baglioni C, Nissen C, Schweinoch A, Riemann D, Spiegelhalder K, Berger M, et al. Polysomnographic characteristics of sleep in stroke: a systematic review and meta-analysis. Romigi A, editor. PLoS ONE. 2016;11(3):e0148496.
49. Bassetti CL, Hermann DM. Sleep and stroke. Sleep Disord. 2011;99:1052–69.
50. Palamarchuk I, Kimpinski K, Lippert C, Hachinski V. Nocturnal deterioration after ischemic stroke and autonomic dysfunction: hypothesis and implications. Cerebrovasc Dis. 2013;36(5–6):454–61.
51. Petrov ME, Howard VJ, Kleindorfer D, Grandner MA, Molano JR, Howard G. Over-the-counter and prescription sleep medication and incident stroke: the reasons for geographic and racial differences in stroke study. J Stroke Cerebrovasc Dis. 2014;23(8):2110–6.
52. Lee S-H, Lim SM. Acupuncture for insomnia after stroke: a systematic review and meta-analysis. BMC Complement Altern Med [Internet]. 2016 [cited 2018 Feb 11];16(1). Available from: http://bmccomplementalternmed.biomedcentral.com/articles/10.1186/s12906-016-1220-z
53. Allan LM, Rowan EN, Thomas AJ, Polvikoski TM, O'Brien JT, et al. Long-term incidence of depression and predictors of depressive symptoms in older stroke survivors. Br J Psychiatry. 2013;203(6):453–60.

15 Sleep in Amyotrophic Lateral Sclerosis

Swapan Dholakia and Saiprakash B. Venkateshiah

Introduction

Amyotrophic lateral sclerosis (ALS) is a progressive neurodegenerative disorder affecting the motor neurons. Upper motor neurons (UMN) are located within the brain and lower motor neurons (LMN) within the brainstem and spinal cord. Clinical findings of UMN involvement are spasticity, hyperreflexia, and pathologic reflexes, whereas those of LMN involvement include weakness, atrophy, and fasciculation (Table 15.1). A combination of UMN and LMN findings raises the suspicion for ALS. In most cases, ALS is a relentlessly progressive disease, culminating in death from respiratory failure. However, a population-based study found that 11.8% patients survived more than 10 years and younger age at onset, male gender, and spinal onset were associated with longer survival [1].

According to the National ALS Registry, the prevalence of ALS in the United States is 5 per 100,000, which is more common in whites, males, and those in the age range of 60–69 years [2]. Both genetic and environmental factors play a role in this disease. About 5–10% of cases are familial, and the remainder are sporadic [3]. Mutations in the gene encoding for the Cu/Zn superoxide dismutase 1 (SOD1) metalloenzyme have been reported in cases of familial ALS [4]. Advances in genetics have identified several other gene mutations and are updated in an ALS online database [5]. The demonstration of ubiquitin-positive neuronal inclusions containing the protein TDP-43 in both ALS and frontotemporal dementia shows that these

S. Dholakia
Department of Neurology, Atlanta VA Medical Center, Emory University School of Medicine, Decatur, GA, USA
e-mail: swapan.a.dholakia@emory.edu

S. B. Venkateshiah (✉)
Division of Pulmonary, Allergy, Critical Care and Sleep Medicine, Emory University School of Medicine, Atlanta VA Medical Center, Decatur, GA, USA

L. M. DelRosso, R. Ferri (eds.), *Sleep Neurology*,
https://doi.org/10.1007/978-3-030-54359-4_15

Table 15.1 Clinical features of ALS

Upper motor neuron features	Lower motor neuron features
Spasticity	Weakness
Hyperreflexia	Atrophy
Pathologic reflexes	Fasciculations

Table 15.2 Causes of sleep disturbance in ALS

Sleep disorder in ALS	Pathophysiology
Sleep-disordered breathing	Respiratory weakness and hypoventilation
Insomnia	Muscle cramps, impaired mobility, excessive secretions, dysphagia, and fasciculations
Restless legs syndrome	Depression and anxiety
Hypersomnia	

neurodegenerative disorders are linked [6]. 50% of patients with ALS have cognitive impairment, while 15% meet the criteria for frontotemporal dementia [7].

The El Escorial criteria were developed in 1994 for the diagnosis of ALS. To improve its sensitivity with the use of electrodiagnostic testing, these criteria were revised in 2000 [8]. The revised El Escorial criteria have four categories. Clinical evidence of UMN and LMN dysfunction is sought in four body regions (bulbar/brainstem, cervical, thoracic, and lumbosacral spinal segments). Clinically definite ALS is defined based on clinical findings of UMN and LMN in the bulbar region plus two spinal regions, or presence of UMN and LMN signs in three spinal regions. Clinically probable ALS is defined by clinical evidence of UMN and LMN signs in two regions with some UMN signs rostral to LMN signs. Clinically probable laboratory-supported ALS is defined by UMN and LMN signs in one region or UMN signs in one region and LMN signs on EMG testing in two regions. When these criteria are not met, clinically possible ALS applies to UMN and LMN signs present in one region, UMN signs alone in two regions or more, or LMN signs present rostral to UMN signs. Other causes should be ruled out by appropriate neuroimaging and laboratory studies.

This chapter will discuss the sleep disorders found in patients with ALS, their pathophysiology (Table 15.2), diagnosis, and treatment options.

Restless Legs Syndrome and Periodic Limb Movements in Sleep

Restless legs syndrome (RLS) and periodic limb movements during sleep (PLMS) have been reported to occur more frequently in patients with ALS than in the general population. A study on 69 ALS patients in a French cohort found a higher frequency of RLS in patients with ALS (20.8%) who were older than 64 years than in the general French population (8.6%) of the same age group [9]. Another study comparing 76 ALS patients (mean age 58.7 ± 12.8 years) with 100 control subjects found a higher frequency of RLS in patients with ALS (25% vs. 8%) and that the duration of RLS symptoms was shorter but the frequency of symptoms was higher in ALS patients. In addition, patients with ALS and RLS had more functional impairment and sleep complaints than ALS patients without RLS [10]. A polysomnography study on 41 patients with ALS showed a high PLMS index of more than 15/hour in 53.6% of patients [11].

Parasomnias

REM sleep without atonia has been observed in polysomnographic studies in patients with ALS and correlates with worsening motor function. However, it is not clear if this represents a risk factor for REM sleep behavior disorder or is an isolated finding [12]. REM sleep behavior disorder has been described in patients with familial ALS [13].

Insomnia

Several disease-specific factors may play a role in impairing sleep onset or sleep continuity. These factors include excessive secretions, dysphagia, impaired mobility, spasticity, muscle cramps, and difficulty with communication. Addressing these and providing symptomatic treatment is an important aspect in the overall management of patients [14]. Patients with ALS have a significantly poor quality of sleep which correlates with daytime sleepiness as well as severity of ALS [15]. Behavioral changes, reduced motivation, depression, and apathy are seen frequently in patients with ALS [16] and may play a role in the development of insomnia. In addition to neuropsychiatric symptoms, sleep disturbances are seen in about 30% of patients with dementia (including frontotemporal dementia) [17]. Thus, ALS patients with frontotemporal dementia and cognitive impairment may exhibit sleep disturbances as a part of the neurocognitive disorder. Polysomnography has revealed increase in awake time and reduced deep and REM sleep in patients with ALS (Table 15.3) [18]. Fragmentation of slow-wave sleep has been described with the development of complete locked-in syndrome in an ALS patient [19].

Table 15.3 Polysomnography findings in ALS

PSG finding	Percent patients affected
Reduced sleep efficiency	55%
Reduced deep sleep	
Reduced REM sleep	
Hypoventilation and hypercapnia	40%
Obstructive and central sleep apnea	45%

Circadian Rhythm Disorders

In addition to reduced sleep efficiency and total sleep time, possibility of a circadian phase delay has been noted (increased activity at night and reduced activity in the morning on actigraphy and sleep diary data) in patients with FTD [20]. Additionally, a reduction in REM sleep has been observed in FTD [21]. The morning plasma cortisol level is elevated in patients with ALS (particularly the spinal form of the disease), and chronic stress may play a role in dysfunction of the hypothalamic-pituitary-adrenal axis [22]. Loss of circadian rhythmicity in cortisol release, with high cortisol levels in the evening, lack of the expected rise in cortisol levels in response to stress [23], and blunting of the normal rise in cortisol level upon awakening [24] have been observed in ALS. Interestingly, a study on ALS mice showed that exposure to stress was associated with higher rate of progression of the disease and reduced survival [25].

Hypersomnia and Fatigue

A common symptom seen in patients with ALS is fatigue [26]. Patients with fatigue have nocturnal complaints such as muscle cramps and nocturia and report more difficulties in staying asleep. They are also more disabled [27]. Modafinil has been shown to have beneficial effects in terms of reducing symptoms of fatigue as well as daytime sleepiness in patients with ALS. A study on 15 ALS patients showed a significant improvement in daytime sleepiness as measured by Epworth Sleepiness Scale (mean score reduction from 8.2 to 4.5) as well as in fatigue as measured by the Fatigue Severity Scale (mean score reduction from 51.3 to 42.8) [28]. There is limited data on the presence of hypersomnias of central origin in ALS. The hypocretin (orexin) neurotransmission system does not seem to be involved in ALS [29].

Sleep-Related Breathing Disorder in ALS

Breathing is controlled by columns of respiratory neurons in the pons and the medulla. The dorsal respiratory group (mainly inspiratory neurons) and the ventral respiratory group (consisting of both inspiratory and expiratory neurons) are located in the medulla [30]. The pontine center is thought to interact with the medullary neurons and play a role in controlling transitions of respiratory phase as well as respiratory reflexes [31]. Neurons from these centers project to the spinal motor neurons which then innervate the inspiratory and expiratory muscles of respiration. During NREM sleep, the activation of GABAergic neurons in the VLPO nucleus of the hypothalamus, with its widespread projections to brainstem arousal systems, results in hypoventilation seen during sleep [32]. During REM sleep, the descending brainstem signals to the spinal motor neurons result in the characteristic muscle atonia [33], leading to a further reduction in ventilation. The diaphragm, which receives its innervation from the C3 to C5 cervical segments via the phrenic nerve, is spared from the motor inhibition in REM sleep. Due to these reasons, neuromuscular disorders such as ALS which result in denervation and weakness of the respiratory muscles manifest with hypoventilation first seen during sleep, particularly in REM sleep. Nocturnal hypoxia may be seen in the absence of significant respiratory weakness or diaphragm denervation and is explained by dysfunction of the central control of breathing [34] due to degeneration of the pre-Botzinger complex (inspiratory rhythm generator neurons in the ventral respiratory group in the medulla) [35].

A high prevalence of obstructive sleep apnea has been noted in patients with ALS (Table 15.3). A polysomnography study on 250 ALS patients found that 45.6% had an AHI >/= 5 and 40% had nocturnal hypoventilation as measured by transcutaneous capnometry [36]. Patients with bulbar symptoms have a higher than predicted prevalence of obstructive events, whereas those without evidence of bulbar signs have a higher prevalence of central events [37]. The presence of obstructive sleep apnea at the onset of ALS has been suggested to play a prognostic role. ALS patients with OSA (AHI >/= 5) had a significantly shorter mean survival as compared to patients without OSA [38]. However, the degree of respiratory muscle weakness relates to quality of life (QoL) and functional status, and not the AHI [39]. Sometimes, ALS may present with unexplained respiratory failure or sleep disturbance, in the absence of significant limb or bulbar symptoms [40].

Symptoms of respiratory muscle weakness and nocturnal hypoventilation are orthopnea, frequent arousals, morning headaches, daytime sleepiness, and cognitive problems. Pulmonary function tests to evaluate for hypoventilation include forced vital capacity (FVC), maximum inspiratory pressure (MIP), nocturnal oximetry, and sniff nasal pressure (SNP). Measuring FVC in the supine position was better at detecting weakness of the diaphragm as compared to measurement in the upright position [41]. MIP and nocturnal oximetry are more sensitive at detecting initial stages of respiratory weakness in comparison with FVC [42]. SNP measured early in the disease was a predictor of mortality or tracheostomy at 1-year follow-up [43]. There may not be a correlation between sleep-disordered breathing and daytime respiratory testing, making the case for nocturnal assessment (with a sleep study or oximetry) of respiratory function as a part of the overall evaluation of patients with ALS [44]. Another study showed that the presence of sleep-disordered breathing correlated with daytime symptoms such as excessive sleepiness, but not with forced vital capacity, again emphasizing the importance of polysomnography when there is a suspicion of a sleep breathing disorder [45]. Nocturnal hypoxia in ALS is associated with cognitive dysfunction on neuropsychological testing [46].

Non-invasive ventilation (NIV) is indicated when PaCO2 > 45 mmHg, MIP <60 cm H2O, FVC < 50% predicted, or SaO2 ≤ 88% for at least 5 continuous minutes (Table 15.4) [47]. The American Academy of Neurology (AAN) practice parameter update also recommends NIV when SNP <40 cm H2O or there is presence of orthopnea [48]. NIV has been shown to improve survival in patients without severe bulbar dysfunction, with a median survival benefit of 205 days [49]. It also positively improves objective sleep architecture (increases slow-wave sleep and

Table 15.4 Indications for initiation of non-invasive ventilation in ALS

Symptom and physical findings	Orthopnea
Oximetry	$SaO2 \leq 88\%$ for at least 5 continuous minutes on nocturnal oximetry
Pulmonary function tests	FVC < 50% of predicted MIP < 60 cm H2O SNP < 40 cm H2O
Blood gases	PaCO2 > 45 mmHg

rapid eye movement sleep) and improves oxygen saturation and nocturnal ventilation (reduction in carbon dioxide levels) [50, 51]. In addition to treating hypoventilation, treatment of obstructive events is also important to confer maximum survival benefit [52].

Early use of NIV is recommended as it improves compliance and results in slower decline in respiratory function [53]. Lower compliance to NIV has been observed in patients with frontotemporal dementia and those with bulbar involvement [48]. Patients with bulbar weakness may be prone to higher mouth leaks and this may affect tolerance to NIV. Early use of non-invasive ventilation is also supported by animal studies. A study on ALS mice showed that intermittent hypoxia led to a more rapid deterioration of motor neurons leading to worsening weakness. It was also associated with a decline in cognitive function [54]. The ability to remotely monitor NIV and make corrective changes reduces the need for office and emergency room visits as well as hospital admissions [55]. While NIV improves QoL of patients with ALS, it does not have a negative impact on the caregivers' QoL or burden [56].

Devices used for NIV include bilevel positive airway pressure therapy (with or without backup rate), portable ventilators, and more advanced volume-targeted devices such as average volume-assured pressure support (AVAPS). A potential advantage of AVAPS in individuals with ALS is that as respiratory function declines and hypoventilation worsens over time, the device can automatically adjust the tidal volume as it is able to increase the pressure support to meet the desired tidal volume [57]. With progression of the disease, inability to clear upper airway secretions may interfere with adequate ventilation. Use of cough assist and suction devices should be considered in such patients.

Both adherence to therapy and the efficacy of NIV in terms of reducing nocturnal hypoxia play a key role in prognosis [58]. Respiratory weakness predisposes to pulmonary infections, with pneumonia being the most common cause of death in ALS [59]. In patients without nocturnal hypoxia, reduced variability in heart rate (possibly due to autonomic dysfunction) is associated with sudden death [60].

Treatment of ALS

There is no effective treatment for ALS and median survival ranges from 20 to 48 months [61]. Treatment is largely symptomatic, and an approach that involves multidisciplinary clinic referral is recommended [48]. Factors associated with

Table 15.5 Treatment of symptoms interfering with sleep in ALS

Spasticity	Stretching exercises, baclofen, tizanidine
Sialorrhea	Amitriptyline, atropine drops, botulinum toxin injection, salivary gland irradiation
Immobility	Use of air mattress or hospital bed
Dyspnea	Non-invasive ventilation
Dysarthria	Communication via writing, electronic devices
Dysphagia	Speech pathology evaluation, enteral nutrition
Depression and anxiety	Antidepressants

reduced survival in ALS include older age, bulbar onset of symptoms, presence of frontotemporal syndrome [62], as well as the rate of progression of symptoms. Riluzole is an antiglutamate agent that has shown to prolong median survival by 2–3 months. It is dosed at 100 mg per day and can lead to mild elevations in aminotransferase levels [63, 64]. Even though symptomatic treatment is the mainstay to improve and maintain QoL in patients with ALS, this is highly underused, especially with symptoms such as fatigue [65]. A survey of neurologists treating ALS patients found little consensus on the treatment of individual symptoms [66]. There is paucity of data in terms of clinical trials for symptomatic treatments in ALS. A number of symptoms can interfere with sleep in patients with ALS and must be addressed and treated (Table 15.5). Amitriptyline, atropine eye drops, and glycopyrrolate are used to treat sialorrhea. If these are not effective, based on the AAN practice guidelines, botulinum toxin or salivary gland irradiation can be used. Baclofen and tizanidine may be used for spasticity. Stretching exercises may help. In individuals with limited mobility, an alternating pressure air mattress or a hospital bed may be used to improve comfort [67].

Invasive Ventilation

Invasive ventilation should be offered when long-term survival is the goal. In view of the incurable and progressive nature of the illness, invasive ventilation via a tracheostomy poses ethically difficult concerns. Invasive ventilation impairs communication and the patient's autonomy. Use of the method implies increasing dependence on caring personnel and/or the next of kin. The patient and family should be fully informed of the risks and benefits of this treatment, and they should be provided the option to stop the ventilation at any time. Patients with ALS who choose invasive ventilation are fewer than 10%. Unfortunately, in most patients, the demands of invasive ventilation are difficult or impossible to manage at home, and nursing home placement results. In individuals who choose to stop invasive ventilation, symptoms of dyspnea, anxiety, and distress should be palliated.

References

1. Al-Chalabi A, Hardiman O. The epidemiology of ALS: a conspiracy of genes, environment and time. Nat Rev Neurol. 2013;9(11):617–28.
2. Mehta P, Kaye W, Bryan L, Larson T, Copeland T, Wu J, et al. Prevalence of amyotrophic lateral sclerosis – United States, 2012–2013. Morb Mortal Wkly Rep Surveill Summ (Washington, DC: 2002). 2016;65(8):1–12.
3. Rowland LP, Shneider NA. Amyotrophic lateral sclerosis. N Engl J Med. 2001;344(22):1688–700.
4. Rosen DR, Siddique T, Patterson D, Figlewicz DA, Sapp P, Hentati A, et al. Mutations in Cu/Zn superoxide dismutase gene are associated with familial amyotrophic lateral sclerosis. Nature. 1993;362(6415):59–62.
5. Abel O, Powell JF, Andersen PM, Al-Chalabi A. ALSoD: a user-friendly online bioinformatics tool for amyotrophic lateral sclerosis genetics. Hum Mutat. 2012;33(9):1345–51.
6. Neumann M, Sampathu DM, Kwong LK, Truax AC, Micsenyi MC, Chou TT, et al. Ubiquitinated TDP-43 in frontotemporal lobar degeneration and amyotrophic lateral sclerosis. Science (New York, NY). 2006;314(5796):130–3.
7. Ringholz GM, Appel SH, Bradshaw M, Cooke NA, Mosnik DM, Schulz PE. Prevalence and patterns of cognitive impairment in sporadic ALS. Neurology. 2005;65(4):586–90.
8. Brooks BR, Miller RG, Swash M, Munsat TL. El Escorial revisited: revised criteria for the diagnosis of amyotrophic lateral sclerosis. Amyotroph Lateral Scler Other Motor Neuron Disord. 2000;1(5):293–9.
9. Limousin N, Blasco H, Corcia P, Arnulf I, Praline J. The high frequency of restless legs syndrome in patients with amyotrophic lateral sclerosis. Amyotrophic Lateral Scler. 2011;12(4):303–6.
10. Lo Coco D, Piccoli F, La Bella V. Restless legs syndrome in patients with amyotrophic lateral sclerosis. Mov Disord. 2010;25(15):2658–61.
11. Lo Coco D, Puligheddu M, Mattaliano P, Congiu P, Borghero G, Fantini ML, et al. REM sleep behavior disorder and periodic leg movements during sleep in ALS. Acta Neurol Scand. 2017;135(2):219–24.
12. Puligheddu M, Congiu P, Arico D, Rundo F, Borghero G, Marrosu F, et al. Isolated rapid eye movement sleep without atonia in amyotrophic lateral sclerosis. Sleep Med. 2016;26:16–22.
13. Ebben MR, Shahbazi M, Lange DJ, Krieger AC. REM behavior disorder associated with familial amyotrophic lateral sclerosis. Amyotrophic Lateral Scler. 2012;13(5):473–4.
14. Jenkins TM, Hollinger H, McDermott CJ. The evidence for symptomatic treatments in amyotrophic lateral sclerosis. Curr Opin Neurol. 2014;27(5):524–31.
15. Lo Coco D, Mattaliano P, Spataro R, Mattaliano A, La Bella V. Sleep-wake disturbances in patients with amyotrophic lateral sclerosis. J Neurol Neurosurg Psychiatry. 2011;82(8):839–42.
16. Lillo P, Mioshi E, Zoing MC, Kiernan MC, Hodges JR. How common are behavioural changes in amyotrophic lateral sclerosis? Amyotrophic Lateral Scler. 2011;12(1):45–51.
17. Fernandez Martinez M, Castro Flores J, Perez de las Heras S, Mandaluniz Lekumberri A, Gordejuela Menocal M, Zarranz Imirizaldu JJ. Prevalence of neuropsychiatric symptoms in elderly patients with dementia in Mungialde County (Basque Country, Spain). Dement Geriatr Cogn Disord. 2008;25(2):103–8.
18. Likhachev SA, Rushkevich YN, Abelskaia IS, Chechik NM, Merkul OV. Polysomnography in patients with amyotrophic lateral sclerosis. Zh Nevrol Psikhiatr Im S S Korsakova. 2016;116(4):37–41.
19. Soekadar SR, Born J, Birbaumer N, Bensch M, Halder S, Murguialday AR, et al. Fragmentation of slow wave sleep after onset of complete locked-in state. J Clin Sleep Med. 2013;9(9):951–3.
20. Anderson KN, Hatfield C, Kipps C, Hastings M, Hodges JR. Disrupted sleep and circadian patterns in frontotemporal dementia. Eur J Neurol. 2009;16(3):317–23.
21. Autret A, Lucas B, Mondon K, Hommet C, Corcia P, Saudeau D, et al. Sleep and brain lesions: a critical review of the literature and additional new cases. Neurophysiol Clin = Clin Neurophysiol. 2001;31(6):356–75.

22. Spataro R, Volanti P, Vitale F, Meli F, Colletti T, Di Natale A, et al. Plasma cortisol level in amyotrophic lateral sclerosis. J Neurol Sci. 2015;358(1–2):282–6.
23. Patacchioli FR, Monnazzi P, Scontrini A, Tremante E, Caridi I, Brunetti E, et al. Adrenal dysregulation in amyotrophic lateral sclerosis. J Endocrinol Invest. 2003;26(12):Rc23–5.
24. Roozendaal B, Kim S, Wolf OT, Kim MS, Sung KK, Lee S. The cortisol awakening response in amyotrophic lateral sclerosis is blunted and correlates with clinical status and depressive mood. Psychoneuroendocrinology. 2012;37(1):20–6.
25. Fidler JA, Treleaven CM, Frakes A, Tamsett TJ, McCrate M, Cheng SH, et al. Disease progression in a mouse model of amyotrophic lateral sclerosis: the influence of chronic stress and corticosterone. FASEB J. 2011;25(12):4369–77.
26. Lou JS. Fatigue in amyotrophic lateral sclerosis. Phys Med Rehabil Clin N Am. 2008;19(3):533–43, ix.
27. Lo Coco D, La Bella V. Fatigue, sleep, and nocturnal complaints in patients with amyotrophic lateral sclerosis. Eur J Neurol. 2012;19(5):760–3.
28. Carter GT, Weiss MD, Lou JS, Jensen MP, Abresch RT, Martin TK, et al. Modafinil to treat fatigue in amyotrophic lateral sclerosis: an open label pilot study. Am J Hosp Palliat Care. 2005;22(1):55–9.
29. Van Rooij FG, Schelhaas HJ, Lammers GJ, Verbeek MM, Overeem S. CSF hypocretin-1 levels are normal in patients with amyotrophic lateral sclerosis. Amyotrophic Lateral Scler. 2009;10(5–6):487–9.
30. Duffin J. Functional organization of respiratory neurons: a brief review of current questions and speculations. Exp Physiol. 2004;89(5):517–29.
31. Ikeda K, Kawakami K, Onimaru H, Okada Y, Yokota S, Koshiya N, et al. The respiratory control mechanisms in the brainstem and spinal cord: integrative views of the neuroanatomy and neurophysiology. J Physiol Sci. 2017;67(1):45–62.
32. Joseph V, Pequignot JM, Van Reeth O. Neurochemical perspectives on the control of breathing during sleep. Respir Physiol Neurobiol. 2002;130(3):253–63.
33. Vetrivelan R, Chang C, Lu J. Muscle tone regulation during REM sleep: neural circuitry and clinical significance. Arch Ital Biol. 2011;149(4):348–66.
34. Atalaia A, De Carvalho M, Evangelista T, Pinto A. Sleep characteristics of amyotrophic lateral sclerosis in patients with preserved diaphragmatic function. Amyotrophic Lateral Scler. 2007;8(2):101–5.
35. Howell BN, Newman DS. Dysfunction of central control of breathing in amyotrophic lateral sclerosis. Muscle Nerve. 2017;56(2):197–201.
36. Boentert M, Glatz C, Helmle C, Okegwo A, Young P. Prevalence of sleep apnoea and capnographic detection of nocturnal hypoventilation in amyotrophic lateral sclerosis. J Neurol Neurosurg Psychiatry. 2018;89(4):418–424. https://doi.org/10.1136/jnnp-2017-316515.
37. Santos C, Braghiroli A, Mazzini L, Pratesi R, Oliveira LV, Mora G. Sleep-related breathing disorders in amyotrophic lateral sclerosis. Monaldi Arch Chest Dis = Arch Monaldi Mal Torace. 2003;59(2):160–5.
38. Quaranta VN, Carratu P, Damiani MF, Dragonieri S, Capozzolo A, Cassano A, et al. The prognostic role of obstructive sleep apnea at the onset of amyotrophic lateral sclerosis. Neurodegener Dis. 2017;17(1):14–21.
39. Bourke SC, Shaw PJ, Gibson GJ. Respiratory function vs sleep-disordered breathing as predictors of QOL in ALS. Neurology. 2001;57(11):2040–4.
40. Scelsa SN, Yakubov B, Salzman SH. Dyspnea-fasciculation syndrome: early respiratory failure in ALS with minimal motor signs. Amyotrophic Lateral Scler Other Motor Neuron Disord. 2002;3(4):239–43.
41. Lechtzin N, Wiener CM, Shade DM, Clawson L, Diette GB. Spirometry in the supine position improves the detection of diaphragmatic weakness in patients with amyotrophic lateral sclerosis. Chest. 2002;121(2):436–42.
42. Jackson CE, Rosenfeld J, Moore DH, Bryan WW, Barohn RJ, Wrench M, et al. A preliminary evaluation of a prospective study of pulmonary function studies and symptoms of hypoventilation in ALS/MND patients. J Neurol Sci. 2001;191(1–2):75–8.

43. Capozzo R, Quaranta VN, Pellegrini F, Fontana A, Copetti M, Carratu P, et al. Sniff nasal inspiratory pressure as a prognostic factor of tracheostomy or death in amyotrophic lateral sclerosis. J Neurol. 2015;262(3):593–603.
44. Tsara V, Serasli E, Steiropoulos P, Tsorova A, Antoniadou M, Zisi P. Respiratory function in amyotrophic lateral sclerosis patients. The role of sleep studies. Hippokratia. 2010;14(1):33–6.
45. David WS, Bundlie SR, Mahdavi Z. Polysomnographic studies in amyotrophic lateral sclerosis. J Neurol Sci. 1997;152(Suppl 1):S29–35.
46. Park SY, Kim SM, Sung JJ, Lee KM, Park KS, Kim SY, et al. Nocturnal hypoxia in ALS is related to cognitive dysfunction and can occur as clusters of desaturations. PLoS One. 2013;8(9):e75324.
47. Clinical indications for noninvasive positive pressure ventilation in chronic respiratory failure due to restrictive lung disease, COPD, and nocturnal hypoventilation—a consensus conference report. Chest. 1999;116(2):521–534. https://doi.org/10.1378/chest.116.2.521.
48. Miller RG, Jackson CE, Kasarskis EJ, England JD, Forshew D, Johnston W, et al. Practice parameter update: the care of the patient with amyotrophic lateral sclerosis: drug, nutritional, and respiratory therapies (an evidence-based review): report of the quality standards subcommittee of the American Academy of Neurology. Neurology. 2009;73(15):1218–26.
49. Bourke SC, Tomlinson M, Williams TL, Bullock RE, Shaw PJ, Gibson GJ. Effects of non-invasive ventilation on survival and quality of life in patients with amyotrophic lateral sclerosis: a randomised controlled trial. Lancet Neurol. 2006;5(2):140–7.
50. Boentert M, Brenscheidt I, Glatz C, Young P. Effects of non-invasive ventilation on objective sleep and nocturnal respiration in patients with amyotrophic lateral sclerosis. J Neurol. 2015;262(9):2073–82.
51. Vrijsen B, Buyse B, Belge C, Robberecht W, Van Damme P, Decramer M, et al. Noninvasive ventilation improves sleep in amyotrophic lateral sclerosis: a prospective polysomnographic study. J Clin Sleep Med. 2015;11(5):559–66.
52. Georges M, Attali V, Golmard JL, Morelot-Panzini C, Crevier-Buchman L, Collet JM, et al. Reduced survival in patients with ALS with upper airway obstructive events on non-invasive ventilation. J Neurol Neurosurg Psychiatry. 2016;87(10):1045–50.
53. Terzano C, Romani S. Early use of non invasive ventilation in patients with amyotrophic lateral sclerosis: what benefits? Eur Rev Med Pharmacol Sci. 2015;19(22):4304–13.
54. Kim SM, Kim H, Lee JS, Park KS, Jeon GS, Shon J, et al. Intermittent hypoxia can aggravate motor neuronal loss and cognitive dysfunction in ALS mice. PLoS One. 2013;8(11):e81808.
55. Pinto A, Almeida JP, Pinto S, Pereira J, Oliveira AG, de Carvalho M. Home telemonitoring of non-invasive ventilation decreases healthcare utilisation in a prospective controlled trial of patients with amyotrophic lateral sclerosis. J Neurol Neurosurg Psychiatry. 2010;81(11):1238–42.
56. Mustfa N, Walsh E, Bryant V, Lyall RA, Addington-Hall J, Goldstein LH, et al. The effect of noninvasive ventilation on ALS patients and their caregivers. Neurology. 2006;66(8):1211–7.
57. Berry RB, Chediak A, Brown LK, Finder J, Gozal D, Iber C, et al. Best clinical practices for the sleep center adjustment of noninvasive positive pressure ventilation (NPPV) in stable chronic alveolar hypoventilation syndromes. J Clin Sleep Med. 2010;6(5):491–509.
58. Gonzalez-Bermejo J, Morelot-Panzini C, Arnol N, Meininger V, Kraoua S, Salachas F, et al. Prognostic value of efficiently correcting nocturnal desaturations after one month of non-invasive ventilation in amyotrophic lateral sclerosis: a retrospective monocentre observational cohort study. Amyotrophic Lateral Scler Frontotemporal Degener. 2013;14(5–6):373–9.
59. Corcia P, Pradat PF, Salachas F, Bruneteau G, Forestier N, Seilhean D, et al. Causes of death in a post-mortem series of ALS patients. Amyotrophic Lateral Scler. 2008;9(1):59–62.
60. Pinto S, Pinto A, De Carvalho M. Decreased heart rate variability predicts death in amyotrophic lateral sclerosis. Muscle Nerve. 2012;46(3):341–5.
61. Chio A, Logroscino G, Hardiman O, Swingler R, Mitchell D, Beghi E, et al. Prognostic factors in ALS: a critical review. Amyotrophic Lateral Scler. 2009;10(5–6):310–23.

62. Govaarts R, Beeldman E, Kampelmacher MJ, van Tol MJ, van den Berg LH, van der Kooi AJ, et al. The frontotemporal syndrome of ALS is associated with poor survival. J Neurol. 2016;263(12):2476–83.
63. Bensimon G, Lacomblez L, Meininger V. A controlled trial of riluzole in amyotrophic lateral sclerosis. ALS/Riluzole study group. N Engl J Med. 1994;330(9):585–91.
64. Miller RG, Mitchell JD, Moore DH. Riluzole for amyotrophic lateral sclerosis (ALS)/motor neuron disease (MND). Cochrane Database Syst Rev. 2012;2012(3):CD001447. Published 2012 Mar 14. https://doi.org/10.1002/14651858.CD001447.pub3.
65. Nicholson K, Murphy A, McDonnell E, Shapiro J, Simpson E, Glass J, et al. Improving symptom management for people with amyotrophic lateral sclerosis. Muscle Nerve. 2018;57(1):20–4.
66. Forshew DA, Bromberg MB. A survey of clinicians' practice in the symptomatic treatment of ALS. Amyotrophic Lateral Scler Other Motor Neuron Disord. 2003;4(4):258–63.
67. Jackson CE, McVey AL, Rudnicki S, Dimachkie MM, Barohn RJ. Symptom management and end-of-life Care in amyotrophic lateral sclerosis. Neurol Clin. 2015;33(4):889–908.

Sleep After Traumatic Brain Injury

16

Patrizia Congiu, Monica Puligheddu, Michela Figorilli, Stefania Mondello, and Raffaele Ferri

Introduction

The most recent consensus definition of traumatic brain injury (TBI) is that proposed by Menon et al. [1] on behalf of the Demographic and Clinical Assessment Working Group of the International and Interagency Initiative toward Common Data Elements for Research on Traumatic Brain Injury and Psychological Health. They define TBI as "an alteration in brain function, or other evidence of brain pathology, caused by an external force." As well specified by Menon et al. [1] in their special communication about TBI, *alteration in brain function* is defined as one of the following clinical signs: (1) *any period of loss of or a decreased level of consciousness*; (2) *any loss of memory for events immediately before (retrograde amnesia) or after the injury* (*PTA*); and (3) *neurologic deficits (weakened, loss of balance, change in vision, dyspraxia paresis/plegia [paralysis], sensory loss, aphasia,* etc.). *Other evidence of brain pathology* may include vis*ual, neuroradiologic, or laboratory confirmation of damage to the brain. Caused by an external force may include any of the following events: the head being struck by an object, the head striking an object, the brain undergoing an acceleration/deceleration movement without direct external trauma to the head, a foreign body penetrating the brain,*

P. Congiu · M. Puligheddu · M. Figorilli
Neurophysiology Unit, Sleep Disorder Centre, Department of Medical Sciences and Public Health, University of Cagliari, Cagliari, Italy
e-mail: puligheddu@unica.it

S. Mondello
Department of Biomedical and Dental Sciences and Morphofunctional Imaging, University of Messina, Messina, Italy
e-mail: smondello@unime.it

R. Ferri (✉)
Sleep Research Centre, Oasi Research Institute – IRCCS, Troina, Italy
e-mail: rferri@oasi.en.it

L. M. DelRosso, R. Ferri (eds.), *Sleep Neurology*,
https://doi.org/10.1007/978-3-030-54359-4_16

forces generated from events such as a blast or explosion, and other force yet to be defined [1].

TBI is an important cause of death and disability at all ages around the world, and it is the leading cause of disability in young adults worldwide [2, 3].

Pathophysiology of TBI

It is important to distinguish between penetrating brain and closed injuries, because neurobiological and neurobehavioral consequences may be different [4]. In fact, in cases of penetrating brain lesions, the brain damage is related to the displacement or destruction of brain tissue consequent to the introduction into the skull of external objects or bone fragments or potential infectious material. The clinical consequences of a penetrating lesion are directly correlated with the location, the trajectory, and the size of the object involved, and these characteristics may largely predict them. Non-penetrating or closed injuries are explained by the complex interaction between the biomechanical forces causing the trauma and the material properties of the brain and its relationship with skull bones. Not rarely, injuries may be due to a combination of these two types [4]. Closed TBI may be due to contact forces or inertial forces. Contact injuries occur when the brain moves inside the skull striking its inner surface. Movements of the brain against some ridges and bone protuberances, especially in the anterior and middle fossae, may be very dangerous to temporal and frontal poles and frontotemporal cortices. Inertial forces are characterized by linear and rotational forces that, producing angular acceleration/deceleration, may cause straining, shearing, and compression of brain tissue. The areas more subject to this type of forces are the junction between the gray and white matter and the rostral brainstem [4].

Furthermore, the brain tissue damage, as well as the spinal cord tissue damage, is the result of the interactions between the primary mechanical injury and the secondary neurodegeneration process. The primary injury is the initial traumatic insult to the brain, the immediate consequences of which are disruption of neuronal and glial cells, vascular damage, and axon shearing. Immediate clinical manifestations of TBI are loss of consciousness, seizure, respiratory depression, ischemia, hypoxia, parenchymal inflammation, and production of nitric oxide. The secondary neurodegeneration process may continue for hours, days, and weeks after the initial trauma. In fact, secondary nonmechanical degeneration is characterized by a complex cascade of biochemical, molecular, metabolic, and cellular events which worsen the initial damage through the activation of the oxidative stress; production of reactive oxygen species (ROS), free radicals, and reactive nitrogen species (RNS); activation of pro-inflammatory cytokines, which in turn may cause mitochondrial dysfunction; neurotransmitter accumulation; blood-brain barrier disruption; activation of inflammatory and immune processes; vascular events; damage to protein and DNA structure and of cell membrane; and excitotoxic damage and consequent apoptosis, which plays an important role in worsening the initial injury (Fig. 16.1) [5, 6].

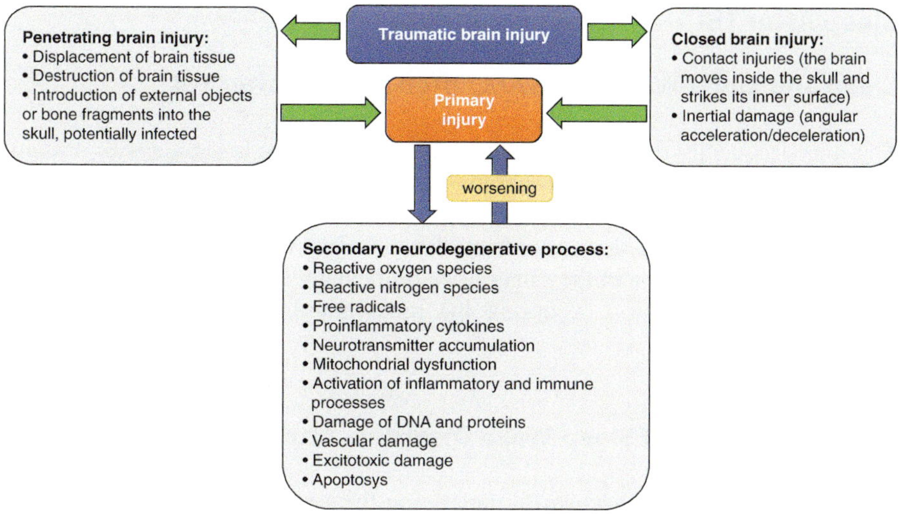

Fig. 16.1 Interactions between the primary mechanical injury and the secondary neurodegeneration process

Table 16.1 Pathophysiological mechanisms underlying sleep disorders after TBI

Pre-TBI mechanisms	Post-TBI mechanisms
Genetics	Hypothalamic damage (decreased hypocretin-1)
Age	Decreased/delayed melatonin
Sex	Lesions in brainstem areas involved in wake regulation
Sleep-wake habits	Pain (e.g., headache)
Sleep disorders	Comorbidities
Psychiatric disorders	New drugs
Cognitive disturbances	Changes in habits
Medical comorbidities	Altered light-dark exposure
Drugs	Hospitalization

The cells of the central nervous system have a low capacity to contrast the effects of oxidative stress, and are highly sensitive to it, which is strongly correlated with the pathogenesis of brain tissue damage due to several brain injuries, including TBI and stroke.

Therefore, the outcome of brain and spinal cord injury strongly depends on the secondary damage, and this seems to be the eligible phase during which to intervene with specific therapies, in order to minimize the effects of TBI.

After TBI, depending on the site of damage and preexisting physical and mental health conditions, regardless of the severity of the trauma, several symptoms and signs may occur, including behavioral alterations; balance disturbances; altered cognition; headache; dizziness; confusion; nausea; difficulties in memory, attention, and concentration; altered mood; and altered sleep. Table 16.1 summarizes the pathophysiological mechanisms underlying sleep disorders after TBI.

Sleep After TBI

Sleep and Sleep-Wake Cycle Are Often Altered After TBI

The most frequently reported sleep disorders after TBI are insomnia with difficulties in falling or staying asleep, excessive daytime sleepiness and increased need of sleep with consequent intrusion of sleep during wakefulness, hypersomnia, and alteration of sleep-wake rhythms. Sleep disorders may be present despite the mild severity of the trauma, and they may arise immediately after the traumatic event, during the hospitalization or during the rehabilitation, and can last for years after TBI.

Pathophysiology of Sleep-Wake Disorders After TBI

The pathophysiological mechanism underlying the sleep-wake disorders after TBI is not completely known, and it is thought that several mechanisms may contribute together.

In particular, a clear relationship between specific brain areas and sleep disturbances has not been established, but several studies have identified hypothalamic dysfunction as a potential cause of sleep-wake alteration [7–9]. Baumann et al. [8, 9] described a decreased secretion of hypocretin-1 following TBI, which returned to normal levels after 6 months after the trauma, but with persistent lower levels in subjects with excessive daytime sleepiness (EDS).

Moreover, Grima et al. [10] and Shekleton et al. [11] debated that TBI is associated with attenuated and delayed melatonin profiles, with reduced evening and overnight melatonin production. Melatonin is a hormone involved in the circadian control of the sleep-wake cycle, and alteration in its profile may contribute to the development of sleep-wake problems following TBI. Moreover, Grima et al. [12] found that melatonin supplementation for a 4-week period is effective and safe in improving subjective sleep quality after TBI.

Besides hypocretin-1 levels, other risk factors for the development of sleep-wake disorders in patients after TBI are pain (especially chronic headache) and its inadequate management; hospitalization with changes in sleep-wake rhythms; psychiatric disorders such as depression, anxiety, and post-traumatic stress disorder; cognitive disturbances; and some drugs. Ouellet et al. [13, 14] proposed a model based on potentially overlapping factors affecting the sleep-wake pattern after TBI, including pre-injury factors, acute factors, and post-acute factors. Pre-injury factors include genetic predisposition, sleep-wake habits, sleep disturbances, health, age, and sex. Acute factors are the brain lesions, hospitalization, changes in habits, drugs, pain, and anxiety. Post-acute factors include pain, health, sleep-related behaviors, drugs, and psychopathology. Ouellet [13] proposed that preexisting factors might contribute to the development of post-acute sleep-wake disturbances; moreover, in the acute phase, lesions of brain areas involved in sleep-wake regulation, psychological distress, changes in behavior, the complaint of pain, and some drugs might contribute to the development of sleep disturbances all together. In the long term, changes of habits, possible neurological and general health complications, and some drugs might interact to perpetuate the sleep-wake disorders.

The brain regions involved in sleep-wake regulation are very vulnerable to trauma, and the prevalence of sleep disorders after TBI is very high, but they are often overlooked; it is estimated that nearly half of subjects with TBI develop some sleep problems early after the traumatic event [2, 15]. Disordered sleep in these patients may be due not only to brain injury itself but also to mood depression, anxiety, irritability, pain, long hospitalization, and cognitive and functional impairment.

Sleep-Wake Disturbances After TBI

Disordered sleep and sleep-wake disturbances are among the most frequent acute and chronic consequences of traumatic brain injury [12, 15–17], often lasting for months to years after the event [7, 13, 18, 19], but their real extent of persistence is not completely clear [15]. In a meta-analysis of 21 clinical studies assessing sleep and sleep disorders following TBI, Mathias and Alvaro [20] showed that about 50% of TBI patients suffered from post-traumatic sleep-wake disturbances and 25–29% had been diagnosed with a sleep disorder such as insomnia (29% vs. 10% in controls), hypersomnia (28% vs. 10%), obstructive sleep apnea (25% vs. 2%), periodic limb movements (8% vs. 4%), and narcolepsy (4% vs. 0.047%), revealing a higher prevalence with respect to the general population. Moreover, these patients had a two- to four-time higher risk to complain specific sleep problems [20], namely, snoring (60% vs. 42%), insomnia (50% vs. 31%), poor sleep efficiency (49% vs. 27%), delayed sleep onset (36% vs. 27%), nightmares (27% vs. 8%), excessive daytime sleepiness (27% vs. 9%), early awakenings (38% vs. 18%), and sleepwalking (9% vs. 2%).

Sleep architecture after TBI is not well explored yet. Polysomnographic recordings in subjects after TBI have been shown to correspond to subjective complaints, confirming that after TBI, the quality of sleep is altered, with increased awakenings and poor sleep efficiency. In particular, PSG studies showed an increased amount of NREM N1 and N2 sleep stages, decreased REM sleep, and shortened REM sleep latency. A few reports have shown an increase in slow-wave sleep, which may reflect its involvement in recovery mechanisms by promoting neurological processes, namely, axonal sprouting and synaptic remodeling [13, 21].

Sleep disorders and their diurnal consequences may have a negative impact on daily activities, and may negatively affect the outcome of rehabilitation, prolonging the duration of hospitalization and delaying the recovery of patients with TBI [20, 22]. Moreover, sleep disorders, especially when they are associated with EDS, are significant risk factors for motor vehicle accidents and can increase the risk of repetitive TBI [17].

We will describe the most frequent sleep disorders following TBI.

Insomnia

Insomnia is defined by the American Academy of Sleep Medicine in the International Classification of Sleep Disorders-Third Edition (ICSD-3) [23] as "a persistent difficulty with sleep initiation, duration, consolidation, or quality that occurs despite adequate opportunity and circumstances of sleep, and results in some form of daytime impairment." Insomnia is reported in 30–60% of individuals after TBI of all degrees of severity [13, 14, 20, 24–26], although it seems to be more frequent following mild injuries, in the presence of pain or depression, maybe because these patients are more aware of their clinical conditions with respect to the more severe cases [3, 13, 24]. Moreover, multiple TBIs are associated with higher risk to develop more severe insomnia and sleep disturbances as shown in veterans [27]. Cognitive-behavioral therapy (CBT) and sleep hygiene strategies seem to determine improvement of insomnia symptoms also in patients suffering from insomnia due to TBI, providing a long-lasting positive effect [13, 14].

Excessive Daytime Sleepiness, Increased Need for Sleep, and Hypersomnia

EDS is reported in a very high percentage of subjects after TBI. EDS is defined as a score >10 at the Epworth Sleepiness Scale (ESS) [28–30] and as an average sleep latency of <10 min at the Multiple Sleep Latency Test (MSLT) [31]. It is characterized by a diurnal increased pressure to fall asleep with sleepiness occurring mostly during sedentary activities and episodes of difficult-to-control need to sleep or unintentional and undesired naps.

Many studies have reported that EDS and increased need of sleep following TBI are very frequent [13, 14, 32, 33]. Subjective complaint of daytime sleepiness is reported in 50–85% of individuals following TBI [16, 18, 34]. Usually EDS remits; however, it may persist in 10–53% of cases [16], and it may be present 6 months after the initial trauma as shown by Imbach et al. [7] and Baumann et al. [9] who evaluated EDS with MSLT and, also years later, as shown by Kempf et al. [19].

Narcolepsy occurs less frequently than the other sleep disorders, but with frequency higher than in the general population (4% vs. 0.47%). Some case reports of narcolepsy after TBI have been described in 1941 [35], but at that time it was impossible to distinguish narcolepsy from other sleep disturbances with hypersomnolence. Moreover, it is not always simple to understand if narcolepsy could have preceded or contributed to the trauma and, in some cases, TBI might have been the trigger event in subjects already at risk to develop narcolepsy [17].

The pathophysiology of EDS and hypersomnia after TBI remains unclear, although damage to hypothalamic neurons and other wake-promoting neuronal population might be associated with EDS, as reported above [8, 36]. Furthermore, short REM latency has been reported following TBI, leading to a narcoleptic-like condition due to other conditions, such as sleep deprivation, withdrawal from REM-suppressing drugs, and changes in sleep-wake schedule [13].

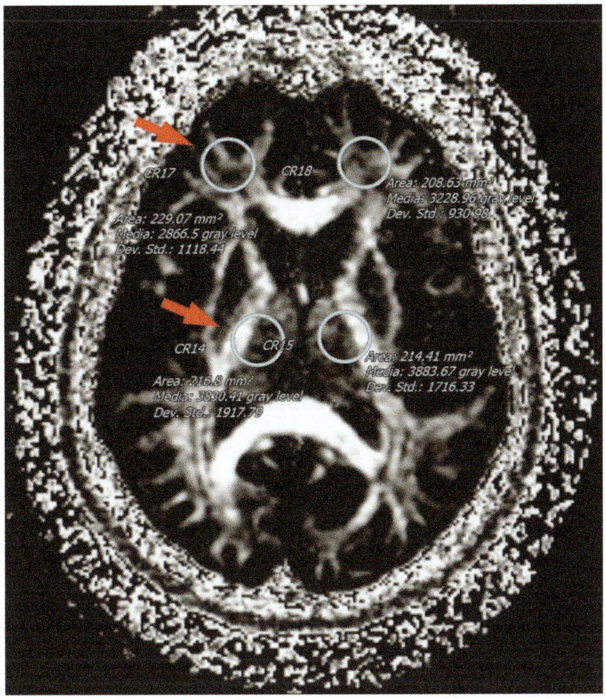

Fig. 16.2 Quantitative assessment using comparative region-of-interest values showing reduced fractional anisotropy in the left frontal region due to relevant left white matter damage (red arrow). The same reduction is evident at the thalamic level (red arrow)

Recently, we have described [37] the case of a 48-year-old healthy woman who had a head trauma with cerebral commotion followed by a 4-month coma. Since the first days after the coma resolution, she reported EDS, with prolonged diurnal non-restorative naps, anterograde and retrograde amnesia, and fronto-bitemporal headache with phonofobia and photophobia, sometimes associated with nausea and vomit. At the time of the visit at our outpatient sleep service, 3 years after the trauma, she complained severe sleep disturbances, such as incoercible daytime sleepiness, not refreshing prolonged naps, prolonged nighttime sleep with snoring, and sleep inertia. EDS was confirmed by PSG and MSLT. A brain MRI with gadolinium and diffusion tensor imaging (DTI) was performed that showed no cerebral lesions, but DTI tractography highlighted reduced neuronal fibers in the thalamo-forebrain circuit; moreover, tractography showed loss of white matter tracts in the left frontal region (Fig. 16.2). Finally, perfusion MRI showed loss of cortical blood flow in the left frontal region due to axonal post-traumatic damage. These findings did suggest us that in this case, EDS and hypersomnia could be explained as results of decreased activity in the cortical wake activation pathways.

Circadian Rhythm Sleep Disorders

There are only few studies assessing the prevalence of circadian rhythm disorders following TBI. Some case reports of post-traumatic delayed phase syndrome

[38–40] and one case of hypernyctohemeral syndrome [41] have been described. Ayalon et al. [40] reported that 36% of patients who were referred for an evaluation of insomnia after a minor TBI were diagnosed also with a circadian rhythm disorder, namely, delayed sleep phase syndrome (52%) and irregular sleep-wake pattern, which were associated with distinct profiles of melatonin and temperature circadian rhythms.

Duclos et al. [40] assessed the rest-activity cycle in 16 hospitalized patients after a moderate/severe TBI by means of a 10-day actigraphy performed 10 days after the injury, and they found that these patients had an altered rest-activity cycle, maybe due to severe sleep fragmentation and wake episodes, with improvement over time. Moreover, a faster return to a normal cycle might predict enhanced brain recovery.

In summary, the real prevalence of circadian rhythm disorder following TBI is poorly studied and needs additional investigation.

Sleep-Breathing Disorders

Both obstructive sleep apnea (OSA) and central sleep apnea (CSA) are reported more frequently in patients after TBI than in the general population. Many studies report a prevalence of 25–35% for OSA following TBI [17, 22]. OSA is the most frequent sleep-disordered breathing (SDB) after TBI. The relationship between SDB and TBI is complex. In fact, preexisting untreated or not-well-treated SDB may cause EDS, predisposing to motor vehicle accidents and consequent TBI. On the other hand, traumas involving the upper airways, associated with TBI, may lead to post-traumatic SDB. Moreover, hospitalization and poor mobility may cause weight gain that is a risk factor for OSA; supine position, loss of muscle tone, and drugs are additional risk factor for SDB [17, 18].

Management of Sleep and Sleep-Wake Disorders After TBI

Although sleep disorders are very frequent following TBI, often they are overlooked and patients drag on EDS without receiving a correct diagnosis and treatment. In fact, there are no guidelines for the management of sleep-wake alterations after TBI.

Diagnosis Because of the high prevalence of disturbed sleep and sleep disorders following TBI, and the fact that they may contribute to worsen clinical condition and outcome, patients after TBI should undergo a careful evaluation of their sleep-wake cycle using sleep questionnaires and sleep logs, also with the help of relatives and caregivers. Other more complex diagnostic procedures, such as actigraphy, cardiorespiratory monitoring, or polysomnographic recording, should be obtained when necessary to define the clinical picture and reach a definite sleep diagnosis [17]. Figure 16.3 shows the schematic representation of the diagnostic process for sleep and sleep-wake disorders in patients after TBI.

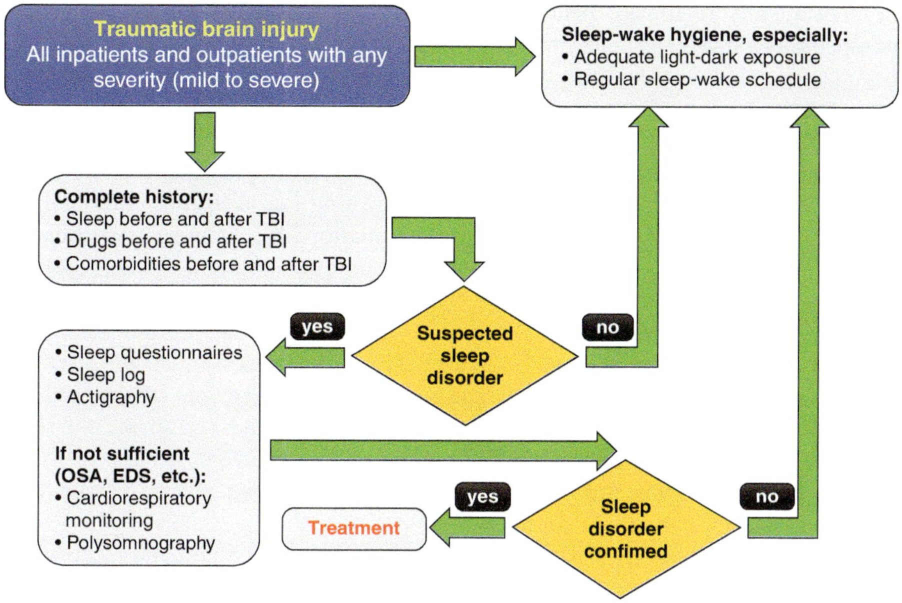

Fig. 16.3 Schematic representation of the diagnostic process for sleep and sleep-wake disorders in patients after TBI

Treatment Options Treatment of sleep disorders strongly depends on which kind of sleep disorders presented by the patients. To our knowledge there are no studies evaluating the safety and efficacy of pharmacological or non-pharmacological treatments in patients suffering from sleep disorders after TBI, except for some studies on the role of melatonin supplementation and the role of CBT and sleep hygiene [6, 12, 42].

There are meta-analyses evaluating the pharmacological treatment of sleep disorders in TBI; clinicians need to refer to evidence from other categories of patients with sleep disorders, paying attention to possible side and adverse effects related to TBI and other concomitant medications.

In all patients, however, regardless of the presence or not of any sleep disturbance, it is fundamental to promote adequate sleep hygiene, with the primary aim to prevent and avoid the development of sleep-wake disorders.

The Role of Sleep in Recovery After TBI

Sleep is an active neurophysiological process characterized by the cyclic succession of complex events. During sleep some cerebral areas are less active than during wakefulness, while others are more active, depending from the sleep stage (light or deep NREM sleep or REM sleep), and the regular alternation and cyclicity of sleep stages are fundamental for the restorative role of sleep. Moreover, sleep is homeostatically regulated; hence, acute sleep deprivation leads to a subsequent compensatory increased amount of sleep (sleep rebound), but chronic sleep deprivation is associated with daytime symptoms and signs such as EDS, fatigue, memory, and concentration dysfunction and often with naps during daytime activities.

Despite the fact that we spend one third of our life sleeping, sleep functions are not completely understood yet, although a solid evidence suggests that sleep has several different functions [43, 44], including resting and restoring brain and the whole body, energy balance [45–47], thermoregulation [48], memory consolidation [49], regulation of metabolism and endocrine, immune and anti-oxidant systems [50–52], and in removing toxic waste substances accumulated through the day, acting as a neuroprotective process [44]. Clinical observations suggest that sleep may be involved also in the modulation of neuronal recovery process and neuroplasticity in brain after an injury [53]. Concerning the neuroprotective role of sleep after a brain injury, Cam et al. [54] demonstrated that in rats with induced stroke, the sleep rebound after pre-stroke sleep deprivation reduces brain damage, while post-stroke sleep deprivation aggravates stroke, confirming the positive effect of sleep, and sleep rebound, on brain tissue damage.

After TBI patients often present a marked alteration of sleep and sleep-wake cycle, which seem to be closely associated with recovery outcomes, both in the acute and post-acute phases of rehabilitation. Often, soon after TBI, patients have a difficulty in staying awake or they miss clinical appointment, showing cognitive alterations [14, 55, 56]. Moreover, the severity of the sleep-wake alteration might predict the duration of post-traumatic amnesia and of hospitalization [57]. Urakami [58] showed, by means of combined EEG-MEG study, a parallel improvement of abnormalities in sleep spindle activity, consciousness, and cognition. In the post-acute phase, sleep problems might contribute to the chronicity of other debilitating symptoms, such as pain (e.g., headache), depressive symptoms, and irritability [13, 59].

As explained above, sleep plays an important role in neurorecovery mechanisms.

In fact, it has been demonstrated in animal models that sleep is involved in the modulation of neural repair mechanisms and, according to it, there is evidence that neurorecovery is slowed or decreased following sleep deprivation [60]. Zunzunegui et al. [53] demonstrated in a rat model of focal ischemic stroke that sleep disorders had *detrimental effects* on functional and morphological/structural outcomes after stroke, suggesting a role for sleep in the modulation of the recovery process and neuroplasticity. The autopsy of the rats demonstrated that sleep-deprived animals had lower amounts of brain repair indices, such as axonal sprouting, synaptogenesis, and vasogenesis. In another study Xie et al. [61] showed that sleep deprivation facilitated the accumulation of beta-amyloid proteins and other neurotoxins that are associated with neurodegenerative disorders, concluding that the restorative function of sleep may be due to the enhanced removal of neurotoxic products accumulated during awake in the central nervous system.

Clinical observations suggest that sleep may be involved also in the modulation of the recovery process and neuroplasticity in the brain after an injury, such as stroke [53].

Only a few studies have been performed in patients with sleep disorders following TBI in order to assess recovery and long-term complications, and it has been shown that subjects with sleep disorders/deprivation may have deficits in memory and executive function, as well as depression, pain, and anxiety, and early treatment may improve the outcome in these patients [62–64].

Conclusions

Disordered sleep and sleep-wake disorders are very frequent acute and chronic consequences of TBI. They are associated with a negative impact on recovery and on long-term outcomes, regardless of the severity of the initial trauma, also because of the important role that sleep plays in neuroplasticity and neurorecovery after a brain injury, indicating that adequate sleep is fundamental after TBI in order to improve the outcome.

Because of the negative impact of sleep disorders and sleep-wake disturbances in subjects after TBI, such as slow recovery, prolonged hospitalization, risk of chronicity, and long-term complications, and its role in improving recovery, the assessment of the sleep-wake cycle, also by means of instrumental tools such as PSG recordings and/or actigraphy, should be considered in all patients after TBI, regardless of its initial severity, in order to detect early and properly treat any sleep-wake disorder. But first, since the early days after TBI, during the hospitalization, sleep quality and quantity should be carefully sought, and all disturbing factors (e.g., noise, altered light-dark exposure, inversion of sleep-wake rhythm) that might favor sleep-wake alterations should be avoided, as much as possible.

Future studies are needed in order to assess the prevalence and the duration of sleep problems following TBI and to establish clear guidelines for their management for both diagnosis and treatment options.

References

1. Menon DK, Schwab K, Wright DW, Maas AI. Position statement: definition of traumatic brain injury. Arch Phys Med Rehabil. 2010;91(11):1637–40. https://doi.org/10.1016/j.apmr.2010.05.017.
2. Singh K, Morse AM, Tkachenko N, Kothare SV. Sleep disorders associated with traumatic brain injury – a review. Pediatr Neurol. 2016;60:30–6. https://doi.org/10.1016/j.pediatrneurol.2016.02.013.
3. Theadom A, Cropley M, Parmar P, et al. Sleep difficulties one year following mild traumatic brain injury in a population-based study. Sleep Med. 2015;16(8):926–32. https://doi.org/10.1016/j.sleep.2015.04.013.
4. McAllister TW. Neurobiological consequences of traumatic brain injury. Dialogues Clin Neurosci. 2011;13(3):287–300.
5. Khalatbary AR, Tiraihi T, Boroujeni MB, Ahmadvand H, Tavafi M, Tamjidipoor A. Effects of epigallocatechin gallate on tissue protection and functional recovery after contusive spinal cord injury in rats. Brain Res. 2010;1306:168–75. https://doi.org/10.1016/j.brainres.2009.09.109.

6. Naseem M, Parvez S. Role of melatonin in traumatic brain injury and spinal cord injury. ScientificWorldJournal. 2014;2014:586270. https://doi.org/10.1155/2014/586270.
7. Imbach LL, Büchele F, Valko PO, et al. Sleep-wake disorders persist 18 months after traumatic brain injury but remain underrecognized. Neurology. 2016;86(21):1945–9. https://doi.org/10.1212/WNL.0000000000002697.
8. Baumann CR, Stocker R, Imhof H-G, et al. Hypocretin-1 (orexin A) deficiency in acute traumatic brain injury. Neurology. 2005;65(1):147–9. https://doi.org/10.1212/01.wnl.0000167605.02541.f2.
9. Baumann CR, Werth E, Stocker R, Ludwig S, Bassetti CL. Sleep-wake disturbances 6 months after traumatic brain injury: a prospective study. Brain. 2007;130(7):1873–83. https://doi.org/10.1093/brain/awm109.
10. Grima NA, Ponsford JL, St Hilaire MA, Mansfield D, Rajaratnam SM. Circadian melatonin rhythm following traumatic brain injury. Neurorehabil Neural Repair. 2016;30(10):972–7. https://doi.org/10.1177/1545968316650279.
11. Shekleton JA, Parcell DL, Redman JR, Phipps-Nelson J, Ponsford JL, Rajaratnam SMW. Sleep disturbance and melatonin levels following traumatic brain injury. Neurology. 2010;74(21):1732–8. https://doi.org/10.1212/WNL.0b013e3181e0438b.
12. Grima NA, Rajaratnam SMW, Mansfield D, Sletten TL, Spitz G, Ponsford JL. Efficacy of melatonin for sleep disturbance following traumatic brain injury: a randomised controlled trial. BMC Med. 2018;16(1):8. https://doi.org/10.1186/s12916-017-0995-1.
13. Ouellet MC, Beaulieu-Bonneau S, Morin CM. Sleep-wake disturbances after traumatic brain injury. Lancet Neurol. 2015;14(7):746–57. https://doi.org/10.1016/S1474-4422(15)00068-X.
14. Ouellet M-C, Beaulieu-Bonneau S, Morin CM. Insomnia in patients with traumatic brain injury: frequency, characteristics, and risk factors. J Head Trauma Rehabil. 2006;21(3):199–212. http://www.ncbi.nlm.nih.gov/pubmed/16717498
15. Mantua J, Grillakis A, Mahfouz SH, et al. A systematic review and meta-analysis of sleep architecture and chronic traumatic brain injury. Sleep Med Rev. 2018:1–17. https://doi.org/10.1016/j.smrv.2018.01.004.
16. Wickwire EM, Williams SG, Roth T, et al. Sleep, sleep disorders, and mild traumatic brain injury. What we know and what we need to know: findings from a National Working Group. Neurotherapeutics. 2016;13(2):403–17. https://doi.org/10.1007/s13311-016-0429-3.
17. Castriotta RJ, Murthy JN. Sleep disorders in patients with traumatic brain injury: a review. CNS Drugs. 2011;25(3):175–85. https://doi.org/10.2165/11584870-000000000-00000.
18. Verma A, Anand V, Verma NP. Sleep disorders in chronic traumatic brain injury. J Clin Sleep Med. 2007;3(4):357–62. http://www.pubmedcentral.nih.gov/articlerender.fcgi?artid=1978305&tool=pmcentrez&rendertype=abstract
19. Kempf J, Werth E, Kaiser PR, Bassetti CL, Baumann CR. Sleep-wake disturbances 3 years after traumatic brain injury. J Neurol Neurosurg Psychiatry. 2010;81(12):1402–5. https://doi.org/10.1136/jnnp.2009.201913.
20. Mathias JL, Alvaro PK. Prevalence of sleep disturbances, disorders, and problems following traumatic brain injury: a meta-analysis. Sleep Med. 2012;13(7):898–905. https://doi.org/10.1016/j.sleep.2012.04.006.
21. Sommerauer M, Valko PO, Werth E, Baumann CR. Excessive sleep need following traumatic brain injury: a case-control study of 36 patients. J Sleep Res. 2013;22(6):634–9. https://doi.org/10.1111/jsr.12068.
22. Grima NA, Ponsford JL, Pase MP. Sleep complications following traumatic brain injury. Curr Opin Pulm Med. 2017;23(6):493–9. https://doi.org/10.1097/MCP.0000000000000429.
23. American Academy of Sleep Medicine. International classification of sleep disorders, 3rd ed. Sateia M, editor. Darien, IL: American Academy of Sleep Medicine. 2014.
24. Sandsmark DK, Elliott JE, Lim MM. Sleep-wake disturbances after traumatic brain injury: synthesis of human and animal studies. Sleep. 2017;40(5):1–18. http://ovidsp.ovid.com/ovidweb.cgi?T=JS&PAGE=reference&D=psyc14&NEWS=N&AN=2017-28449-001
25. Fichtenberg NL, Zafonte RD, Putnam S, Mann NR, Millard AE. Insomnia in a post-acute brain injury sample. Brain Inj. 2002;16(3):197–206. https://doi.org/10.1080/02699050110103940.

26. Guilleminault C, Yuen KM, Gulevich MG, Karadeniz D, Leger D, Philip P. Hypersomnia after head-neck trauma: a medicolegal dilemma. Neurology. 2000;54(3):653–9. http://www.ncbi.nlm.nih.gov/pubmed/10680799
27. Bryan CJ. Repetitive traumatic brain injury (or concussion) increases severity of sleep disturbance among deployed military personnel. Sleep. 2013;36(6):941–6. https://doi.org/10.5665/sleep.2730.
28. Johns MW. Daytime sleepiness, snoring, and obstructive sleep apnea. The Epworth sleepiness scale. Chest. 1993;103(1):30–6. http://www.ncbi.nlm.nih.gov/pubmed/8417909
29. Johns MW. A new method for measuring daytime sleepiness: the Epworth sleepiness scale. Sleep. 1991;14(6):540–5. http://www.ncbi.nlm.nih.gov/pubmed/1798888
30. Vignatelli L, Plazzi G, Barbato A, et al. Italian version of the Epworth sleepiness scale: external validity. Neurol Sci. 2003;23(6):295–300. https://doi.org/10.1007/s100720300004.
31. Carskadon MA, Dement WC, Mitler MM, Roth T, Westbrook PR, Keenan S. Guidelines for the multiple sleep latency test (MSLT): a standard measure of sleepiness. Sleep. 1986;9(4):519–24. http://www.ncbi.nlm.nih.gov/pubmed/3809866
32. Guilleminault C, Faull KF, Miles L, van den Hoed J. Posttraumatic excessive daytime sleepiness: a review of 20 patients. Neurology. 1983;33(12):1584–9. http://www.ncbi.nlm.nih.gov/pubmed/6685831
33. Masel BE, Scheibel RS, Kimbark T, Kuna ST. Excessive daytime sleepiness in adults with brain injuries. Arch Phys Med Rehabil. 2001;82(11):1526–32. https://doi.org/10.1053/apmr.2001.26093.
34. Collen J, Orr N, Lettieri CJ, Carter K, Holley AB. Sleep disturbances among soldiers with combat-related traumatic brain injury. Chest. 2012;142(3):622–30. https://doi.org/10.1378/chest.11-1603.
35. Wilson Gill A. Idiopathic and traumatic narcolepsy. Lancet. 1941;237(6137):474–6. https://doi.org/10.1016/S0140-6736(00)61098-X.
36. Ripley B, Overeem S, Fujiki N, et al. CSF hypocretin/orexin levels in narcolepsy and other neurological conditions. Neurology. 2001;57(12):2253–8. http://www.ncbi.nlm.nih.gov/pubmed/11756606
37. Puligheddu M, Laccu I, Gioi G, Congiu P, Figorilli M, Peltz MT. Diffusion tensor imaging tractography and MRI perfusion in post traumatic brain injury hypersomnia. Sleep Med. 2017;34:96–8. https://doi.org/10.1016/j.sleep.2017.03.011.
38. Nagtegaal JE, Kerkhof GA, Smits MG, Swart AC, van der Meer YG. Traumatic brain injury-associated delayed sleep phase syndrome. Funct Neurol. 12(6):345–8. http://www.ncbi.nlm.nih.gov/pubmed/9503198
39. Smits MG, Nagtegaal JE. Post-traumatic delayed sleep phase syndrome. Neurology. 2000;55(6):902–3. http://www.ncbi.nlm.nih.gov/pubmed/10994031
40. Patten SB, Lauderdale WM. Delayed sleep phase disorder after traumatic brain injury. J Am Acad Child Adolesc Psychiatry. 1992;31(1):100–2. https://doi.org/10.1097/00004583-199201000-00015.
41. Boivin DB, Caliyurt O, James FO, Chalk C. Association between delayed sleep phase and hypernyctohemeral syndromes: a case study. Sleep. 2004;27(3):417–21. http://www.ncbi.nlm.nih.gov/pubmed/15164893
42. Bogdanov S, Naismith S, Lah S. Sleep outcomes following sleep-hygiene-related interventions for individuals with traumatic brain injury: a systematic review. Brain Inj. 2017;31(4):422–33. https://doi.org/10.1080/02699052.2017.1282042.
43. Vassalli A, Dijk D-J. Sleep function: current questions and new approaches. Eur J Neurosci. 2009;29(9):1830–41. https://doi.org/10.1111/j.1460-9568.2009.06767.x.
44. Underwood E. Neuroscience. Sleep: the brain's housekeeper? Science. 2013;342(6156):301. https://doi.org/10.1126/science.342.6156.301.
45. Netchiporouk L, Shram N, Salvert D, Cespuglio R. Brain extracellular glucose assessed by voltammetry throughout the rat sleep-wake cycle. Eur J Neurosci. 2001;13(7):1429–34. http://www.ncbi.nlm.nih.gov/pubmed/11298804

46. Kong J, Shepel PN, Holden CP, Mackiewicz M, Pack AI, Geiger JD. Brain glycogen decreases with increased periods of wakefulness: implications for homeostatic drive to sleep. J Neurosci. 2002;22(13):5581–7.
47. Kennedy C, Gillin JC, Mendelson W, et al. Local cerebral glucose utilization in slow-wave sleep. Trans Am Neurol Assoc. 1981;106:25–8. http://www.ncbi.nlm.nih.gov/pubmed/7349002
48. Lack LC, Gradisar M, Van Someren EJW, Wright HR, Lushington K. The relationship between insomnia and body temperatures. Sleep Med Rev. 2008;12(4):307–17. https://doi.org/10.1016/j.smrv.2008.02.003.
49. Ackermann S, Rasch B. Differential effects of non-REM and REM sleep on memory consolidation? Curr Neurol Neurosci Rep. 2014;14(2):430. https://doi.org/10.1007/s11910-013-0430-8.
50. Singh R, Kiloung J, Singh S, Sharma D. Effect of paradoxical sleep deprivation on oxidative stress parameters in brain regions of adult and old rats. Biogerontology. 2008;9(3):153–62. https://doi.org/10.1007/s10522-008-9124-z.
51. Ramanathan L, Hu S, Frautschy SA, Siegel JM. Short-term total sleep deprivation in the rat increases antioxidant responses in multiple brain regions without impairing spontaneous alternation behavior. Behav Brain Res. 2010;207(2):305–9. https://doi.org/10.1016/j.bbr.2009.10.014.
52. Ramanathan L, Gulyani S, Nienhuis R, Siegel JM. Sleep deprivation decreases superoxide dismutase activity in rat hippocampus and brainstem. Neuroreport. 2002;13(11):1387–90. http://www.ncbi.nlm.nih.gov/pubmed/12167758
53. Zunzunegui C, Gao B, Cam E, Hodor A, Bassetti CL. Sleep disturbance impairs stroke recovery in the rat. Sleep. 2011;34(9):1261–9. https://doi.org/10.5665/SLEEP.1252.
54. Cam E, Gao B, Imbach L, Hodor A, Bassetti CL. Sleep deprivation before stroke is neuroprotective: a pre-ischemic conditioning related to sleep rebound. Exp Neurol. 2013;247:673–9. https://doi.org/10.1016/j.expneurol.2013.03.003.
55. Worthington AD, Melia Y. Rehabilitation is compromised by arousal and sleep disorders: results of a survey of rehabilitation centres. Brain Inj. 2006;20(3):327–32. https://doi.org/10.1080/02699050500488249.
56. Silva MA, Nakase-Richardson R, Sherer M, Barnett SD, Evans CC, Yablon SA. Posttraumatic confusion predicts patient cooperation during traumatic brain injury rehabilitation. Am J Phys Med Rehabil. 2012;91(10):890–3. https://doi.org/10.1097/PHM.0b013e31825a1648.
57. Nakase-Richardson R, Sherer M, Barnett SD, et al. Prospective evaluation of the nature, course, and impact of acute sleep abnormality after traumatic brain injury. Arch Phys Med Rehabil. 2013;94(5):875–82. https://doi.org/10.1016/j.apmr.2013.01.001.
58. Urakami Y. Relationship between, sleep spindles and clinical recovery in patients with traumatic brain injury: a simultaneous EEG and MEG study. Clin EEG Neurosci. 2012;43(1):39–47. https://doi.org/10.1177/1550059411428718.
59. Chaput G, Giguère J-F, Chauny J-M, Denis R, Lavigne G. Relationship among subjective sleep complaints, headaches, and mood alterations following a mild traumatic brain injury. Sleep Med. 2009;10(7):713–6. https://doi.org/10.1016/j.sleep.2008.07.015.
60. Barshikar S, Bell KR. Sleep disturbance after TBI. Curr Neurol Neurosci Rep. 2017;17(11):87. https://doi.org/10.1007/s11910-017-0792-4.
61. Xie L, Kang H, Xu Q, et al. Sleep drives metabolite clearance from the adult brain. Science. 2013;342(6156):373–7. https://doi.org/10.1126/science.1241224.
62. Wilde MC, Castriotta RJ, Lai JM, Atanasov S, Masel BE, Kuna ST. Cognitive impairment in patients with traumatic brain injury and obstructive sleep apnea. Arch Phys Med Rehabil. 2007;88(10):1284–8. https://doi.org/10.1016/j.apmr.2007.07.012.
63. Duclos C, Beauregard M-P, Bottari C, Ouellet M-C, Gosselin N. The impact of poor sleep on cognition and activities of daily living after traumatic brain injury: a review. Aust Occup Ther J. 2015;62(1):2–12. https://doi.org/10.1111/1440-1630.12164.
64. Fogelberg DJ, Hoffman JM, Dikmen S, Temkin NR, Bell KR. Association of sleep and co-occurring psychological conditions at 1 year after traumatic brain injury. Arch Phys Med Rehabil. 2012;93(8):1313–8. https://doi.org/10.1016/j.apmr.2012.04.031.

Disorders of Sleep and Wakefulness in Parkinson's Disease and Other Movement Disorders

17

John C. Carter and Vishesh K. Kapur

Introduction

The term "movement disorders" refers to a large group of neurological conditions characterized by abnormal voluntary or involuntary movements. These include a number of different conditions such as Parkinson's disease and the parkinsonian spectrum, restless legs syndrome, essential tremor, Tourette syndrome, dystonia, Huntington's disease, ataxia, etc. These diseases can have very different pathophysiologic underpinnings and, for this reason, may manifest with different sleep problems. Sleep complaints are common in patients with movement disorders, ranging from fatigue and sleepiness to sleep attacks, insomnia, and motor activity during sleep. A variety of sleep disorders may underlie these complaints including obstructive sleep apnea (OSA), circadian dysrhythmias, and rapid eye movement sleep (REM) behavior disorder. Sleep issues often significantly impact quality of life and thus represent an important focus for clinical care. Some sleep and movement disorders share common pathogenetic mechanisms, particularly via dopaminergic neurotransmission. In this chapter we review the disorders of sleep and wakefulness that commonly occur in patients with movement disorders, with an emphasis on Parkinson's disease.

J. C. Carter
Department of Medicine, Division of Pulmonary, Critical Care, and Sleep Medicine, Case Western Reserve University School of Medicine and MetroHealth Medical Cente, Cleveland, OH, USA
e-mail: jcarter5@metrohealth.org

V. K. Kapur (✉)
Department of Medicine, Division of Pulmonary, Critical Care and Sleep Medicine, University of Washington, Harborview Medical Center, Seattle, WA, USA
e-mail: vkapur@uw.edu

L. M. DelRosso, R. Ferri (eds.), *Sleep Neurology*,
https://doi.org/10.1007/978-3-030-54359-4_17

Parkinson's Disease

Parkinson's disease (PD), which affects roughly 300 per 100,000 people, is characterized by motor symptoms (tremor, rigidity, and bradykinesia) and non-motor symptoms such as neuropsychiatric impairment, depression, autonomic instability, and sleep disruption. Though non-motor symptoms may have a greater impact on patients' quality of life [1–3], recognition of and directed therapies for these symptoms are relatively recent developments [4].

Sleep problems, among the most common non-motor symptoms in PD, were recognized by Parkinson in his original 1817 manuscript [5]. Sleep disturbances such as insomnia, nocturia, frequent awakenings, restless legs syndrome (RLS), REM sleep behavior disorder (RBD), OSA, and excessive daytime sleepiness (including sleep attacks) affect up to 90% of PD patients and may occur concurrently [6–9]. Sleep problems are important to recognize, evaluate, and manage as they can cause significant impairment to quality of life.

Insomnia in Parkinson's Disease

Insomnia, including difficulty falling asleep and maintaining sleep, is the most common sleep complaint in PD, with sleep maintenance problems (frequent or prolonged awakenings) being the most frequent manifestation [7]. Both disease-specific and non-disease-specific factors contribute to the development of insomnia in PD. In addition to insomnia related to poor sleep hygiene or psychophysiologic factors (non-disease-specific factors), individuals with PD experience sleep disruption secondary to their disease. For example, motor symptoms may be triggered by an arousal from sleep [10], leading to prolonged awakenings and difficulty returning to sleep, particularly when dopaminergic medications are at low serum levels. Nocturia (present in 80% of PD patients) [6] may also result in frequent nighttime awakenings and tends to worsen as PD severity progresses. Nocturia may be secondary to autonomic dysfunction in PD but can also be caused by non-disease-specific factors such as obstructive sleep apnea. Additionally, sleep-associated hallucinations, reported by 15–20% of PD patients, have been associated with difficulty falling asleep and prolonged sleep latency [11]. Both daytime and sleep-related hallucinations may be exacerbated by dopaminergic medications.

Depression may be related to sleep onset and maintenance insomnia in PD, although the direction of causation is unclear. In one study, depression severity (Zung scores) positively correlated with severity of sleep initiation and maintenance complaints in PD [12]. Depression may arise from or be exacerbated by sleep disruption, though sleep disruption can also be a consequence of neurophysiologic changes in depression [13]. Both directions of causation are likely relevant, and the neurodegenerative processes in PD may represent a common underlying pathophysiology of sleep disruption and depression in some patients [14].

Insomnia in PD is best addressed through a combined approach of treating PD as well as the insomnia directly using pharmacologic [15] and non-pharmacologic (behavioral and psychological) [16] methods. Long-acting L-dopa at night [17, 18], catechol-O-methyl transferase (COMT) inhibitors [19], and cabergoline have been shown to increase sleep efficiency and amount of deep (slow-wave) sleep as well as subjective sleep quality [20]. Symptomatic treatment of insomnia may be accomplished using non-benzodiazepine agonists of the benzodiazepine receptor (e.g., eszopiclone 2 or 3 mg, associated with improved subjective sleep quality but not total sleep time) [21], melatonin (3 mg, associated with improved Pittsburg Sleep Quality Index and subjective sleep quality but not polysomnogram-derived measures of sleep or motor function in one large randomized controlled trial) [22], or doxepin (10 mg, associated with improved Insomnia Severity Index scores and Scales for Outcomes in PD scores) [23]. Cognitive behavioral therapy for insomnia (CBTI) improves Insomnia Severity Index scores compared to placebo, although in one large study of CBTI in PD, there was no demonstrable improvement in quality of life measures [23].

REM Behavior Disorder in Parkinson's Disease

REM behavior disorder (RBD) is seen in 15–60% of PD cases [24–26], in contrast to an overall prevalence of 0.5% in the general population [27, 28]. RBD results from a lack of normal skeletal muscle atonia in REM sleep (see Fig. 17.1), leading to dream-enactment behavior [28, 29]. Typical behaviors include kicking, punching, jumping, talking, screaming, laughing, and singing [30]. Some such behaviors may

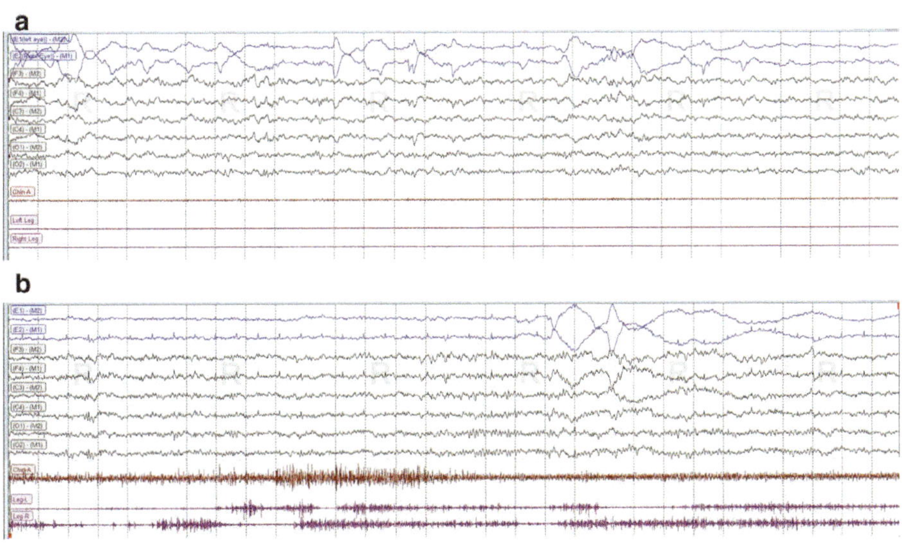

Fig. 17.1 Polysomnographic morphology of REM sleep in the normal case (**a.**) and in a patient with RBD (**b.**). In the normal case, note flat chin and leg electromyogram (EMG) tone, which typify skeletal muscle atonia seen in normal REM. The patient shown in part B. is a 75-year-old man with parkinsonism and dream enactment behavior. Note high chin and leg EMG tone

result in self-injury or injury to bedpartners. Dream content frequently involves a sense of being pursued or defending oneself. The diagnosis of RBD requires a consistent clinical history and evidence of REM without atonia on polysomnography [28]. Polysomnography is also helpful to differentiate RBD from parasomnias arising from non-REM sleep or arousals in REM due to sleep apnea (termed pseudo-RBD).

In about 20% of patients, RBD precedes the onset of motor symptoms in PD by years or decades [30, 31]. In patients diagnosed with idiopathic RBD, the 5-year risk of developing PD or another neurodegenerative disorder (typically another synucleinopathy or Alzheimer's disease) is 18%; the 10-year risk is 40% and the 12-year risk is 52% [32]. Thus, in patients diagnosed with idiopathic RBD, it is incumbent upon the diagnosing provider to carefully screen for signs of a neurodegenerative disease, maintain vigilant follow-up, and have a frank discussion with patients about the likelihood of developing a neurodegenerative condition. In patients with PD and RBD, the RBD symptoms tend to subside as PD severity progresses [33].

The typical time course of RBD symptoms predating onset of motor symptoms of PD may be explained by early involvement of the lower brainstem followed by extension of the synucleinopathy to the upper pons [34], ultimately affecting the locus subcoeruleus and pedunculopontine regions and cholinergic, serotonergic, and noradrenergic neurons [35, 36] (see Fig. 17.2). However, there is preliminary evidence that patients who develop RBD prior to onset of PD tend to have milder symptoms and take lower doses of dopaminergic medications compared to those who develop RBD concomitantly with PD or after onset of PD [37].

The pathophysiologic mechanisms underlying RBD in PD and other neurodegenerative conditions have not been fully elucidated. However, recently a "dual pathway" hypothesis was proposed, wherein involvement of the dorsal midbrain leads to REM sleep without atonia and involvement of the pons is responsible for generating complex locomotor patterns [38]. Further investigation is likely to lead to a better understanding of the neuropathology of RBD and its associated conditions.

Management of RBD consists first and foremost of environmental safety measures to minimize the risk of injury to the patient and bedpartners. Consider removing sharp and breakable objects from the bedroom, covering sharp corners, or placing the mattress on the floor in case of falls [33]. Unlike somnambulism, RBD rarely involves significant displacement from the bed. A careful review of medications is also indicated, as RBD may be precipitated by the use of certain tricyclic antidepressants and selective serotonin reuptake inhibitors [39], as well as withdrawal from alcohol and barbiturates. Pharmacologic treatment of RBD includes administration of clonazepam used at doses of 0.25–1 mg [40] and/or melatonin (often at high doses up to 10–15 mg) [41]. Generally, clonazepam is quite effective but should be considered carefully in patients with dementia, gait/balance issues, or obstructive sleep apnea. Evidence for the use of dopamine agonists is contradictory [42], and likewise there is limited evidence for rivastigmine [43] and donepezil [44].

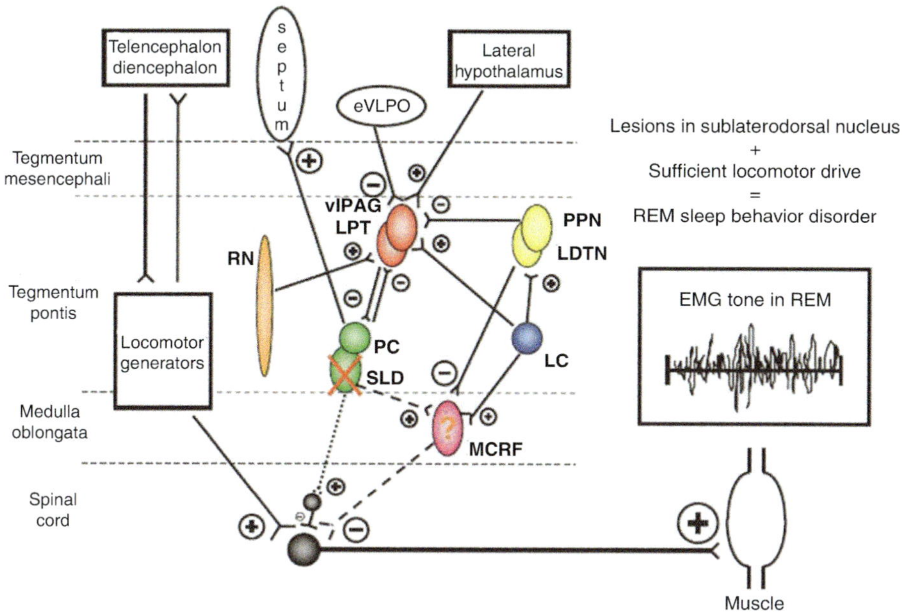

Fig. 17.2 Theoretical pathophysiology of REM behavior disorder. The REM-off region is represented by the vlPAG and LPT in red, and the REM-on region is represented by the PC and SLD in green. The "direct pathway" of skeletal muscle tone inhibition in REM is represented by the projection of the SLD onto spinal interneurons. Cell death in the SLD or lesions to the SLD result in decreased activation of spinal interneurons, resulting in decreased inhibition of spinal motoneurons and thus increased muscle activity and EMG tone. The "indirect pathway," denoted by the dashed line between the SLD and MCRF and then to the spinal interneurons, is thought to also contribute to REM-related skeletal muscle atonia. Decreased signaling from the SLD to the MCRF results in loss of inhibition from the MCRF onto spinal motoneurons, again resulting in increased muscle activity and EMG tone. *EMG* electromyographic, *eVLPO* extended part of the ventrolateral preoptic nucleus, *LC* locus coeruleus, *LDTN* laterodorsal tegmental nucleus, *LPT* lateral pontine tegmentum, *MCRF* magnocellular reticular formation, *PC* pre-coeruleus, *PPN* pedunculopontine nucleus, *REM* rapid eye movement, *RN* raphe nucleus, *SLD* sublaterodorsal nucleus, *vlPAG* ventrolateral part of the periaqueductal grey matter. (Adapted from Boeve BF, Silber MH, Saper CB, Ferman TJ, Dickson DW, Parisi JE, et al. [38])

Restless Legs Syndrome in Parkinson's Disease

Restless legs syndrome (RLS) is characterized by a bothersome urge to move the limbs (usually legs), typically occurring in the evening or before bed, worsened by inactivity or suggested immobility, and temporarily relieved with movement [28]. RLS is common in the general population (prevalence of 2.5–10%) and in PD patients (prevalence of 8–50%) [6, 45–49]. The reported prevalence of RLS in PD may be confounded by symptoms of dystonia, akathisia, painful neuropathy, and biphasic dyskinesia seen in PD patients, and an increased prevalence of RLS in PD has not been clearly demonstrated [50].

Further, the exact relationship between RLS and PD is not established. In one cross-sectional study, men with RLS had a threefold higher rate of PD compared to the general population [51], while in a prospective study of 22,999 health professionals, men with RLS were more likely to develop PD than men without RLS (adjusted relative risk of 1.47) [52]. Although hypofunction of dopaminergic pathways underlies both PD and RLS, RLS is not generally considered an early manifestation of PD (unlike RBD), and the upstream mechanisms of dopaminergic dysfunction appear to differ in the two conditions. Neurodegeneration is seen in the substantia nigra in PD but not in RLS [53], and normal substantia nigra iron deposition is observed in PD but not RLS [54, 55]. Similarly, neuroimaging via dopamine transporter single-photon emission computerized tomography (DaT-SPECT or DaT-Scan) shows striatal dopamine transporter levels are abnormal in PD [56], with mixed results in RLS [57–59]. Further, there are no reports that the disease-specific genetic loci of one disease are associated with those of the other [60, 61]. In summary, the association between PD and RLS is uncertain, and most patients with RLS are not prone to develop PD during their lifetime.

Management of RLS should include a review of medications and substances known to exacerbate these conditions, including nighttime use of selective serotonin reuptake inhibitors, tricyclic antidepressants, antihistamines, alcohol, and nicotine. Importantly, akathisia, which is common in PD, should be differentiated from RLS – the diurnal pattern of RLS and the subjective sensory complaints in RLS are useful distinguishers [62]. Screening for and correcting deficiencies in iron (goal ferritin levels above 50–75 ng/mL) [63] and vitamin D (goal levels not definitively established but generally recommended to correct deficiency and insufficiency) [64] can be helpful. Pharmacologic treatment with levodopa or dopamine agonists, gabapentin enacarbil [65], or opioids [64] should be considered if symptoms are problematic. Notably, levodopa and dopamine agonists can be associated with augmentation of RLS symptoms over time, and close monitoring for augmentation is recommended, with discontinuation of the dopaminergic agent (if possible) when augmentation is noted [66]. This phenomenon has not been well studied in the PD population, in which doses of dopaminergic agents tend to increase over time.

Periodic limb movements in sleep (PLMS) are frequently present in RLS patients but may occur independently of subjective leg discomfort. PLMS are characterized by repetitive involuntary limb movements during sleep which occur periodically and may result in cortical arousal. PLMS are common in PD, with prevalence reports of 30–80% [67], and may be more common in more advanced PD [68, 69]. The clinical significance of PLMS (occurring independently of RLS) has not been established, though there is epidemiologic evidence to suggest that sleep disruption from PLMD and RLS may be associated with hypertension, cardiovascular disease, and cognitive decline [70, 71]. Evidence supporting the benefits of pharmacologic therapy for PLMS outside the context of RLS is lacking.

Sleep Apnea in Parkinson's Disease

OSA is characterized by repetitive narrowing or collapse of the upper airway during sleep, leading to sleep disruption and/or oxyhemoglobin desaturations. The association of OSA with PD has not been adequately evaluated, as reflected by the significant limitations of studies in this area including single-center design, small sample sizes, use of patients clinically referred for polysomnography and/or use of inadequate control groups, as well as the contradictory findings reported. While some studies report a rate at least equivalent to the general population (estimates range from 6% to 17% and increase with age) [72], others suggest an increased prevalence of OSA in PD [73, 74]. At least one study showed a lower prevalence of OSA in PD compared to age-matched hospitalized non-PD patients with suspected venous thromboembolism, with no difference in daytime sleepiness between PD patients with OSA and those without [75]. Factors associated with PD that may predispose to OSA include open mouth associated with advanced-stage disease that results in oral breathing, jaw movement dysfunction, orofacial sensory-motor function deficit, and mandibular dystonia [76].

There is some evidence that OSA may promote the development of PD. A recent large longitudinal population study suggested an increased risk of developing PD in individuals with OSA compared to those without (this relationship was driven primarily by men and older individuals >60 years) [77]. If this association is confirmed, one conceivable mechanism is via oxidative stress from chronic intermittent hypoxemia [78], which can promote neurodegeneration and accumulation of alpha-synuclein plaques [79, 80]. An important unanswered question is whether treatment of OSA would reduce the incidence of PD and other neurodegenerative conditions.

OSA is most effectively treated with continuous positive airway pressure (CPAP), which stabilizes the upper airway and prevents repetitive narrowing and collapse. An abundance of evidence supports the use of CPAP in treating OSA, although CPAP has not been extensively studied in the context of PD specifically, where its use may be challenging in terms of compliance and tolerance [81]. At least one randomized controlled trial showed benefit of CPAP in improving excessive daytime sleepiness in PD [82]. Cognition and neuropsychiatric function are worse in patients with PD and concomitant OSA, but CPAP therapy has not been shown to definitively improve these domains [83]. Inadequate adherence to CPAP therapy is an important factor complicating such studies. In other neurodegenerative conditions such as Alzheimer's disease [84], there is evidence that consistent usage of CPAP for OSA can improve cognitive function. Alternative treatments for OSA, such as dental appliances for mandibular advancement, positional therapy (avoidance of supine sleep), and surgical procedures, have not been systematically evaluated in PD.

Central sleep apnea, characterized by repetitive cessation of respiratory effort in the presence of a patent upper airway, is uncommon in PD and the general population. Very limited data suggest that the distribution of apneas in patients with idiopathic RBD referred for PSG skewed toward a greater proportion of central and mixed events than in age-, gender-, and AHI-matched sleep controls [85], perhaps related to the autonomic instability seen in advanced PD.

Circadian Rhythm Disturbances in Parkinson's Disease

The timing and variation of sleepiness and wakefulness during a 24-hour period are governed in large part by the circadian rhythm, disruption of which is associated with deficits in sleep quality, daytime alertness, and neurocognitive function [86, 87].

Circadian rhythmicity is manifested in a multitude of physiologic parameters, including heart rate, blood pressure, immune function, and gene expression. Changes to the circadian rhythm are seen with normal aging, including decreased circadian amplitude and stability of rest-activity cycles [88, 89]. In PD, fluctuations in symptoms throughout the day are characteristic, including motor symptoms [90], autonomic function [91], sleep-wake schedules [10, 92, 93], and responsiveness to treatment [90, 93]. PD patients tend to have lower peak activity levels during the day [10] and increased nighttime activity [92], as well as more fragmented patterns of rest-activity [94] compared to healthy older adults. Continuous blood pressure monitoring in PD patients shows reversal of the typical day-night pattern of blood pressure (with the blood pressure nadir occurring at night) and increased diurnal blood pressure variability [95]. Similarly, loss of morning peak sympathetic tone has been described in PD [96].

Taken together, these findings suggest a dysregulation of circadian rhythmicity in PD, though the etiology of this dysregulation in unclear. Small studies implicate blunted secretion of melatonin in PD [97], as well as downregulation of melatonin receptors in relevant brain regions including the substantia nigra and amygdala [98]. It is not well understood whether structural or functional changes in the suprachiasmatic nucleus (the hypothalamic region mainly responsible for circadian regulation) or a downstream signaling pathway is responsible for the circadian dysregulation seen in PD, but dopaminergic neurons in the hypothalamus do not appear to be affected in PD [99].

Treatment of circadian dysrhythmia in PD has not been rigorously evaluated, though two important influencers of the circadian rhythm, melatonin and light, have been studied with respect to their effect on sleep and other symptoms in PD. Use of melatonin at bedtime is associated with improvement in subjective sleep quality, but not polysomnographic measures of sleep [22], and no improvement in PD motor symptoms has been associated with melatonin. Importantly, studies of melatonin in PD typically evaluate relatively large doses of melatonin administered at bedtime (3–6 mg), as opposed to lower doses (0.5–1 mg) timed to shift the circadian rhythm forward or backwards. Morning bright light therapy has been successfully used to improve depression and core motor symptoms in PD [100], although evening light therapy has also been noted to improve primary and secondary features of PD, including significant improvement in motor symptoms to the point of reducing required doses of levodopa [101]. It remains unclear whether and how modifying the circadian rhythm in PD affects core features of the disorder and by what mechanisms agents such as melatonin and light therapy exert a therapeutic benefit.

Excessive Daytime Sleepiness and Sleep Attacks in Parkinson's Disease

Excessive daytime sleepiness (EDS), defined as an increased tendency to fall asleep during the typical wakeful period, is very common in PD, both as a primary feature and as a result of medications. Though it is uncommon in newly diagnosed patients

[102], EDS tends to emerge or worsen with dopaminergic treatment [103] and with disease progression. Interestingly, data do not show a clear link between nocturnal symptoms (awakenings, perceived sleep quality) and severity of EDS in PD [73, 104, 105], but severity of PD tends to be associated with worse EDS. Sleep attacks are rarer but more concerning sudden overwhelming bouts of sleepiness that were first described in patients with PD taking pramipexole and ropinirole and occur with most dopamine agonists [106]. EDS affects as many as 50% of PD patients, and prevalence estimates of sleep attacks range from <5% to 43% [107], with one large study of 2952 patients with PD reporting a 6% prevalence [108]. In another study of 638 patients with relatively mild PD, 3.8% had sleep attacks while driving [109].

The etiology of EDS in PD is likely multifactorial. Despite the relatively low rate of EDS in newly diagnosed PD patients [102], at least one population study demonstrated a threefold increased risk of developing PD in men with EDS [110], and another large study showed a correlation between EDS and subsequent diagnosis of PD, but no effect of pre-diagnosis total sleep time [111]. Thus, EDS may represent more than a core non-motor feature of PD and may be an important precursor to the onset of motor symptoms. Loss of hypothalamic neurons that secrete the wakefulness-promoting neuropeptide hypocretin and lower levels of hypocretin in CSF have been noted in PD patients [112, 113]. Autopsy studies indicate that loss of hypocretin-secreting neurons seems to parallel PD disease progression [113, 114]. Dopaminergic medications also play a significant role in EDS in PD. Major clinical trials first indicated a 13–36% rate of EDS with use of dopaminergic medications [115] and a lower but significant rate of sleep attacks. The incidence of EDS appears to be highest with levodopa plus a dopamine agonist and lowest for levodopa monotherapy [108, 115].

Management of EDS begins with ensuring an adequate sleep duration of 7–9 hours per 24-hour period, a review of soporific medications, and screening for sleep disorders. Identification of EDS can be accomplished using any of a number of validated instruments, including the Epworth Sleepiness Scale (ESS), which is validated in PD patients. Interestingly, many patients with elevated ESS scores do not recognize that they are sleepy [116]. The Parkinson's Disease Sleep Scale [117] and its revised version [118] have been validated and may be useful for screening for sleep problems in PD, and the SCOPA-SLEEP is designed specifically for PD and correlates well with ESS and PSQI [119]. The ESS is the most commonly used of these scales in clinical practice but is limited to one manifestation of subjective sleepiness (the tendency to doze) and does not capture other valid complaints such as feeling tired or fatigued. Although evaluation for concomitant sleep disorders, namely, obstructive sleep apnea, via polysomnography is advisable in PD patients with symptoms, daytime testing via the multiple sleep latency test (MSLT) is not recommended as a routine test [119].

Evidence supports the treatment of EDS in PD using wakefulness-promoting agents, melatonin, and reduction of soporific medications. Modafinil at doses of 100–200 mg demonstrated improved EDS in PD in RCTs [120, 121], though at least one trial failed to show a benefit in improving subjective EDS [122]. A double-blind crossover trial showed benefit of melatonin over placebo in total sleep time and subjective sleep quality and daytime alertness with melatonin doses as low as 5 mg [123]. Reduction of soporific medications to their minimum effective dose and optimizing timing of medications to limit their effect on daytime alertness are also recommended. In patients with residual sleepiness and especially those with sleep attacks, avoidance of driving and other situations in which loss of alertness could be catastrophic should be recommended.

Sleep in Other Movement Disorders

Although the majority of research on sleep in movement disorders has focused on sleep disorders in PD, there is some literature regarding sleep problems in other movement disorders. As with sleep problems in PD, identifying and addressing sleep concerns in other movement disorders may lead to improved quality of life.

Essential Tremor

Despite the significant prevalence of essential tremor (ET), there are relatively few studies examining comorbid sleep disorders in this condition. Patients with ET report decreased subjective sleep quality compared to controls [124], and RLS may be more common in ET [125], but sleep disorders such as RBD which are common in PD are observed at equal frequencies in essential tremor and the general population [126]. One polysomnographic study of PD versus ET found RBD and REM without atonia were less common in ET than PD, even after adjusting for use of antidepressant medications, although a small number of the ET patients did have REM without atonia (which is classically seen in RBD) [127]. In this study, patients with ET occupied a "middle ground" between PD and controls with regard to subjective sleepiness and sleep disruption, as assessed by the Epworth Sleepiness Scale and Pittsburgh Sleep Quality Index. However, there was no significant difference with respect to clinical insomnia, polysomnogram-based arousals from sleep, RLS/PLMS, or OSA in PD versus ET.

Tourette Syndrome

Children and adolescents with Tourette syndrome (TS) commonly report difficulty with sleep initiation and maintenance [128], with up to 75% of those with Tourette syndrome plus attention deficit hyperactivity disorder reporting significant insomnia [129]. Insomnia prevalence appears to be lower in children not treated with stimulant medications. Polysomnographic studies show prolonged sleep latency and decreased sleep efficiency with increased awake time after sleep onset in patients with Tourette syndrome, though sleep architecture, distribution of NREM and REM sleep, and sleep state transitions do not significantly differ between TS and healthy age-matched controls [130, 131]. Periodic limb movements in TS [131] can be difficult to differentiate from motor tics, and there is discordant evidence regarding their prevalence. Although classically tics diminish during sleep, some may persist, especially during arousals and sleep state transitions [132], suggesting that inappropriate activation of central motor pattern generators may play a role in the pathogenesis of tics.

Dystonia

Sleep problems in patients with dystonia are poorly understood. Patients with focal cervical dystonia report poor sleep quality, though they do not report more daytime somnolence than controls [133]. Polysomnographic evidence suggests that patients with focal cervical dystonia have longer sleep latency and lower sleep efficiency than controls, along with more subjective complaints about sleep quality [134]. One study of primary dystonia found that dystonic symptoms during sleep were uncommon at baseline [135] and though motor symptoms improved after treatment with botulinum toxin injection, subjective sleep problems remained unchanged. Patients with dopa-responsive dystonia also tend to report more subjective sleep complaints (particularly reduced sleep quality) than controls, but small polysomnographic studies of dopa-responsive dystonia have not revealed differences in sleep onset latency, total sleep time, sleep architecture, or number of arousals [135, 136].

Huntington's Disease

In addition to its well-recognized motor, cognitive, and psychiatric symptoms, Huntington's disease (HD) is also frequently complicated by circadian rhythm and sleep disturbances. Nearly 90% of HD patients or their caregivers reported sleep problems, which were rated as important over 60% of the time [137]. As with PD, sleep symptoms often predate the development of motor and psychiatric symptoms in HD [138], raising the question of whether sleep disturbance is part of the HD gene-related pathology. Typical sleep problems in HD include prolonged sleep latency and insomnia, advanced sleep phase, excessive daytime sleepiness, and lower CSF hypocretin levels [139, 140], but not frank narcolepsy [141]. Polysomnography and actigraphy studies in small groups of HD patients are notable for reduced sleep efficiency, delayed and reduced REM sleep, frequent arousals, and increased periodic limb movements [141–143]. Circadian dysregulation, irregular sleep-wake patterns, and day-night reversal appear to be fairly common in advanced HD and particularly problematic for caregivers [144, 145]. Sleep-disordered breathing does not appear to be significantly more frequent in HD compared to the general population [146]. Interventions aimed at improving sleep quality in HD have not been systematically evaluated, though melatonin and melatonin receptor agonists are commonly used in clinical practice. Further, there is increasing scientific interest in the antiapoptotic role of melatonin for neuroprotection in neurodegenerative disorders [147].

Cerebellar Ataxias

As with other movement disorders, there is growing awareness of sleep disruption in the cerebellar ataxias. In the autosomal dominant cerebellar ataxias, RLS is common, with prevalence estimates ranging between 20% and 55% [148, 149]. The most robust data exist for spinocerebellar ataxia type 3 (SCA3; Machado-Joseph disease), in which insomnia, RLS, and RBD appear to be fairly common. RBD is present in up to 50% of cases of SCA3, suggesting a common pathway linked to dysfunction of pontine noradrenergic and midbrain cholinergic function. Like RBD in PD, onset of RBD symptoms may predate motor symptoms/ataxia in SCA3 by up to a decade [150]. Obstructive sleep apnea is reported at rates of 20–25% in SCA3 [151], though affected individuals tend to be older and with more advanced disease. EDS is reported in nearly 50% of SCA3 patients [152], though it is not clear whether this is due to comorbid sleep disorders or a fundamental feature of the neurodegenerative process.

In the recessive cerebellar ataxias, disordered nocturnal breathing appears to be the predominant sleep finding, though data are generally limited to small case series. Both obstructive and central sleep apnea have been described in Friedrich's ataxia [153]. A polysomnographic case series of 12 adolescent patients with ataxia telangiectasia showed that 8 (67%) had OSA and 2 (16%) had borderline or mild nocturnal hypoventilation [154]. In Joubert syndrome, the well-described daytime respiratory abnormalities can also extend into sleep, manifesting mainly as isolated tachypnea or repetitive tachypnea followed by central apneas [155]. Although robust data are lacking, patients with recessive cerebellar ataxias should be screened for symptoms of sleep-disordered breathing and referred for polysomnography if symptoms are present.

Conclusion

Sleep complaints are common in patients with movement disorders and significantly impact quality of life. Some sleep and movement disorders may share pathophysiologic mechanisms, offering insight into the pathogenesis of these conditions. Polysomnographic evaluation should be considered for patients with frequent awakenings, dream enactment behavior, and symptoms of sleep-disordered breathing. In many cases, sleep symptoms may precede the onset of core motor symptoms. Treatment of sleep disorders can lead to improved quality of life and may benefit core symptoms of movement disorders.

References

1. Karlsen KH, Larsen JP, Tandberg E, Maeland JG. Influence of clinical and demographic variables on quality of life in patients with Parkinson's disease. J Neurol Neurosurg Psychiatry. 1999;66(4):431–5.

2. Aarsland D, Alves G, Larsen JP. Disorders of motivation, sexual conduct, and sleep in Parkinson's disease. Adv Neurol. 2005;96:56–64.
3. Shulman LM, Taback RL, Bean J, Weiner WJ. Comorbidity of the nonmotor symptoms of Parkinson's disease. Mov Disord. 2001;16(3):507–10.
4. Wulff K, Gatti S, Wettstein JG, Foster RG. Sleep and circadian rhythm disruption in psychiatric and neurodegenerative disease. Nat Rev Neurosci. 2010;11(8):589–99.
5. Parkinson J. An essay on the shaking palsy. 1817. J Neuropsychiatry Clin Neurosci. 2002;14(2):223–36; discussion 2
6. Lees AJ, Blackburn NA, Campbell VL. The nighttime problems of Parkinson's disease. Clin Neuropharmacol. 1988;11(6):512–9.
7. Factor SA, McAlarney T, Sanchez-Ramos JR, Weiner WJ. Sleep disorders and sleep effect in Parkinson's disease. Mov Disord. 1990;5(4):280–5.
8. Tandberg E, Larsen JP, Karlsen K. A community-based study of sleep disorders in patients with Parkinson's disease. Mov Disord. 1998;13(6):895–9.
9. Comella C. Sleep episodes in Parkinson's disease: more questions remain. Sleep Med. 2003;4(4):267–8.
10. van Hilten B, Hoff JI, Middelkoop HA, van der Velde EA, Kerkhof GA, Wauquier A, et al. Sleep disruption in Parkinson's disease. Assessment by continuous activity monitoring. Arch Neurol. 1994;51(9):922–8.
11. Ylikoski A, Martikainen K, Partinen M. Parasomnias and isolated sleep symptoms in Parkinson's disease: a questionnaire study on 661 patients. J Neurol Sci. 2014;346(1–2):204–8.
12. Happe S, Schrodl B, Faltl M, Muller C, Auff E, Zeitlhofer J. Sleep disorders and depression in patients with Parkinson's disease. Acta Neurol Scand. 2001;104(5):275–80.
13. Taylor DJ, Lichstein KL, Durrence HH, Reidel BW, Bush AJ. Epidemiology of insomnia, depression, and anxiety. Sleep. 2005;28(11):1457–64.
14. Ehmann TS, Beninger RJ, Gawel MJ, Riopelle RJ. Depressive symptoms in Parkinson's disease: a comparison with disabled control subjects. J Geriatr Psychiatry Neurol. 1990;3(1):3–9.
15. Sateia MJ, Buysse DJ, Krystal AD, Neubauer DN, Heald JL. Clinical practice guideline for the pharmacologic treatment of chronic insomnia in adults: an American Academy of Sleep Medicine clinical practice guideline. J Clin Sleep Med. 2017;13(2):307–49.
16. Morgenthaler T, Kramer M, Alessi C, Friedman L, Boehlecke B, Brown T, et al. Practice parameters for the psychological and behavioral treatment of insomnia: an update. An American academy of sleep medicine report. Sleep. 2006;29(11):1415–9.
17. Pahwa R, Busenbark K, Huber SJ, Michalek D, Hubble JP, Koller WC. Clinical experience with controlled-release carbidopa/levodopa in Parkinson's disease. Neurology. 1993;43(4):677–81.
18. Van den Kerchove M, Jacquy J, Gonce M, De Deyn PP. Sustained-release levodopa in parkinsonian patients with nocturnal disabilities. Acta Neurol Belg. 1993;93(1):32–9.
19. Ebersbach G, Hahn K, Lorrain M, Storch A. Tolcapone improves sleep in patients with advanced Parkinson's disease (PD). Arch Gerontol Geriatr. 2010;51(3):e125–8.
20. Romigi A, Stanzione P, Marciani MG, Izzi F, Placidi F, Cervellino A, et al. Effect of cabergoline added to levodopa treatment on sleep-wake cycle in idiopathic Parkinson's disease: an open label 24-hour polysomnographic study. J Neural Transm (Vienna). 2006;113(12):1909–13.
21. Menza M, Dobkin RD, Marin H, Gara M, Bienfait K, Dicke A, et al. Treatment of insomnia in Parkinson's disease: a controlled trial of eszopiclone and placebo. Mov Disord. 2010;25(11):1708–14.
22. Medeiros CA, Carvalhedo de Bruin PF, Lopes LA, Magalhaes MC, de Lourdes Seabra M, de Bruin VM. Effect of exogenous melatonin on sleep and motor dysfunction in Parkinson's disease. A randomized, double blind, placebo-controlled study. J Neurol. 2007;254(4):459–64.
23. Rios Romenets S, Creti L, Fichten C, Bailes S, Libman E, Pelletier A, et al. Doxepin and cognitive behavioural therapy for insomnia in patients with Parkinson's disease – a randomized study. Parkinsonism Relat Disord. 2013;19(7):670–5.
24. Comella CL, Nardine TM, Diederich NJ, Stebbins GT. Sleep-related violence, injury, and REM sleep behavior disorder in Parkinson's disease. Neurology. 1998;51(2):526–9.

25. Gagnon JF, Bedard MA, Fantini ML, Petit D, Panisset M, Rompre S, et al. REM sleep behavior disorder and REM sleep without atonia in Parkinson's disease. Neurology. 2002;59(4):585–9.
26. Scaglione C, Vignatelli L, Plazzi G, Marchese R, Negrotti A, Rizzo G, et al. REM sleep behaviour disorder in Parkinson's disease: a questionnaire-based study. Neurol Sci. 2005;25(6):316–21.
27. Ohayon MM, Caulet M, Priest RG. Violent behavior during sleep. J Clin Psychiatry. 1997;58(8):369–76; quiz 77
28. American Academy of Sleep Medicine. International classification of sleep disorders. 3rd ed. Darien: American Academy of Sleep Medicine; 2014.
29. Schenck CH, Bundlie SR, Ettinger MG, Mahowald MW. Chronic behavioral disorders of human REM sleep: a new category of parasomnia. Sleep. 1986;9(2):293–308.
30. Iranzo A, Santamaria J, Tolosa E. The clinical and pathophysiological relevance of REM sleep behavior disorder in neurodegenerative diseases. Sleep Med Rev. 2009;13(6):385–401.
31. Claassen DO, Josephs KA, Ahlskog JE, Silber MH, Tippmann-Peikert M, Boeve BF. REM sleep behavior disorder preceding other aspects of synucleinopathies by up to half a century. Neurology. 2010;75(6):494–9.
32. Postuma RB, Gagnon JF, Vendette M, Fantini ML, Massicotte-Marquez J, Montplaisir J. Quantifying the risk of neurodegenerative disease in idiopathic REM sleep behavior disorder. Neurology. 2009;72(15):1296–300.
33. Schenck CH, Mahowald MW. REM sleep behavior disorder: clinical, developmental, and neuroscience perspectives 16 years after its formal identification in SLEEP. Sleep. 2002;25(2):120–38.
34. Braak H, Del Tredici K, Rub U, de Vos RA, Jansen Steur EN, Braak E. Staging of brain pathology related to sporadic Parkinson's disease. Neurobiol Aging. 2003;24(2):197–211.
35. Kimura K, Tachibana N, Kohyama J, Otsuka Y, Fukazawa S, Waki R. A discrete pontine ischemic lesion could cause REM sleep behavior disorder. Neurology. 2000;55(6):894–5.
36. Plazzi G, Corsini R, Provini F, Pierangeli G, Martinelli P, Montagna P, et al. REM sleep behavior disorders in multiple system atrophy. Neurology. 1997;48(4):1094–7.
37. Ferri R, Cosentino FI, Pizza F, Arico D, Plazzi G. The timing between REM sleep behavior disorder and Parkinson's disease. Sleep Breath. 2014;18(2):319–23.
38. Boeve BF, Silber MH, Saper CB, Ferman TJ, Dickson DW, Parisi JE, et al. Pathophysiology of REM sleep behaviour disorder and relevance to neurodegenerative disease. Brain. 2007;130(Pt 11):2770–88.
39. Mahowald MW, Schenck CH, Bornemann MA. Pathophysiologic mechanisms in REM sleep behavior disorder. Curr Neurol Neurosci Rep. 2007;7(2):167–72.
40. Schenck CH, Mahowald MW. Rapid eye movement sleep parasomnias. Neurol Clin. 2005;23(4):1107–26.
41. Kunz D, Bes F. Melatonin as a therapy in REM sleep behavior disorder patients: an open-labeled pilot study on the possible influence of melatonin on REM-sleep regulation. Mov Disord. 1999;14(3):507–11.
42. Fantini ML, Gagnon JF, Filipini D, Montplaisir J. The effects of pramipexole in REM sleep behavior disorder. Neurology. 2003;61(10):1418–20.
43. Aurora RN, Zak RS, Maganti RK, Auerbach SH, Casey KR, Chowdhuri S, et al. Best practice guide for the treatment of REM sleep behavior disorder (RBD). J Clin Sleep Med. 2010;6(1):85–95.
44. Ringman JM, Simmons JH. Treatment of REM sleep behavior disorder with donepezil: a report of three cases. Neurology. 2000;55(6):870–1.
45. Tan EK, Lum SY, Wong MC. Restless legs syndrome in Parkinson's disease. J Neurol Sci. 2002;196(1–2):33–6.
46. Krishnan PR, Bhatia M, Behari M. Restless legs syndrome in Parkinson's disease: a case-controlled study. Mov Disord. 2003;18(2):181–5.
47. Ondo WG, Vuong KD, Jankovic J. Exploring the relationship between Parkinson disease and restless legs syndrome. Arch Neurol. 2002;59(3):421–4.

48. Nomura T, Inoue Y, Miyake M, Yasui K, Nakashima K. Prevalence and clinical characteristics of restless legs syndrome in Japanese patients with Parkinson's disease. Mov Disord. 2006;21(3):380–4.
49. Garcia-Borreguero D, Egatz R, Winkelmann J, Berger K. Epidemiology of restless legs syndrome: the current status. Sleep Med Rev. 2006;10(3):153–67.
50. Hogl B, Stefani A. Restless legs syndrome and periodic leg movements in patients with movement disorders: specific considerations. Mov Disord. 2017;32(5):669–81.
51. Gao X, Schwarzschild MA, O'Reilly EJ, Wang H, Ascherio A. Restless legs syndrome and Parkinson's disease in men. Mov Disord. 2010;25(15):2654–7.
52. Wong JC, Li Y, Schwarzschild MA, Ascherio A, Gao X. Restless legs syndrome: an early clinical feature of Parkinson disease in men. Sleep. 2014;37(2):369–72.
53. Pittock SJ, Parrett T, Adler CH, Parisi JE, Dickson DW, Ahlskog JE. Neuropathology of primary restless leg syndrome: absence of specific tau- and alpha-synuclein pathology. Mov Disord. 2004;19(6):695–9.
54. Allen RP, Barker PB, Wehrl FW, Song HK, Earley CJ. MRI measurement of brain iron in patients with restless legs syndrome. Neurology. 2001;56(2):263–5.
55. Godau J, Schweitzer KJ, Liepelt I, Gerloff C, Berg D. Substantia nigra hypoechogenicity: definition and findings in restless legs syndrome. Mov Disord. 2007;22(2):187–92.
56. Bajaj N, Hauser RA, Grachev ID. Clinical utility of dopamine transporter single photon emission CT (DaT-SPECT) with (123I) ioflupane in diagnosis of parkinsonian syndromes. J Neurol Neurosurg Psychiatry. 2013;84(11):1288–95.
57. Earley CJ, Kuwabara H, Wong DF, Gamaldo C, Salas R, Brasic J, et al. The dopamine transporter is decreased in the striatum of subjects with restless legs syndrome. Sleep. 2011;34(3):341–7.
58. Kim KW, Jhoo JH, Lee SB, Lee SD, Kim TH, Kim SE, et al. Increased striatal dopamine transporter density in moderately severe old restless legs syndrome patients. Eur J Neurol. 2012;19(9):1213–8.
59. Eisensehr I, Wetter TC, Linke R, Noachtar S, von Lindeiner H, Gildehaus FJ, et al. Normal IPT and IBZM SPECT in drug-naive and levodopa-treated idiopathic restless legs syndrome. Neurology. 2001;57(7):1307–9.
60. Moller JC, Unger M, Stiasny-Kolster K, Oertel WH. Restless legs syndrome (RLS) and Parkinson's disease (PD)-related disorders or different entities? J Neurol Sci. 2010;289(1–2):135–7.
61. Kumar KR, Lohmann K, Klein C. Genetics of Parkinson disease and other movement disorders. Curr Opin Neurol. 2012;25(4):466–74.
62. Walters AS, Hening W, Rubinstein M, Chokroverty S. A clinical and polysomnographic comparison of neuroleptic-induced akathisia and the idiopathic restless legs syndrome. Sleep. 1991;14(4):339–45.
63. Allen RP, Picchietti DL, Auerbach M, Cho YW, Connor JR, Earley CJ, et al. Evidence-based and consensus clinical practice guidelines for the iron treatment of restless legs syndrome/ Willis-Ekbom disease in adults and children: an IRLSSG task force report. Sleep Med. 2018;41:27–44.
64. Silber MH, Becker PM, Earley C, Garcia-Borreguero D, Ondo WG. Medical Advisory Board of the Willis-Ekbom Disease F. Willis-Ekbom Disease Foundation revised consensus statement on the management of restless legs syndrome. Mayo Clin Proc. 2013;88(9):977–86.
65. Agarwal P, Griffith A, Costantino HR, Vaish N. Gabapentin enacarbil – clinical efficacy in restless legs syndrome. Neuropsychiatr Dis Treat. 2010;6:151–8.
66. Garcia-Borreguero D, Silber MH, Winkelman JW, Hogl B, Bainbridge J, Buchfuhrer M, et al. Guidelines for the first-line treatment of restless legs syndrome/Willis-Ekbom disease, prevention and treatment of dopaminergic augmentation: a combined task force of the IRLSSG, EURLSSG, and the RLS-foundation. Sleep Med. 2016;21:1–11.
67. Poewe W, Hogl B. Akathisia, restless legs and periodic limb movements in sleep in Parkinson's disease. Neurology. 2004;63(8 Suppl 3):S12–6.

68. Wetter TC, Collado-Seidel V, Pollmacher T, Yassouridis A, Trenkwalder C. Sleep and periodic leg movement patterns in drug-free patients with Parkinson's disease and multiple system atrophy. Sleep. 2000;23(3):361–7.
69. Young A, Home M, Churchward T, Freezer N, Holmes P, Ho M. Comparison of sleep disturbance in mild versus severe Parkinson's disease. Sleep. 2002;25(5):573–7.
70. Walters AS, Rye DB. Review of the relationship of restless legs syndrome and periodic limb movements in sleep to hypertension, heart disease, and stroke. Sleep. 2009;32(5):589–97.
71. Leng Y, Blackwell T, Stone KL, Hoang TD, Redline S, Yaffe K. Periodic limb movements in sleep are associated with greater cognitive decline in older men without dementia. Sleep. 2016;39(10):1807–10.
72. Senaratna CV, Perret JL, Lodge CJ, Lowe AJ, Campbell BE, Matheson MC, et al. Prevalence of obstructive sleep apnea in the general population: a systematic review. Sleep Med Rev. 2017;34:70–81.
73. Arnulf I, Konofal E, Merino-Andreu M, Houeto JL, Mesnage V, Welter ML, et al. Parkinson's disease and sleepiness: an integral part of PD. Neurology. 2002;58(7):1019–24.
74. Norlinah MI, Afidah KN, Noradina AT, Shamsul AS, Hamidon BB, Sahathevan R, et al. Sleep disturbances in Malaysian patients with Parkinson's disease using polysomnography and PDSS. Parkinsonism Relat Disord. 2009;15(9):670–4.
75. Cochen De Cock V, Abouda M, Leu S, Oudiette D, Roze E, Vidailhet M, et al. Is obstructive sleep apnea a problem in Parkinson's disease? Sleep Med. 2010;11(3):247–52.
76. Raggi A, Bella R, Pennisi G, Neri W, Ferri R. Sleep disorders in Parkinson's disease: a narrative review of the literature. Rev Neurosci. 2013;24(3):279–91.
77. Chou PS, Lai CL, Chou YH, Chang WP. Sleep apnea and the subsequent risk of Parkinson's disease: a 3-year nationwide population-based study. Neuropsychiatr Dis Treat. 2017;13:959–65.
78. Lurie A. Inflammation, oxidative stress, and procoagulant and thrombotic activity in adults with obstructive sleep apnea. Adv Cardiol. 2011;46:43–66.
79. Xiang W, Schlachetzki JC, Helling S, Bussmann JC, Berlinghof M, Schaffer TE, et al. Oxidative stress-induced posttranslational modifications of alpha-synuclein: specific modification of alpha-synuclein by 4-hydroxy-2-nonenal increases dopaminergic toxicity. Mol Cell Neurosci. 2013;54:71–83.
80. Guo Y, Scarlata S. A loss in cellular protein partners promotes alpha-synuclein aggregation in cells resulting from oxidative stress. Biochemistry. 2013;52(22):3913–20.
81. Terzaghi M, Spelta L, Minafra B, Rustioni V, Zangaglia R, Pacchetti C, et al. Treating sleep apnea in Parkinson's disease with C-PAP: feasibility concerns and effects on cognition and alertness. Sleep Med. 2017;33:114–8.
82. Neikrug AB, Liu L, Avanzino JA, Maglione JE, Natarajan L, Bradley L, et al. Continuous positive airway pressure improves sleep and daytime sleepiness in patients with Parkinson disease and sleep apnea. Sleep. 2014;37(1):177–85.
83. Harmell AL, Neikrug AB, Palmer BW, Avanzino JA, Liu L, Maglione JE, et al. Obstructive sleep apnea and cognition in Parkinson's disease. Sleep Med. 2016;21:28–34.
84. Cooke JR, Ayalon L, Palmer BW, Loredo JS, Corey-Bloom J, Natarajan L, et al. Sustained use of CPAP slows deterioration of cognition, sleep, and mood in patients with Alzheimer's disease and obstructive sleep apnea: a preliminary study. J Clin Sleep Med. 2009;5(4):305–9.
85. Diederich NJ, Vaillant M, Leischen M, Mancuso G, Golinval S, Nati R, et al. Sleep apnea syndrome in Parkinson's disease. A case-control study in 49 patients. Mov Disord. 2005;20(11):1413–8.
86. Buysse DJ, Monk TH, Carrier J, Begley A. Circadian patterns of sleep, sleepiness, and performance in older and younger adults. Sleep. 2005;28(11):1365–76.
87. Silva EJ, Wang W, Ronda JM, Wyatt JK, Duffy JF. Circadian and wake-dependent influences on subjective sleepiness, cognitive throughput, and reaction time performance in older and young adults. Sleep. 2010;33(4):481–90.
88. Czeisler CA, Dumont M, Duffy JF, Steinberg JD, Richardson GS, Brown EN, et al. Association of sleep-wake habits in older people with changes in output of circadian pacemaker. Lancet. 1992;340(8825):933–6.

89. Hofman MA. The human circadian clock and aging. Chronobiol Int. 2000;17(3):245–59.
90. Bonuccelli U, Del Dotto P, Lucetti C, Petrozzi L, Bernardini S, Gambaccini G, et al. Diurnal motor variations to repeated doses of levodopa in Parkinson's disease. Clin Neuropharmacol. 2000;23(1):28–33.
91. Pursiainen V, Haapaniemi TH, Korpelainen JT, Huikuri HV, Sotaniemi KA, Myllyla VV. Circadian heart rate variability in Parkinson's disease. J Neurol. 2002;249(11):1535–40.
92. van Hilten JJ, Hoogland G, van der Velde EA, Middelkoop HA, Kerkhof GA, Roos RA. Diurnal effects of motor activity and fatigue in Parkinson's disease. J Neurol Neurosurg Psychiatry. 1993;56(8):874–7.
93. Placidi F, Izzi F, Romigi A, Stanzione P, Marciani MG, Brusa L, et al. Sleep-wake cycle and effects of cabergoline monotherapy in de novo Parkinson's disease patients. An ambulatory polysomnographic study. J Neurol. 2008;255(7):1032–7.
94. Whitehead DL, Davies AD, Playfer JR, Turnbull CJ. Circadian rest-activity rhythm is altered in Parkinson's disease patients with hallucinations. Mov Disord. 2008;23(8):1137–45.
95. Ejaz AA, Sekhon IS, Munjal S. Characteristic findings on 24-h ambulatory blood pressure monitoring in a series of patients with Parkinson's disease. Eur J Intern Med. 2006;17(6):417–20.
96. Devos D, Kroumova M, Bordet R, Vodougnon H, Guieu JD, Libersa C, et al. Heart rate variability and Parkinson's disease severity. J Neural Transm (Vienna). 2003;110(9):997–1011.
97. Bordet R, Devos D, Brique S, Touitou Y, Guieu JD, Libersa C, et al. Study of circadian melatonin secretion pattern at different stages of Parkinson's disease. Clin Neuropharmacol. 2003;26(2):65–72.
98. Adi N, Mash DC, Ali Y, Singer C, Shehadeh L, Papapetropoulos S. Melatonin MT1 and MT2 receptor expression in Parkinson's disease. Med Sci Monit. 2010;16(2):BR61–7.
99. Matzuk MM, Saper CB. Preservation of hypothalamic dopaminergic neurons in Parkinson's disease. Ann Neurol. 1985;18(5):552–5.
100. Paus S, Schmitz-Hubsch T, Wullner U, Vogel A, Klockgether T, Abele M. Bright light therapy in Parkinson's disease: a pilot study. Mov Disord. 2007;22(10):1495–8.
101. Willis GL, Turner EJ. Primary and secondary features of Parkinson's disease improve with strategic exposure to bright light: a case series study. Chronobiol Int. 2007;24(3):521–37.
102. Fabbrini G, Barbanti P, Aurilia C, Vanacore N, Pauletti C, Meco G. Excessive daytime sleepiness in de novo and treated Parkinson's disease. Mov Disord. 2002;17(5):1026–30.
103. Kaynak D, Kiziltan G, Kaynak H, Benbir G, Uysal O. Sleep and sleepiness in patients with Parkinson's disease before and after dopaminergic treatment. Eur J Neurol. 2005;12(3):199–207.
104. Rye DB, Bliwise DL, Dihenia B, Gurecki P. FAST TRACK: daytime sleepiness in Parkinson's disease. J Sleep Res. 2000;9(1):63–9.
105. Roth T, Rye DB, Borchert LD, Bartlett C, Bliwise DL, Cantor C, et al. Assessment of sleepiness and unintended sleep in Parkinson's disease patients taking dopamine agonists. Sleep Med. 2003;4(4):275–80.
106. Frucht S, Rogers JD, Greene PE, Gordon MF, Fahn S. Falling asleep at the wheel: motor vehicle mishaps in persons taking pramipexole and ropinirole. Neurology. 1999;52(9):1908–10.
107. Montastruc JL, Brefel-Courbon C, Senard JM, Bagheri H, Ferreira J, Rascol O, et al. Sleep attacks and antiparkinsonian drugs: a pilot prospective pharmacoepidemiologic study. Clin Neuropharmacol. 2001;24(3):181–3.
108. Paus S, Brecht HM, Koster J, Seeger G, Klockgether T, Wullner U. Sleep attacks, daytime sleepiness, and dopamine agonists in Parkinson's disease. Mov Disord. 2003;18(6):659–67.
109. Hobson DE, Lang AE, Martin WR, Razmy A, Rivest J, Fleming J. Excessive daytime sleepiness and sudden-onset sleep in Parkinson disease: a survey by the Canadian Movement Disorders group. JAMA. 2002;287(4):455–63.
110. Abbott RD, Ross GW, White LR, Tanner CM, Masaki KH, Nelson JS, et al. Excessive daytime sleepiness and subsequent development of Parkinson disease. Neurology. 2005;65(9):1442–6.
111. Gao J, Huang X, Park Y, Hollenbeck A, Blair A, Schatzkin A, et al. Daytime napping, nighttime sleeping, and Parkinson disease. Am J Epidemiol. 2011;173(9):1032–8.

112. Drouot X, Moutereau S, Nguyen JP, Lefaucheur JP, Creange A, Remy P, et al. Low levels of ventricular CSF orexin/hypocretin in advanced PD. Neurology. 2003;61(4):540–3.
113. Fronczek R, Overeem S, Lee SY, Hegeman IM, van Pelt J, van Duinen SG, et al. Hypocretin (orexin) loss in Parkinson's disease. Brain. 2007;130(Pt 6):1577–85.
114. Thannickal TC, Lai YY, Siegel JM. Hypocretin (orexin) cell loss in Parkinson's disease. Brain. 2007;130(Pt 6):1586–95.
115. Parkinson Study G. Pramipexole vs levodopa as initial treatment for Parkinson disease: a randomized controlled trial. Parkinson study group. JAMA. 2000;284(15):1931–8.
116. Brodsky MA, Godbold J, Roth T, Olanow CW. Sleepiness in Parkinson's disease: a controlled study. Mov Disord. 2003;18(6):668–72.
117. Chaudhuri KR, Pal S, DiMarco A, Whately-Smith C, Bridgman K, Mathew R, et al. The Parkinson's disease sleep scale: a new instrument for assessing sleep and nocturnal disability in Parkinson's disease. J Neurol Neurosurg Psychiatry. 2002;73(6):629–35.
118. Trenkwalder C, Kohnen R, Hogl B, Metta V, Sixel-Doring F, Frauscher B, et al. Parkinson's disease sleep scale – validation of the revised version PDSS-2. Mov Disord. 2011;26(4):644–52.
119. Marinus J, Visser M, van Hilten JJ, Lammers GJ, Stiggelbout AM. Assessment of sleep and sleepiness in Parkinson disease. Sleep. 2003;26(8):1049–54.
120. Adler CH, Caviness JN, Hentz JG, Lind M, Tiede J. Randomized trial of modafinil for treating subjective daytime sleepiness in patients with Parkinson's disease. Mov Disord. 2003;18(3):287–93.
121. Hogl B, Saletu M, Brandauer E, Glatzl S, Frauscher B, Seppi K, et al. Modafinil for the treatment of daytime sleepiness in Parkinson's disease: a double-blind, randomized, crossover, placebo-controlled polygraphic trial. Sleep. 2002;25(8):905–9.
122. Ondo WG, Fayle R, Atassi F, Jankovic J. Modafinil for daytime somnolence in Parkinson's disease: double blind, placebo controlled parallel trial. J Neurol Neurosurg Psychiatry. 2005;76(12):1636–9.
123. Dowling GA, Mastick J, Colling E, Carter JH, Singer CM, Aminoff MJ. Melatonin for sleep disturbances in Parkinson's disease. Sleep Med. 2005;6(5):459–66.
124. Chandran V, Pal PK, Reddy JY, Thennarasu K, Yadav R, Shivashankar N. Non-motor features in essential tremor. Acta Neurol Scand. 2012;125(5):332–7.
125. Ondo WG, Lai D. Association between restless legs syndrome and essential tremor. Mov Disord. 2006;21(4):515–8.
126. Adler CH, Hentz JG, Shill HA, Sabbagh MN, Driver-Dunckley E, Evidente VG, et al. Probable RBD is increased in Parkinson's disease but not in essential tremor or restless legs syndrome. Parkinsonism Relat Disord. 2011;17(6):456–8.
127. Barut BO, Tascilar N, Varo A. Sleep disturbances in essential tremor and Parkinson disease: a polysomnographic study. J Clin Sleep Med. 2015;11(6):655–62.
128. Modafferi S, Stornelli M, Chiarotti F, Cardona F, Bruni O. Sleep, anxiety and psychiatric symptoms in children with Tourette syndrome and tic disorders. Eur J Paediatr Neurol. 2016;20(5):696–703.
129. Ghosh D, Rajan PV, Das D, Datta P, Rothner AD, Erenberg G. Sleep disorders in children with Tourette syndrome. Pediatr Neurol. 2014;51(1):31–5.
130. Kostanecka-Endress T, Banaschewski T, Kinkelbur J, Wullner I, Lichtblau S, Cohrs S, et al. Disturbed sleep in children with Tourette syndrome: a polysomnographic study. J Psychosom Res. 2003;55(1):23–9.
131. Cohrs S, Rasch T, Altmeyer S, Kinkelbur J, Kostanecka T, Rothenberger A, et al. Decreased sleep quality and increased sleep related movements in patients with Tourette's syndrome. J Neurol Neurosurg Psychiatry. 2001;70(2):192–7.
132. Fish DR, Sawyers D, Allen PJ, Blackie JD, Lees AJ, Marsden CD. The effect of sleep on the dyskinetic movements of Parkinson's disease, Gilles de la Tourette syndrome, Huntington's disease, and torsion dystonia. Arch Neurol. 1991;48(2):210–4.
133. Eichenseer SR, Stebbins GT, Comella CL. Beyond a motor disorder: a prospective evaluation of sleep quality in cervical dystonia. Parkinsonism Relat Disord. 2014;20(4):405–8.

134. Antelmi E, Ferri R, Provini F, Scaglione CML, Mignani F, Rundo F, et al. Modulation of the muscle activity during sleep in cervical dystonia. Sleep. 2017;40(7):zsx088.
135. Hertenstein E, Tang NK, Bernstein CJ, Nissen C, Underwood MR, Sandhu HK. Sleep in patients with primary dystonia: a systematic review on the state of research and perspectives. Sleep Med Rev. 2016;26:95–107.
136. Bruggemann N, Stiller S, Tadic V, Kasten M, Munchau A, Graf J, et al. Non-motor phenotype of dopa-responsive dystonia and quality of life assessment. Parkinsonism Relat Disord. 2014;20(4):428–31.
137. Taylor N, Bramble D. Sleep disturbance and Huntingdon's disease. Br J Psychiatry. 1997;171:393.
138. van Wamelen DJ, Roos RA, Aziz NA. Therapeutic strategies for circadian rhythm and sleep disturbances in Huntington disease. Neurodegener Dis Manag. 2015;5(6):549–59.
139. Aziz A, Fronczek R, Maat-Schieman M, Unmehopa U, Roelandse F, Overeem S, et al. Hypocretin and melanin-concentrating hormone in patients with Huntington disease. Brain Pathol. 2008;18(4):474–83.
140. Roos RA, Aziz NA. Hypocretin-1 and secondary signs in Huntington's disease. Parkinsonism Relat Disord. 2007;13(Suppl 3):S387–90.
141. Arnulf I, Nielsen J, Lohmann E, Schiefer J, Wild E, Jennum P, et al. Rapid eye movement sleep disturbances in Huntington disease. Arch Neurol. 2008;65(4):482–8.
142. Wiegand M, Moller AA, Schreiber W, Lauer C, Krieg JC. Brain morphology and sleep EEG in patients with Huntington's disease. Eur Arch Psychiatry Clin Neurosci. 1991;240(3):148–52.
143. Videnovic A, Leurgans S, Fan W, Jaglin J, Shannon KM. Daytime somnolence and nocturnal sleep disturbances in Huntington disease. Parkinsonism Relat Disord. 2009;15(6):471–4.
144. Morton AJ, Wood NI, Hastings MH, Hurelbrink C, Barker RA, Maywood ES. Disintegration of the sleep-wake cycle and circadian timing in Huntington's disease. J Neurosci. 2005;25(1):157–63.
145. Hurelbrink CB, Lewis SJ, Barker RA. The use of the Actiwatch-Neurologica system to objectively assess the involuntary movements and sleep-wake activity in patients with mild-moderate Huntington's disease. J Neurol. 2005;252(6):642–7.
146. Piano C, Losurdo A, Della Marca G, Solito M, Calandra-Buonaura G, Provini F, et al. Polysomnographic findings and clinical correlates in Huntington disease: a cross-sectional cohort study. Sleep. 2015;38(9):1489–95.
147. Wang X. The antiapoptotic activity of melatonin in neurodegenerative diseases. CNS Neurosci Ther. 2009;15(4):345–57.
148. Schols L, Bauer P, Schmidt T, Schulte T, Riess O. Autosomal dominant cerebellar ataxias: clinical features, genetics, and pathogenesis. Lancet Neurol. 2004;3(5):291–304.
149. Abele M, Burk K, Laccone F, Dichgans J, Klockgether T. Restless legs syndrome in spinocerebellar ataxia types 1, 2, and 3. J Neurol. 2001;248(4):311–4.
150. D'Abreu A, Friedman J, Coskun J. Non-movement disorder heralds symptoms of Machado-Joseph disease years before ataxia. Mov Disord. 2005;20(6):739–41.
151. D'Abreu A, Franca M Jr, Conz L, Friedman JH, Nucci AM, Cendes F, et al. Sleep symptoms and their clinical correlates in Machado-Joseph disease. Acta Neurol Scand. 2009;119(4):277–80.
152. Friedman JH, Fernandez HH, Sudarsky LR. REM behavior disorder and excessive daytime somnolence in Machado-Joseph disease (SCA-3). Mov Disord. 2003;18(12):1520–2.
153. Reddy PL, Grewal RP. Friedreich's ataxia: a clinical and genetic analysis. Clin Neurol Neurosurg. 2007;109(2):200–2.
154. McGrath-Morrow SA, Sterni L, McGinley B, Lefton-Greif MA, Rosquist K, Lederman H. Polysomnographic values in adolescents with ataxia telangiectasia. Pediatr Pulmonol. 2008;43(7):674–9.
155. Boltshauser E, Herdan M, Dumermuth G, Isler W. Joubert syndrome: clinical and polygraphic observations in a further case. Neuropediatrics. 1981;12(2):181–91.

Sleep in Elderly Adults and in Subjects with Dementia

18

Helmut Frohnhofen and Dirk M. Hermann

Introduction

Sleep is a circadian behavior that is regulated autonomously. The normal 24-hour rhythm in adult humans is characterized by a stable change between daytime wakefulness and sleep at night. Sufficient and refreshing sleep is an important factor for well-being, self-management, and quality of life. Disturbed sleep compromises the individual affected, his relatives, and caregivers. However, the diagnosis and management of sleep disorders are challenges in an aging society with a large percentage of subjects exhibiting multimorbidity, need of care, nursing home residency, and dementia. Approximately half of elderly subjects are dissatisfied with their sleep, complaining of early morning awakenings, problems with initiating or maintaining sleep, frequent nocturnal awakenings, non-restorative sleep, or daytime sleepiness [1]. Despite their high prevalence and relevance, sleep disturbances in elderly people are hardly taken into account. Up to now, sleep disorders are not routinely assessed in geriatric medicine [2].

Regulation of Sleep and Wakefulness

Sleep and wakefulness are regulated by widespread neuronal networks involving multiple brain regions. The most important brain structures are the ventrolateral preoptic nucleus (VLPO) in the anterior hypothalamus, the lateral and posterior

H. Frohnhofen (✉)
Department of Internal Medicine and Geriatrics, Alfried Krupp Krankenhaus, Essen, Germany
e-mail: helmut.frohnhofen@krupp-Krankenhaus.de

D. M. Hermann
Department of Neurology, Universitätsklinikum Essen, Essen, Germany
e-mail: dirk.hermann@uk-essen.de

L. M. DelRosso, R. Ferri (eds.), *Sleep Neurology*,
https://doi.org/10.1007/978-3-030-54359-4_18

hypothalamus, raphe nuclei, and the locus coeruleus (LC) in the pons. The VLPO contains inhibitory galaninergic and GABAergic neurons that are active during sleep [3]. Lesions of the VLPO result in disrupted sleep [4]. Hypocretinergic neurons in the lateral hypothalamus send excitatory projections to multiple nuclei and promote wakefulness [5]. Monoaminergic and noradrenergic neurons in the posterior hypothalamus, raphe nuclei, and locus coeruleus promote wakefulness via ascending arousal systems. Sleep-promoting systems in the VLPO and wakefulness-promoting systems in the lateral and posterior hypothalamus, raphe nuclei, and locus coeruleus mutually inhibit each other and control the transition between sleep and wakefulness [6, 7].

The suprachiasmatic nucleus (SCN) is the central pacemaker of the mammalian circadian system, and circadian rhythm is mainly determined by the genotype [8]. The interaction between the abovementioned nuclei that regulate the sleep-wake rhythm is complex and polysynaptic [9]. Furthermore, the SCN is connected to the retina. This connection enables the entrainment of endogenous rhythms to the periodic change of day and night. Blindness interrupts this connection. Therefore, blind people are prone to be no longer entrained in circadian rhythms.

The role of the SCN can be demonstrated experimentally in animal models, in which the abolition of the SCN leads to a disruption of the circadian rhythm [10] (Fig. 18.1).

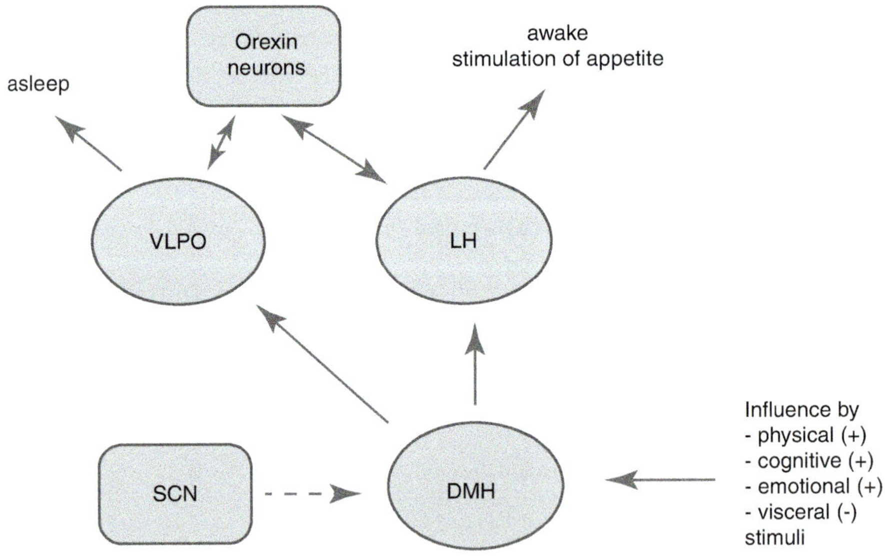

Fig. 18.1 Simplified scheme of the neuronal centers involved in the regulation of sleep and wakefulness. *SCN* suprachiasmatic nucleus, *VLPO* ventrolateral preoptic nucleus, *LH* lateral hypothalamus, *DMH* dorsomedial hypothalamus

In view of the involvement of multiple neuronal systems and neurotransmitters in the regulation of sleep and wakefulness, sleep is susceptible to neurodegenerative diseases and to the impact of many drugs. The initiation and maintenance of sleep may be a desired effect of drugs in patients suffering from insomnia (these drugs are then called hypnotics) or an undesired side effect that induces sleepiness (antihistaminics are a typical example). However, the sleep induced by drugs (hypnotics) differs substantially from normal sleep with a negative impact on the restorative function of sleep [11].

The Functions of Sleep

Sleep has well-known restorative functions. After a good night sleep, we feel awake, fresh, and alert. When a full night sleep is not achieved, the sleep deficit may induce micro-sleep during the day. In addition, cognition, alertness, verbal fluency, attention, and memory are impaired [12–15]. Restricting sleep even to 5 hours a night over several nights has a cumulative negative effect on cognitive performance in a dose-dependent fashion [16]. Since these decrements are state dependent, sufficient sleep will remove these impairments. In particular, slow-wave sleep is a requirement for memory consolidation [17, 18].

Sleep in the Healthy Elderly

Aging has an impact on sleep timing, sleep duration, and sleep consolidation. In healthy older adults, the duration of sleep – which is also genetically influenced [19] – decreases by about 10 minutes per decade from the fourth to the seventh decade and hardly changes afterwards. In a large population-based study, deep non-REM sleep (N3) decreased by about 2% and sleeping efficiency by about 3% per decade. From the age of 30 years to the age of 70 years, wake-after sleep onset (WASO) time increases for about 10 minutes per decade and hardly changed afterwards. Less clear changes were seen by the onset of sleep and the proportion of sleep stages N1 and N2. The sleep latency of 20-year-old subjects was about 5% shorter than that of over 70-year-old subjects [20]. Overall sleep in the elderly decreases and is more disrupted compared to younger people [21–23].

Anatomical and physiological studies have shown that age-related changes occur within the suprachiasmatic nucleus (SCN) [24], which is the circadian pacemaker in the human brain that becomes progressively disturbed during aging [25, 26]. Additionally, several biological rhythms are affected by aging. There are a reduction in nocturnal melatonin and pineal N-acetyltransferase rhythm, a decline in body temperature, and a reduction in the amplitude of light-induced phase response curve [27]. With age hypocretin receptor expression declines [28] and projections of these neurons degrade [29]. There is a link between the functional impairment of the master clock (SCN) and observed circadian patterns [30].

As a clinical consequence of these morphological changes, alternations in circadian patterns are observed in older subjects. These alternations encompass a reduction in the amplitude of rhythms and disorganization of their temporal order, loss of entrainment stability and responsiveness to zeitgebers, and changes in clock period length and stability [31, 32]. Furthermore, at the macro level of sleep, older persons show an advanced sleep timing, with earlier bedtimes and earlier risetimes; a longer sleep onset latency; a shorter overall sleep time during the night; increased sleep fragmentation with less consolidation; more awakenings, or transitions to lighter sleep, and a more fragile sleep; a reduced amount of deeper sleep; and an increased time spent awake during the night [33, 34].

However, there is a huge variability in the degree of sleep disruption. Some older subjects have little sleep changes, whereas others show dramatic alterations despite being in the same chronological age. Of note, disrupted sleep is correlated with cognitive performance [35]. Since aging is associated with changes in the brain, it is difficult to estimate the specific effects of sleep disturbance to cognitive performance and circadian rhythmicity.

Polysomnographic Measurement of Sleep

Human sleep can be characterized by three polysomnographic measures, electroencephalography (EEG), electrooculography (EOG), and electromyography (EMG) [36]. Based on these measures, sleep is classified into rapid-eye-movement (REM) sleep and non-REM sleep [37]. Non-REM sleep is further subdivided into three stages of progressively deeper sleep, which are called N1, N2, and N3. Although the functions of the different sleep stages remain unknown, both REM sleep and non-REM sleep are involved in memory consolidation [17].

During the night REM sleep and non-REM sleep alternate within cycles with a duration of about 90 minutes. Such a cycle begins typically with the non-REM sleep stage N1, progresses further to stages N2 and N3, and ends with a REM sleep period. A whole night's sleep encompasses three to four of such cycles.

During the first part of the night, deep non-REM sleep (N3) predominates and decreases as the night progresses. Conversely, the amount of REM sleep is relatively small at the beginning of the night and increases during the night. The average proportions of sleep stages in healthy young adults are 2–5% for stage N1, 45–55% for stage N2, 30–50% for stage N3, and 25% for REM sleep. Furthermore, non-REM stage N2 is characterized by K-complexes and so-called sleep spindles in the EEG [38].

In normal adults sleep is a nocturnal event. Sleep initiation – although regulated autonomously – occurs at a relatively predictable time, and sleep usually continues through the night with minimal interruption. The total time of sleep needed is relatively constant in older adults and encompasses about 6–8 hours. Very few subjects need less sleep to be refreshed, and about 15% of the population usually needs more sleep. The amount of sleep needed to be refreshed is genetically determined [2]. Therefore, each person should know about the individual need of sleep. The main parameter to evaluate sufficient sleep is the subjective feeling during the day. Subjects with a sufficient amount of sleep usually feel refreshed with vigor and without tiredness or sleepiness during the day.

Sleep in Subjects with Dementia

The sensitivity and vulnerability of sleep are of utmost importance for understanding the changes in sleep, cognition, and behavior in subjects with dementia. Dementia is associated with sleep and circadian disturbances. More than 50% of patients with dementia are estimated to have some kind of sleep disturbances [39–41].

Sleep disturbances are attributed to morphological changes in the brain that occur with a predictable pattern [42]. Neuronal degeneration correlates clinically with disruption of sleep and circadian rhythms, and sleep disturbances increase with the severity of dementia [43–47]. The loss of neurons in the SCN in subjects with Alzheimer's disease correlates with circadian rhythm and motor activity in these patients [48]. Unlike the typical advance in circadian phase with aging, subjects with Alzheimer's disease demonstrate a delay in circadian phase [49].

Sleep disturbances can have a debilitating impact on cognition, vigilance, and functional status. Disturbed sleep is one of the main reasons for nursing home placement in subjects with dementia [50].

Sleep studies with polysomnography have been conducted also in patients with dementia. The majority of sleep architecture changes that occur in dementia are more pronounced than changes seen in an elderly population without dementia. But there are some important differences between subjects with dementia and non-demented older persons. For example, the reduction in slow-wave sleep that accompanies the aging process is more intense in dementia [51]. Conversely, sleep spindles, indicative of sleep state N2 and associated with memory consolidation, are reduced in subjects with dementia [22]. REM sleep – which depends on acetylcholine – duration and latency are shorter in subjects with dementia compared to non-demented older persons [52].

Of note, since there is an overlap of diseases in older people with and without dementia (multimorbidity), it is difficult to determine in many patients how much of sleep disturbance can be attributed to the neurodegenerative process per se. Yet, there is no doubt that neurodegenerative diseases affect the complex brain networks responsible for sleep and circadian rhythm regulation [53–57].

In addition, the different types of dementia differ in the patterns of sleep disturbances they present [39]. To evaluate the prevalence of specific sleep disorders in subjects with different types of dementia, 431 community-dwelling older subjects with mild to moderate dementia or mild cognitive impairment were evaluated extensively via questionnaires [39]. The main result of that pivotal study was a high prevalence of any sleep disorder in demented patients, ranging from 65% in subjects with mild cognitive impairment to 90% in subjects with Lewy body dementia (LBD). Furthermore, most individuals had more than two different sleep disorders at the same time. Only 10% of participants showed isolated insomnia and only 8% had isolated daytime sleepiness. Furthermore, frequencies of specific sleep disorders differed between the types of dementia. Compared to patients with Alzheimer's dementia, subjects with LBD had a 2.6 times higher frequency of REM sleep-behavior disorders (RBDs), and subjects with vascular dementia had a 2.4 times higher frequency of insomnia and a 2.5 times higher frequency of sleep-disordered breathing [39]. This study underscores the need for a careful and comprehensive evaluation of sleep and sleep disorders in subjects with dementia.

However, correlation does not prove causality, and it does not clarify the direction of the causative link in individual patients. The structural injury to neuronal circuits controlling sleep and circadian rhythms makes a role of neurodegenerative diseases in the manifestation of sleep disturbances likely. Is there any evidence in the opposite direction that sleep disturbances may promote neurodegeneration processes?

Sleep Disturbances and Risk of Dementia

The relationship between sleep and cognitive symptoms is mutual in dementia. On the one hand, neuronal degeneration affects the systems responsible for sleep regulation, and on the other hand, disturbed sleep causes symptoms that superimpose and possibly exacerbate the dementia-related cognitive deficits. For example, sleep disturbances strongly impair memory consolidation, since it depends on continued non-REM sleep. Therefore, sleep disturbance may aggravate cognitive dysfunction in subjects with dementia [11].

But might sleep disruption also directly induce dementia? This is an important question since there could be the unique opportunity to prevent or even slow down the development of dementia by treatment of sleep disorders.

The main hypothesis in the pathology of Alzheimer's disease is the accumulation of β-amyloid that destroys neurons. Even in non-demented persons, β-amyloid deposition in the brain impacts sleep architecture [58]. Diurnal variation of β-amyloid levels suggests that this may be cleared from the brain during sleep. Since normal sleep promotes the clearance of brain β-amyloid, disturbed sleep predisposes to accumulation of β-amyloid. Indeed, cerebrospinal fluid (CSF) β-amyloid levels correlate inversely with the amount of sleep [59]. Sleep deprivation leads to a significant elevation of interstitial fluid levels of β-amyloid [60, 61]. Chronic sleep restriction increased cerebral β-amyloid plaque formation in amyloid precursor protein transgenic mice, whereas enhancing sleep with an orexin receptor antagonist reduced β-amyloid deposition [60] (Fig. 18.2).

In a recent meta-analysis that included 27 studies, sleep problems including short and long sleep duration, poor sleep quality, circadian rhythm abnormalities, insomnia, and obstructive sleep apnea were associated with a significantly increased relative risk (RR) for the combined outcome of cognitive impairment or Alzheimer's disease (RR, 1.68; 95% CI, 1.51–1.87) [11].

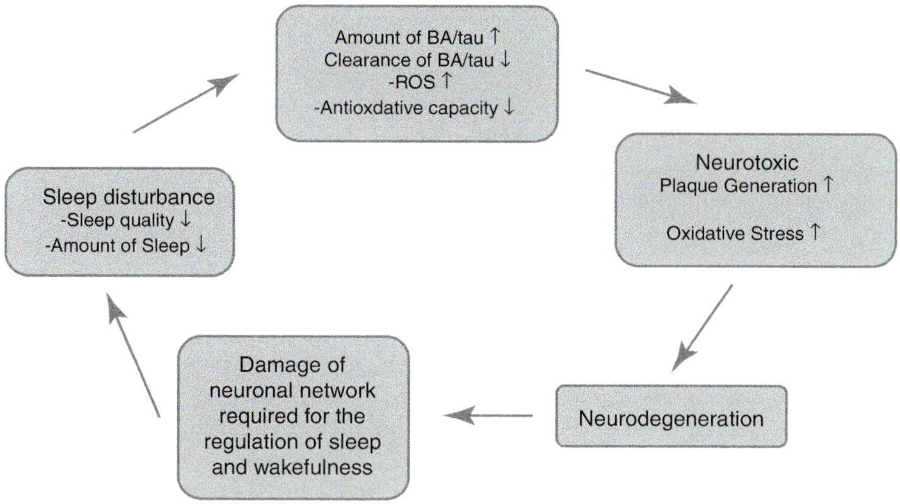

Fig. 18.2 Mutual relationship between sleep disturbances and neurodegeneration processes

In a subgroup analysis, untreated obstructive sleep apnea had the highest impact with a RR of 2.37 (95% CI: 1.82–3.08). But other sleep disturbances like sleep quantity (RR, 1.86; 95% CI, 1.34–2.57), bad sleep quality (RR, 1.62; 95% CI, 1.34–1.97), and insomnia (RR, 1.38; 95% CI, 1.13–1.67) were significantly associated with an increased dementia risk. The calculated population attributable risk (PAR) percentage suggested that about 15% of Alzheimer's disease may be caused by sleep problems. And obstructive sleep apnea appeared to be a strong and modifiable risk factor [11].

Results of the population-based Rotterdam study revealed that elimination of risk factors for dementia like overweight, hypertension, diabetes mellitus, cholesterol, smoking, stroke, coronary heart disease, heart failure, and atrial fibrillation would lead to a 30% reduction in dementia incidence [62]. However, sleep and sleep disturbances were not evaluated in this study.

Two population-based studies specifically evaluated the effect of sleep apnea on the development of cognitive impairment in elderly subjects. In a cohort of 298 women with a mean age of 82 years, which were followed up over 4.7 years, the presence of sleep apnea (SA), which was detected in 105 subjects, promoted the incidence of mild cognitive impairment or dementia [63]. In a multivariable regression, which was adjusted for age, education, race, vascular risk factors, and prescription of psychotropic drugs, a high percentage of sleep exhibiting apneas or hypopneas elevated the risk of mild cognitive impairment or dementia with an odds ratio (OR) of 2.04 (95% CI: 1.10–3.78) [63]. This finding was replicated in a second cohort of 1414 subjects with SA and 7070 nested controls, which is matched control subjects without SA obtained from the same population-based study: The presence of SA again increased the risk of dementia in a multivariable regression analysis which was adjusted for age, sex, social status, vascular risk factors, and history of stroke events, over a follow-up period of 5 years with a hazard ratio [HR] of 1.70 (95% CI: 1.26–2.31) [64].

A single prospective controlled randomized study evaluated the effect of continuous positive airway pressure (CPAP) therapy on sleep and cognitive performance in patients with dementia. In 52 patients with mild or moderate dementia and a mean age of 78 years, which were randomized for CPAP therapy for 6 weeks or sham-CPAP therapy for 3 weeks followed by CPAP therapy for 3 weeks, CPAP therapy reduced SA-associated sleepiness [65], improved sleep architecture [66], and increased cognitive performance evaluated by a neuropsychological composite score [67]. Improvements in cognitive performance particularly affected the neuropsychological domains verbal learning and cognitive flexibility [67]. These data were corroborated by an open observation study in 23 patients with mild or moderate dementia (mean age 75 years), which were followed up over 3.3 years. In this cohort, the adherence to CPAP therapy in 14 patients was associated with a significantly enhanced course of cognitive performance, evaluated by the Mini-Mental-State Examination (MMSE) score, when compared with 9 patients not tolerating CPAP therapy [65]. Larger randomized studies, including multicenter trials, evaluating the effect of CPAP therapy on the development and course of mild cognitive impairment and dementia are urgently needed. In view of the failure of disease-modifying pharmacological treatments in randomized trials, the detection of significant beneficial effects of CPAP therapy in the aforementioned two studies is an impressive, possibly ground-shaking finding, which, in case of replication in larger studies, might open fascinating perspectives for dementia care.

Treatment of Sleep Disorders in Subjects with Dementia

Patients with dementia show an overlap of behavior disorders affecting sleep and circadian rhythms and primary sleep disorders [39].

Given the substantial impact of disturbances of sleep and circadian rhythm in subjects with dementia, there is a huge interest in identifying effective treatments. The aims of treatments are reducing the caregiver burden, improvement of patient's quality of life, postponing institutionalization, and slowing cognitive decline [68].

The first step in treating sleep disturbances even in subjects with dementia is to identify and treat primary sleep disorders. SA affects more than half of subjects with dementia [69, 70]. SA has a detrimental impact on cognition and increases the risk of incident dementia [63]. Based on a controlled randomized study [67], treatment of SA with CPAP therapy is indicated and effective even in dementia patients.

Restless legs syndrome (RLS) is common in subjects with α-synucleinopathies like Lewy body dementia and Parkinson's disease. The treatment of RLS is the same as in idiopathic RLS [71].

REM sleep-behavior disorders (RBDs) are often seen in subjects with α-synucleinopathies like Parkinson syndrome and Lewy body dementia. RBD can precede the clinical manifestation of these diseases for years [72]. Clonazepam and melatonin have been shown to be effective [73].

Non-pharmacological Treatment

The basic treatment for any sleep disorder is sleep hygiene [74]. In a small study in subjects with dementia, sleep hygiene resulted in a longer total sleep time as compared to controls [75]. However, sleep hygiene as a single intervention is often not sufficient [74].

Increased social and physical activities during the day have a moderate positive impact on daytime alertness and nighttime sleep [76]. Of note, up to now there are no formal or standard recommendations for frequency, quantity, or quality of such interventions [68]. But, with the intention to minimize harm, such an intervention should be applied to demented patients prior to any pharmacological treatment.

Bright light therapy is applied to treat circadian disorders. Light therapy yields a benefit for sleep onset latency, total sleep time, time in bed, and sleep efficiency [77].

Aromatherapy with lavender, but not with melissa, was effective as an adjunct therapy in alleviating agitation in patients with dementia [78]. Music therapy reduced agitation [79] and anxiety [80]. Therapeutic touch therapy with sessions two times daily of 5- to 7-minute duration and for 3 days reduced behavioral symptoms compared to a group without any touch intervention [81].

Summary

This brief review emphasizes the importance of sleep disturbances both as a consequence and as an additional risk factor for dementia. Consequently, sleep history should be taken in any subject, and obvious sleep disturbances – especially sleep apnea – should be managed stringently. Future studies on the incidence and management of dementia should more systematically take into account sleep and sleep disturbances. Some important future research questions are how to diagnose and treat sleep disturbances in subjects with dementia, how such a treatment of sleep disturbances may influence the course of the disease, and whether prevention of sleep disorders may reduce the incidence of dementia. Further studies are warranted and necessary.

References

1. Cooke JR, Ancoli-Israel S. Normal and abnormal sleep in the elderly. Handb Clin Neurol. 2011;98:653–65.
2. Roepke SK, Ancoli-Israel S. Sleep disorders in the elderly. Indian J Med Res. 2010;131:302–10.
3. Sherin JE, Shiromani PJ, McCarley RW, Saper CB. Activation of ventrolateral preoptic neurons during sleep. Science (New York, N.Y.). 1996;271(5246):216–9.
4. Lu J, Greco MA, Shiromani P, Saper CB. Effect of lesions of the ventrolateral preoptic nucleus on NREM and REM sleep. J Neurosci. 2000;20(10):3830–42.
5. Peyron C, Tighe DK, van den Pol AN, et al. Neurons containing hypocretin (orexin) project to multiple neuronal systems. J Neurosci. 1998;18(23):9996–10015.
6. Saper CB. The central circadian timing system. Curr Opin Neurobiol. 2013;23(5):747–51.

7. Saper CB, Scammell TE, Lu J. Hypothalamic regulation of sleep and circadian rhythms. Nature. 2005;437(7063):1257–63.
8. Ralph MR, Hurd MW. Pacemaker interactions in the mammalian circadian system. Braz J Med Biol Res = Rev Bras Pesqui Med Biol. 1996;29(1):77–85.
9. Aston-Jones G, Chen S, Zhu Y, Oshinsky ML. A neural circuit for circadian regulation of arousal. Nat Neurosci. 2001;4(7):732–8.
10. Ralph MR, Hurd MW. Circadian pacemakers in vertebrates. Ciba Found Symp. 1995;183:67–81; discussion 81–7
11. Bubu OM, Brannick M, Mortimer J, et al. Sleep, cognitive impairment, and Alzheimer's disease: a systematic review and meta-analysis. Sleep. 2017;40(1):zsw032.
12. Williams HL, Gieseking CF, Lubin A. Some effects of sleep loss on memory. Percept Mot Skills. 1966;23(3):1287–93.
13. Rogers NL, Dorrian J, Dinges DF. Sleep, waking and neurobehavioural performance. Frontiers Biosci. 2003;8:67.
14. Polzella DJ. Effects of sleep deprivation on short-term recognition memory. J Exp Psychol Hum Learn Mem. 1975;104(2):194–200.
15. Drummond SP, Brown GG, Gillin JC, et al. Altered brain response to verbal learning following sleep deprivation. Nature. 2000;403(6770):655–7.
16. Van Dongen HPA, Maislin G, Mullington JM, Dinges DF. The cumulative cost of additional wakefulness: dose-response effects on neurobehavioral functions and sleep physiology from chronic sleep restriction and total sleep deprivation. Sleep. 2003;26(2):117–26.
17. Stickgold R. Sleep-dependent memory consolidation. Nature. 2005;437(7063):1272–8.
18. Walker MP, Stickgold R. Sleep, memory, and plasticity. Annu Rev Psychol. 2006;57:139–66.
19. Jones SE, Tyrrell J, Wood AR, et al. Genome-wide association analyses in 128,266 individuals identifies new morningness and sleep duration loci. PLoS Genet. 2016;12(8):e1006125.
20. Ohayon MM, Carskadon MA, Guilleminault C, Vitiello MV. Meta-analysis of quantitative sleep parameters from childhood to old age in healthy individuals: developing normative sleep values across the human lifespan. Sleep. 2004;27(7):1255–73.
21. Bliwise DL. Sleep in normal aging and dementia. Sleep. 1993;16(1):40–81.
22. Reynolds CF3, Kupfer DJ, Taska LS, et al. EEG sleep in elderly depressed, demented, and healthy subjects. Biol Psychiatry. 1985;20(4):431–42.
23. Bloom HG, Ahmed I, Alessi CA, et al. Evidence-based recommendations for the assessment and management of sleep disorders in older persons. J Am Geriatr Soc. 2009;57(5):761–89.
24. Hofman MA. Lifespan changes in the human hypothalamus. Exp Gerontol. 1997;32(4–5):559–75.
25. Hofman MA, Swaab DF. Influence of aging on the seasonal rhythm of the vasopressin-expressing neurons in the human suprachiasmatic nucleus. Neurobiol Aging. 1995;16(6):965–71.
26. Hofman MA, Swaab DF. Living by the clock: the circadian pacemaker in older people. Ageing Res Rev. 2006;5(1):33–51.
27. Brock MA. Chronobiology and aging. J Am Geriatr Soc. 1991;39(1):74–91.
28. Terao A, Apte-Deshpande A, Morairty S, et al. Age-related decline in hypocretin (orexin) receptor 2 messenger RNA levels in the mouse brain. Neurosci Lett. 2002;332(3):190–4.
29. Zhang JH, Sampogna S, Morales FR, Chase MH. Age-related changes in hypocretin (orexin) immunoreactivity in the cat brainstem. Brain Res. 2002;930(1–2):206–11.
30. Zee PC, Rosenberg RS, Turek FW. Effects of aging on entrainment and rate of resynchronization of circadian locomotor activity. Am J Phys. 1992;263(5 Pt 2):103.
31. Pittendrigh CS, Daan S. Circadian oscillations in rodents: a systematic increase of their frequency with age. Science (New York, N.Y.). 1974;186(4163):548–50.
32. Sloan EP, Flint AJ, Reinish L, Shapiro CM. Circadian rhythms and psychiatric disorders in the elderly. J Geriatr Psychiatry Neurol. 1996;9(4):164–70.
33. Conte F, Arzilli C, Errico BM, et al. Sleep measures expressing 'functional uncertainty' in elderlies' sleep. Gerontology. 2014;60(5):448–57.
34. Zhdanova IV, Masuda K, Quasarano-Kourkoulis C, et al. Aging of intrinsic circadian rhythms and sleep in a diurnal nonhuman primate, Macaca mulatta. J Biol Rhythm. 2011;26(2):149–59.

35. Miyata S, Noda A, Iwamoto K, et al. Poor sleep quality impairs cognitive performance in older adults. J Sleep Res. 2013;22(5):535–41.
36. Rechtschaffen A, editor. Ein Manual der standardisierten Terminologie, Techniken und Auswertung der Schlafstadien beim Menschen. Landsberg/Lech: ecomed; 2002.
37. Grigg-Damberger MM. The AASM scoring manual four years later. J Clin Sleep Med. 2012;8(3):323–32.
38. Ohayon MM. Interactions between sleep normative data and sociocultural characteristics in the elderly. J Psychosom Res. 2004;56(5):479–86.
39. Guarnieri B, Adorni F, Musicco M, et al. Prevalence of sleep disturbances in mild cognitive impairment and dementing disorders: a multicenter Italian clinical cross-sectional study on 431 patients. Dement Geriatr Cogn Disord. 2012;33(1):50–8.
40. Moran M, Lynch CA, Walsh C, et al. Sleep disturbance in mild to moderate Alzheimer's disease. Sleep Med. 2005;6(4):347–52.
41. Moran M, Walsh C, Lynch A, et al. Syndromes of behavioural and psychological symptoms in mild Alzheimer's disease. Int J Geriatr Psychiatry. 2004;19(4):359–64.
42. Braak H, Braak E. Neuropathological stageing of Alzheimer-related changes. Acta Neuropathol. 1991;82(4):239–59.
43. Bliwise DL, Hughes M, McMahon PM, Kutner N. Observed sleep/wakefulness and severity of dementia in an Alzheimer's disease special care unit. J Geront A Biol Sci Med Sci. 1995;50(6):6.
44. Volicer L, Harper DG, Manning BC, et al. Sundowning and circadian rhythms in Alzheimer's disease. Am J Psychiatry. 2001;158(5):704–11.
45. Harper DG, Stopa EG, McKee AC, et al. Differential circadian rhythm disturbances in men with Alzheimer disease and frontotemporal degeneration. Arch Gen Psychiatry. 2001;58(4):353–60.
46. Harper DG, Stopa EG, McKee AC, et al. Dementia severity and Lewy bodies affect circadian rhythms in Alzheimer disease. Neurobiol Aging. 2004;25(6):771–81.
47. Boddy F, Rowan EN, Lett D, et al. Subjectively reported sleep quality and excessive daytime somnolence in Parkinson's disease with and without dementia, dementia with Lewy bodies and Alzheimer's disease. Int J Geriatr Psychiatry. 2007;22(6):529–35.
48. Wang JL, Lim AS, Chiang W-Y, et al. Suprachiasmatic neuron numbers and rest-activity circadian rhythms in older humans. Ann Neurol. 2015;78(2):317–22.
49. Monk TH. Sleep and circadian rhythms. Exp Gerontol. 1991;26(2–3):233–43.
50. Pollak CP, Perlick D. Sleep problems and institutionalization of the elderly. J Geriatr Psychiatry Neurol. 1991;4(4):204–10.
51. Feinberg I, Koresko RL, Heller N. EEG sleep patterns as a function of normal and pathological aging in man. J Psychiatr Res. 1967;5(2):107–44.
52. Bliwise DL, Tinklenberg J, Yesavage JA, et al. REM latency in Alzheimer's disease. Biol Psychiatry. 1989;25(3):320–8.
53. Bliwise DL, Watts RL, Watts N, et al. Disruptive nocturnal behavior in Parkinson's disease and Alzheimer's disease. J Geriatr Psychiatry Neurol. 1995;8(2):107–10.
54. Bliwise DL. Sleep disorders in Alzheimer's disease and other dementias. Clin Cornerstone. 2004;6(Suppl 1A):28.
55. Bliwise DL, Mercaldo ND, Avidan AY, et al. Sleep disturbance in dementia with Lewy bodies and Alzheimer's disease: a multicenter analysis. Dement Geriatr Cogn Disord. 2011;31(3):239–46.
56. Bombois S, Derambure P, Pasquier F, Monaca C. Sleep disorders in aging and dementia. J Nutr Health Aging. 2010;14(3):212–7.
57. Bonanni E, Maestri M, Tognoni G, et al. Daytime sleepiness in mild and moderate Alzheimer's disease and its relationship with cognitive impairment. J Sleep Res. 2005;14(3):311–7.
58. Ju Y-ES, McLeland JS, Toedebusch CD, et al. Sleep quality and preclinical Alzheimer disease. JAMA Neurol. 2013;70(5):587–93.
59. Huang Y, Potter R, Sigurdson W, et al. Effects of age and amyloid deposition on Abeta dynamics in the human central nervous system. Arch Neurol. 2012;69(1):51–8.

60. Kang J-E, Lim MM, Bateman RJ, et al. Amyloid-beta dynamics are regulated by orexin and the sleep-wake cycle. Science (New York, N.Y.). 2009;326(5955):1005–7.
61. Xie L, Kang H, Xu Q, et al. Sleep drives metabolite clearance from the adult brain. Science (New York, N.Y.). 2013;342(6156):373–7.
62. de Bruijn RFAG, Bos MJ, Portegies MLP, et al. The potential for prevention of dementia across two decades: the prospective, population-based Rotterdam Study. BMC Med. 2015;13:132.
63. Yaffe K, Laffan AM, Harrison SL, et al. Sleep-disordered breathing, hypoxia, and risk of mild cognitive impairment and dementia in older women. JAMA. 2011;306(6):613–9.
64. Chong MS, Ayalon L, Marler M, et al. Continuous positive airway pressure reduces subjective daytime sleepiness in patients with mild to moderate Alzheimer's disease with sleep disordered breathing. J Am Geriatr Soc. 2006;54(5):777–81.
65. Troussière A-C, Charley CM, Salleron J, et al. Treatment of sleep apnoea syndrome decreases cognitive decline in patients with Alzheimer's disease. J Neurol Neurosurg Psychiatry. 2014;85(12):1405–8.
66. Cooke JR, Ancoli-Israel S, Liu L, et al. Continuous positive airway pressure deepens sleep in patients with Alzheimer's disease and obstructive sleep apnea. Sleep Med. 2009;10(10):1101–6.
67. Ancoli-Israel S, Palmer BW, Cooke JR, et al. Cognitive effects of treating obstructive sleep apnea in Alzheimer's disease: a randomized controlled study. J Am Geriatr Soc. 2008;56(11):2076–81.
68. Ooms S, Ju Y-E. Treatment of sleep disorders in dementia. Curr Treat Options Neurol. 2016;18(9):40.
69. Ancoli-Israel S, Klauber MR, Butters N, et al. Dementia in institutionalized elderly: relation to sleep apnea. J Am Geriatr Soc. 1991;39(3):258–63.
70. Frohnhofen H, Roffe C. Intermittent nocturnal hypoxemia in individuals with dementia: prevalence and relationship with functional status. J Am Geriatr Soc. 2012;60(10):1997–9.
71. Frohnhofen H, Schlitzer J. CME Zertifizierte Fortbildung: Schlaf und Schlafstörungen beim alten Menschen; Teil 3: Restless-legs-Syndrom. Zeitschrift Gerontologie Geriatrie. 2015;48(4):379–87.
72. McCann H, Stevens CH, Cartwright H, Halliday GM. alpha-Synucleinopathy phenotypes. Parkinsonism Relat Disord. 2014;20(Suppl 1):7.
73. Aurora RN, Zak RS, Maganti RK, et al. Best practice guide for the treatment of REM sleep behavior disorder (RBD). J Clin Sleep Med. 2010;6(1):85–95.
74. Schlitzer J, Heubaum S, Frohnhofen H. CME Zertifizierte Fortbildung: Schlaf und Schlafstörungen beim alten Menschen; Teil 2: Therapie. Z Gerontol Geriatr Eur J Geriatr. 2014;47(7):611–20.
75. McCurry SM, LaFazia DM, Pike KC, et al. Development and evaluation of a sleep education program for older adults with dementia living in adult family homes. Am J Geriatr Psychiatry. 2012;20(6):494–504.
76. McCurry SM, Pike KC, Vitiello MV, et al. Increasing walking and bright light exposure to improve sleep in community-dwelling persons with Alzheimer's disease: results of a randomized, controlled trial. J Am Geriatr Soc. 2011;59(8):1393–402.
77. van Maanen A, Meijer AM, van der Heijden KB, Oort FJ. The effects of light therapy on sleep problems: a systematic review and meta-analysis. Sleep Med Rev. 2016;29:52–62.
78. Lin PW-k, Chan W-c, Ng BF-l, Lam LC-w. Efficacy of aromatherapy (Lavandula angustifolia) as an intervention for agitated behaviours in Chinese older persons with dementia: a cross-over randomized trial. Int J Geriatr Psychiatry. 2007;22(5):405–10.
79. Garland K, Beer E, Eppingstall B, O'Connor DW. A comparison of two treatments of agitated behavior in nursing home residents with dementia: simulated family presence and preferred music. Am J Geriatr Psychiatry. 2007;15(6):514–21.
80. Sung H-c, Lee W-l, Li T-l, Watson R. A group music intervention using percussion instruments with familiar music to reduce anxiety and agitation of institutionalized older adults with dementia. Int J Geriatr Psychiatry. 2012;27(6):621–7.
81. Woods DL, Dimond M. The effect of therapeutic touch on agitated behavior and cortisol in persons with Alzheimer's disease. Biol Res Nurs. 2002;4(2):104–14.

Index

L. M. DelRosso, R. Ferri (eds.), *Sleep Neurology*,
https://doi.org/10.1007/978-3-030-54359-4

W